Lecture Notes in Physics

The Editorial Policy for Proceedings

The series Lecture Notes in Physics reports new developments in physical research and teaching – quickly, informally, and at a high level. The proceedings to be considered for publication in this series should be limited to only a few areas of research, and these should be closely related to each other. The contributions should be of a high standard and should avoid lengthy redraftings of papers already published or about to be published elsewhere. As a whole, the proceedings should aim for a balanced presentation of the theme of the conference including a description of the techniques used and enough motivation for a broad readership. It should not be assumed that the published proceedings must reflect the conference in its entirety. (A listing or abstracts of papers presented at the meeting but not included in the proceedings could be added as an appendix.)

When applying for publication in the series Lecture Notes in Physics the volume's editor(s) should submit sufficient material to enable the series editors and their referees to make a fairly accurate evaluation (e.g. a complete list of speakers and titles of papers to be presented and abstracts). If, based on this information, the proceedings are (tentatively) accepted, the volume's editor(s), whose name(s) will appear on the title pages, should select the papers suitable for publication and have them refereed (as for a journal) when appropriate. As a rule discussions will not be accepted. The series editors and Springer-Verlag will normally not interfere with the detailed editing except in fairly obvious cases or on technical matters.

Final acceptance is expressed by the series editor in charge, in consultation with Springer-Verlag only after receiving the complete manuscript. It might help to send a copy of the authors' manuscripts in advance to the editor in charge to discuss possible revisions with him. As a general rule, the series editor will confirm his tentative acceptance if the final manuscript corresponds to the original concept discussed, if the quality of the contribution meets the requirements of the series, and if the final size of the manuscript does not greatly exceed the number of pages originally agreed upon. The manuscript should be forwarded to Springer-Verlag shortly after the meeting. In cases of extreme delay (more than six months after the conference) the series editors will check once more the timeliness of the papers. Therefore, the volume's editor(s) should establish strict deadlines, or collect the articles during the conference and have them revised on the spot. If a delay is unavoidable, one should encourage the authors to update their contributions if appropriate. The editors of proceedings are strongly advised to inform contributors about these points at an early stage.

The final manuscript should contain a table of contents and an informative introduction accessible also to readers not particularly familiar with the topic of the conference. The contributions should be in English. The volume's editor(s) should check the contributions for the correct use of language. At Springer-Verlag only the prefaces will be checked by a copy-editor for language and style. Grave linguistic or technical shortcomings may lead to the rejection of contributions by the series editors. A conference report should not exceed a total of 500 pages. Keeping the size within this bound should be achieved by a stricter selection of articles and not by imposing an upper limit to the length of the individual papers. Editors receive jointly 30 complimentary copies of their book. They are entitled to purchase further copies of their book at a reduced rate. As a rule no reprints of individual contributions can be supplied. No royalty is paid on Lecture Notes in Physics volumes. Commitment to publish is made by letter of interest rather than by signing a formal contract. Springer-Verlag secures the copyright for each volume.

The Production Process

The books are hardbound, and the publisher will select quality paper appropriate to the needs of the author(s). Publication time is about ten weeks. More than twenty years of experience guarantee authors the best possible service. To reach the goal of rapid publication at a low price the technique of photographic reproduction from a camera-ready manuscript was chosen. This process shifts the main responsibility for the technical quality considerably from the publisher to the authors. We therefore urge all authors and editors of proceedings to observe very carefully the essentials for the preparation of camera-ready manuscripts, which we will supply on request. This applies especially to the quality of figures and halftones submitted for publication. In addition, it might be useful to look at some of the volumes already published. As a special service, we offer free of charge LaTeX and TeX macro packages to format the text according to Springer-Verlag's quality requirements. We strongly recommend that you make use of this offer, since the result will be a book of considerably improved technical quality. To avoid mistakes and time-consuming correspondence during the production period the conference editors should request special instructions from the publisher well before the beginning of the conference. Manuscripts not meeting the technical standard of the series will have to be returned for improvement.

For further information please contact Springer-Verlag, Physics Editorial Department II, Tiergartenstrasse 17, D-69121 Heidelberg, Germany

Pedro L. Garrido Joaquín Marro (Eds.)

Third Granada Lectures in Computational Physics

Proceedings of the III Granada Seminar
on Computational Physics
Held at Granada, Spain, 5-10 September 1994

Springer

Editors

Pedro L. Garrido
Joaquín Marro
Instituto Carlos I de Física Teórica y Computacional
Facultad de Ciencias, Universidad de Granada
Campus de Fuentenueva
E-18071 Granada, Spain

ISBN 978-3-662-14046-8 ISBN 978-3-540-49221-4 (eBook)
DOI 10.1007/978-3-540-49221-4

CIP data applied for

Originally published by Springer-Verlag Berlin Heidelberg New York in 1995
Softcover reprint of the hardcover 1st edition 1995

Typesetting: Camera-ready by the editors
SPIN: 10481101 55/3142-543210 - Printed on acid-free paper

GRANADA SEMINAR on COMPUTATIONAL PHYSICS

STEERING COMMITTEE

Preface

The methods developed to deal with the computational aspects of physical problems are useful in an increasing number of situations, from chemistry, biology and geology to engineering, communications and economics. In fact, *Computational Physics* has evolved into a trans-disciplinary field now concerned with the creative use of computers in scientific research. Moreover, computational methods often help students to develop a deeper understanding of key concepts, and enhance their problem-solving ability. Therefore, computational physics is recognized to have an important educational value, and educators face the task of outlining appropriate curricula to take advantage of these unique features.

This is an important motivation for the publication of the contents of the Seminar on Computational Physics which is held in Granada every two years. The Seminar aims at bringing together small groups of students and active researchers on different aspects of computational physics. It is part of the doctoral programme of the University of Granada. The proceedings of the previous edition were published as *II Granada Lectures in Computational Physics* (World Scientific, Singapore 1993).

This book contains the invited lecture notes and a very brief account of contributions by participants at the *III Granada Seminar on Computational Physics* (Granada, Spain, 5-10 September 1994). The lecturers this year have covered the basics and some generalizations of Monte Carlo methods, and its application to discrete and field theoretic models; techniques to generate pseudo-random numbers; the study of nonequilibrium models of granular media, half way between fluids and solids, by computer simulation and other techniques; the description of sandpile models and phenomena such as self-organized criticality; the diffusion of a Brownian particle in a *large* bath of moving disks; some computational aspects of nonlinear systems with disorder that can be formulated in terms of wave equations; the application of Fourier analysis to the numerical solution of time-dependent partial differential equations with study of stability, consistency, and numerical dispersion and dissipation; statistical properties of dynamical systems, including notions of Lyapunov exponents, nonequilibrium ensembles and ergodic theory. Furthermore, the lectures deal with details of phenomena such as chaos, segregation, pattern formation and phase transitions, convection, fluidification, density waves, surface reaction and growth, spread of epidemics, acoustics, deformation, etc.

The *Granada Lectures* series aims to provide a smooth transition from 'one-semester' course books to the current literature in specialized journals, which we believe is an important gap to be filled. Therefore, an effort has been made to write the notes pedagogically for the lowest graduate student level, and to give a comprehensive description of each topic. Moreover, some exercises are proposed and some practical computer sessions are illustrated. In this way, the book may introduce the beginner to the world of computational science. In addition, it may serve as a work of reference for teachers, students and researchers looking for a state-of-the-art brief description of specific areas of computational physics. In fact, the series is concerned not only with methods and techniques but also with problems at the frontier that are specially amenable to investigations by computational means.

We wish to express our gratitude here to all those who have collaborated in making the 1994 edition of the Seminar a success. We would first like to mention the invited lecturers who established such a good rapport with the participants and whose personal effort enabled us to accomplish the goals of the Seminar, and the other members of the Steering Committee who have effectively helped us to design the format and contents. The smoothness of the organization was ensured by the collaboration of the students C. López, I. Polo, F. de los Santos, J. Torres, and J. A. Vacas. Financial support given by the *Dirección General de Investigación Científica y Técnica* of the Spanish Ministry of Education and Science, as well as the sponsorship of the *Real Sociedad Española de Física*, of the *Consejería de Educación y Ciencia* of the Andalusian administration *Junta de Andalucía,* of the *Vicerrectorados de Investigación y Relaciones Internacionales y de Ordenación Académica,* and of the *Facultad de Ciencias* of the University of Granada is acknowledged.

Granada, January 1995

The Editors

Contents

Part I: INVITED LECTURES

Computational Field Theory and Pattern Formation
R. Toral 3

1. Monte Carlo Methods 3
   1.1 Introduction 3
   1.2 Uniform Sampling 5
   1.3 Importance Sampling for N=1 6
   1.4 Rejection Method for N=1 10
   1.5 Rejection with Repetition 12
   1.6 The Algorithm of Metropolis et al. 14
   1.7 Rejection Method in the N-dimensional Case 18
   1.8 Heat Bath 19
2. The Φ^4 Model 20
   2.1 Introduction and Basic Definitions 20
   2.2 Monte Carlo Methods 24
   2.3 Histogram Extrapolation 28
   2.4 Finite Size Scaling 30
3. Field Dynamics 33
   3.1 Introduction 33
   3.2 The Langevin Equation 34
   3.3 Numerical Solutions of Stochastic Differential Equations 37
4. Hybrid Monte Carlo 41
   4.1 Introduction 41
   4.2 Molecular Dynamics 42
   4.3 Hybrid Monte Carlo 44
   4.4 Generalized Hybrid Monte Carlo 46
   4.5 Fourier Acceleration 48
5. Applications in Domain Growth 50
   5.1 Dynamics of First Order Phase Transitions 50
   5.2 Growth of Rough Surfaces 57
   References 63

Simulating Granular Media on the Computer
H.J. Herrmann 67

1. Introduction 67
2. Molecular Dynamics Technique 70
3. Angle of Repose and Angle of Marginal Stability 72
4. Simulating Granular Media on a Vibrating Plate 73
5. Simulating Size Segregation 80

6. Simulating the Flow Through Hoppers and Pipes 85
7. Traffic Models 89
8. Lattice Gas and Boltzmann Lattice Models 91
9. A Thermodynamic Approach to Granular Media 100
10. Sound Propagation in Granular Assemblies 102
11. Plastic Deformation and Shear Bands in Dense Packings 106
12. Discussion 109
References 110

Some Physical and Computational Aspects of Self–organized Criticality
A. Díaz-Guilera 115

1. Introduction 115
2. Sandpiles, Cellular Automata and Self–organized Criticality 116
   2.1 The Bak, Tang and Wiesenfeld Sandpile Model 116
   2.2 Critical Behavior 119
   2.3 What is New in Self–organized Criticality 120
   2.4 Relationship to Real Sandpiles 121
3. Experiments, Computers and Theories 121
   3.1 Physical Systems Exhibiting Self–organized Criticality 121
   3.2 Simple Numerical Models 122
   3.3 Cellular Automata and Coupled Map Lattice Models 123
   3.4 Theoretical Studies 124
4. Computational Aspects of Self–Organized Criticality 124
   4.1 Preparing the System 125
   4.2 Time Evolution 125
   4.3 Extracting Information 128
References 137

Critical Percolation with Moving Disks
I.M. Polo and J.F. Fernández 141

1. Introduction 141
2. The Model 142
3. Results 143
References 143

Fourier Techniques in Numerical Methods for Evolutionary Problems
J.M. Sanz-Serna 145

1. Introduction 145
   1.1 Scope 145
   1.2 Some Mathematical Preliminaries 146
2. Review of Fourier Series 147
   2.1 The L^2 Theory 147
   2.2 The Trigonometric Version 149

2.3 Fourier Series and Derivatives 151
3. Fourier Analysis of Initial Value Problems 153
3.1 Formal Solution 153
3.2 Well-Posed Problems 154
3.3 Dissipation and Dispersion 156
4. Discrete Fourier Analysis 159
4.1 The Discrete Fourier Transform 159
4.2 An Application: Systems of Ordinary Differential Equations with Circulant Matrices 160
4.3 The Fast Fourier Transform 162
5. Discrete Fourier Transform vs. Fourier Series 165
5.1 A First Look at Aliasing 165
5.2 Trigonometric Interpolation 167
5.3 More Aliasing: Discrete Fourier Coefficients vs. Fourier Coefficients 171
6. Fourier Analysis of Finite-Difference Algorithms: the Time-Continuous Case 173
6.1 Spatial Discretizations of Initial Value Problems 173
6.2 The Von Neumann Stability Analysis 178
6.3 The Roles of Stability and Consistency from a Fourier Viewpoint 179
6.4 Numerical Dissipation and Numerical Dispersion 183
7. Fourier Analysis of Finite-Difference Algorithms: the Fully-Discrete Case 184
7.1 Full Discretization of Initial Value Problems 184
7.2 The Von Neumann Stability Analysis 187
7.3 The Roles of Stability and Consistency from a Fourier Viewpoint 187
7.4 Numerical Dissipation and Numerical Dispersion 188
8. The Practical Relevance of Fourier Analysis of Difference Methods 189
9. Spectral Methods 191
9.1 Spectral Methods for Periodic Linear Constant Coefficient Problems 191
9.2 Pseudospectral Difference Matrices 193
9.3 The Pseudospectral Method for Periodic Nonlinear Problems 196
References 198

A Fast Algorithm for the Generation of Random Numbers with Exponential and Normal Distributions
J.F. Fernández and J. Rivero 201

1. Introduction 201
2. Physical Motivation 203
3. The Algorithms 204
3.1 Exponential Random Deviates 204
3.2 Normal Random Deviates 205

4. Derivation 206
5. Performance 207
References 209

Numerical Simulations of a Nonlinear Klein–Gordon Model. Applications
P.J. Pascual, S. Jiménez and L. Vázquez 211

1. Introduction 211
2. Solitary Waves 212
3. Numerical Treatment of the Sine–Gordon Equation 213
   3.1 Numerical Schemes 213
   3.2 A stochastic Sine–Gordon model 216
4. Numerical Schemes for Euler–Lagrange Equations 217
   4.1 Schemes for the Collective Coordinates Study 217
   4.2 Extension to a General Lagrangian 218
5. Sine–Gordon Equation Under Multiplicative Noise 221
   5.1 The Soliton 221
   5.2 The Breather 227
   5.3 Soliton Under Dissipation 235
   5.4 Breather Under Dissipation 238
6. Sine–Gordon Equation Under Additive Noise and Periodic Perturbation 239
   6.1 The Soliton 239
   6.2 The Breather 244
   6.3 Soliton with Dissipation 248
   6.4 Breather with Dissipation 252
7. Charge Density Waves Modelling 257
   7.1 General Aspects 257
   7.2 Models 259
   7.3 CDW Overdamped Model 261
   7.4 Underlying Elastic Lattice Model 264
References 269

Topics on Chaotic Dynamics
G. Gallavotti 271

1. Dynamical Systems and Their Statistics 271
2. Dynamical Bases and Lyapunov Exponents 275
3. Chaotic Motions 282
4. Time Reversible Systems. Ruelle's Principle 292
5. Summary and Outlook 300
Appendix: Philosophical Questions; and a Few Concrete Ones 302
References 310

Part II: CONTRIBUTIONS

Multistable Steady States and Oscillations in a Surface Reaction Model
M.C. Lemos, J.J. Luque and F. Jiménez-Morales 315

Comparison of Two Fokker–Planck Modelings for a Master Equation
M.A. Muñoz and P.L. Garrido 316

Computational Studies of the Complex Ginzburg–Landau Equation and Its Non-equilibrium Potential
R. Montagne, E. Hernández-García and M. San Miguel 317

Deterministic Simulation of Domain Pattern Formation Under Instantaneous Nucleation Conditions
C.H. Wörner 319

Morphological Instabilities of a Nematic-Smectic B Interface
R. González-Cinca, L. Ramírez-Piscina, J. Casademunt and A. Hernández-Machado 321

Dynamic Time Rescaled Monte Carlo Algorithm
L. Bitar, P.A. Serena, P. García-Mochales and N. García 322

Pattern Formation in Convection of Rotating Fluids
J. Millán-Rodriguez, C. Pérez-García and M. Bestehorn 323

Codimension Two Bifurcation in Unidimensional Heating
A.M. Mancho-Sánchez and C. Pérez-García 325

Cylinders and Wheels Rolling into Chaos
C.M. Arizmendi, H.N. Núñez-Yépez and A.L. Salas-Brito 326

On the Transition to Columnar Convection
A. Alonso and M. Net 327

Finite Temperature dc Conductance of Random Dimer Superlattices
E. Díez, A. Sánchez and F. Domínguez-Adame 329

A Multigrid Approach to Parallel Simulations in Nonlinear Optics
I. Martín, V.M. Pérez-García, F. Tirado and L. Vázquez 331

Surface Electromagnetic Waves on Index-Matched Interfaces
M. Lester and R. Depine 333

Conservative Numerical Schemes for Euler–Lagrange Equations
S. Jiménez 335

Suppression of Chaotic Behaviour in Models of Chemical Chaos
J. Güémez, J.M. Gutiérrez, A. Iglesias and M.A. Matías 337

Perturbing Variables in Poincaré Sections to Suppress Chaos
J.M. Gutiérrez, A. Iglesias, J. Güémez and M.A. Matías 339

Ergodic Theory for Unidimensional Jacobi Matrices
A.I. Alonso and C. Nuñez 341

Order in Binary Sequences and the Properties of the Routes to Chaos
R. López-Ruiz 342

List of Participants 343

I

Invited Lectures

Computational Field Theory and Pattern Formation

Raúl Toral

Departament de Física,
Universitat de les Illes Balears,
07071-Palma de Mallorca, Spain

1. Monte Carlo Methods

1.1. Introduction

In many occasions we need to compute the value of a high-dimensional integral of the form:

$$\int_{-\infty}^{\infty} dx_1 \ldots \int_{-\infty}^{\infty} dx_N f_{\hat{\mathbf{x}}}(x_1,\ldots,x_N) G(x_1,\ldots,x_N) \equiv \int dX f_{\hat{\mathbf{x}}}(X) G(X) \quad (1.1)$$

here N is a *large* number and we have denoted by X the set $(x_1,\ldots,x_N)$ of real variables. The function $f_{\hat{\mathbf{x}}}(X)$ is non-negative and normalized, i.e.:

$$\begin{aligned} &(i) \qquad f_{\hat{\mathbf{x}}}(X) \geq 0 \\ &(ii) \qquad \int f_{\hat{\mathbf{x}}}(X)\, dX \equiv \int_{-\infty}^{\infty} dx_1 \ldots \int_{-\infty}^{\infty} dx_N f_{\hat{\mathbf{x}}}(x_1,\ldots,x_N) = 1 \end{aligned} \quad (1.2)$$

In next chapters we will give specific examples of when one needs to perform this type of calculation. If the above integral cannot be computed by analytical methods (which happens more often than desired), one has to resource to numerical methods. One must realize, however, that a simple extension of the standard methods used in small–N numerical integration (Simpson rules, Gauss integration, etc. (Faires and Burden, 1993)) to compute numerically the value of the integral will fail for very high–dimensional integrals. For instance, let us take $N = 100$ (and this is a rather small number for typical applications) and suppose we want to use Simpson methods. We need to generate a grid in the X space and sum up all the values of $G(X)$ on the points of this grid. If we choose, say, a grid of 10 points per every coordinate, we will have 10^{100} terms to add in the Simpson rule. If every addition takes 10^{-12} s (and today's computers take much longer than that) the time needed for the calculation exceeds any astronomical unit of time (including the age of the universe). The situation does not improve by diminishing the number of points in the integration grid. Even 2 points per integration variable is beyond the capability of any present (or future) computer.

The solution to this problem is given by application of the Monte Carlo techniques (Kalos and Whitlock, 1986; Binder, 1992; Heermann, 1986). For our particular example, they consist in interpreting $f_{\hat{\mathbf{x}}}(X)$ as a probability density function of some random variable $\hat{\mathbf{x}}$ and, consequently, the integral is nothing but the average value of the function $G(X)$,

$$\langle G\rangle = \int dX\, f_{\hat{\mathbf{x}}}(X)G(X) \tag{1.3}$$

The key idea behind Monte Carlo integration is to approximate the previous average by a sample average:

$$\langle G\rangle \approx \mu_M[G] = \frac{1}{M}\sum_{k=1}^{M} G(X^{(k)}) \tag{1.4}$$

where $X^{(k)}$, $k = 1,\dots,M$ are values of the N–dimensional variable X distributed according to the probability density function $f_{\hat{\mathbf{x}}}(X)$. This is called the technique of **importance sampling**. The basic principle of sampling is the same one that allows politicians to have an accurate knowledge of the election results before election day by making random polls amongst a representative sample of the electors.

A more precise statement on the relation between the average $\langle G\rangle$ and the sample average $\mu_M[G]$ is given by the *central limit theorem* (Feller, 1971). This theorem states that, if the values of $X^{(k)}$ are statistically independent, then the sample average μ_M tends in the limit $M \to \infty$ to a Gaussian distribution of mean $\langle G\rangle$ and variance $\sigma^2[\mu_M]$ given by:

$$\sigma^2[\mu_M] = \frac{1}{M}\sigma^2[G] = \frac{1}{M}\left[\langle G^2\rangle - \langle G\rangle^2\right] \tag{1.5}$$

It is usual to express the relation between $\mu_M[G]$ and $\langle G\rangle$ as:

$$\langle G\rangle = \mu_M[G] \pm \sigma[\mu_M] \tag{1.6}$$

which is to be interpreted in the usual statistical sense for the Gaussian distribution, i.e. that there is a 95.45% probability that $\langle G\rangle$ lies in the interval $(\mu_M[G] - 2\sigma[\mu_M], \mu_M[G] + 2\sigma[\mu_M])$, 99.73% in the interval $(\mu_M[G] - 3\sigma[\mu_M], \mu_M[G] + 3\sigma[\mu_M])$, etc. In practice, the unknown variance $\sigma^2[G]$ can be replaced by the sample variance:

$$\sigma^2[G] \approx \frac{1}{M}\sum_{k=1}^{M} G(X^{(k)})^2 - \left(\frac{1}{M}\sum_{k=1}^{M} G(X^{(k)})\right)^2 \tag{1.7}$$

According to relation (1.5) the error in a Monte Carlo integration decreases as the inverse square root of the number of samplings M. In the case that the values $X^{(k)}$ are not independent, two major modifications are needed:

(a) The variance of the sample average is given by:

$$\sigma^2[\mu_M] = \frac{\sigma^2[G]}{M}(2\tau_G + 1) \tag{1.8}$$

where the *autocorrelation time* τ_G is given (in the limit of large M) by:

$$\tau_G = \sum_{k=1}^{\infty} \rho_G(k) \tag{1.9}$$

here the *normalized autocorrelation function* $\rho_G(k)$ is defined as:

$$\rho_G(k) = \frac{\langle G(X^{(k')})G(X^{(k+k')})\rangle - \langle G\rangle^2}{\sigma^2[G]} \tag{1.10}$$

The situation for non zero autocorrelation time can arise, for instance, when the value $X^{(k)}$ depends on the value $X^{(k-1)}$. In our poll example, this can occur if the k–th person to be questioned lives in the neighbourhood of the $(k-1)$–th person, such that their social classes are likely to be similar. Intuitively, τ_G measures the number of values of the sample $X^{(k)}, X^{(k+1)}, \ldots, X^{(k+\tau_G)}$ that we have to discard in order to consider that $X^{(k)}$ and $X^{(k+\tau_G)}$ are independent of each other. A problem of Monte Carlo methods is that, in many cases of interest, τ_G becomes very large and grows with the number of variables N.

(b) The relation $\langle G\rangle = \mu_M[G] \pm \sigma[\mu_M]$ has to be interpreted now according to Chebichev's theorem (Feller, 1971):

$$P(|\mu_M[G] - \langle G\rangle| > k\sigma[\mu_M]) > 1 - \frac{1}{k^2} \tag{1.11}$$

i.e. the probability that $\langle G\rangle$ lies in the interval $(\mu_M[G] - 2\sigma[\mu_M], \mu_M[G] + 2\sigma[\mu_M])$ is *at least* 75%, that it lies in $(\mu_M[G] - 3\sigma[\mu_M], \mu_M[G] + 3\sigma[\mu_M])$, *at least* 88.89%, etc.

The above Monte Carlo importance sampling procedure is one of the most powerful methods available to compute high–dimensional integrals. In order to completely specify the method, though, we need a procedure to generate the values of the samples $X^{(k)}$ that appear in the above formulae. These should be values of the N–dimensional variable $(x_1, \ldots, x_N)$ distributed according to the probability density function $f_{\hat{\mathbf{x}}}(X)$. There are several methods devised to generate the required values $X^{(k)}$. Before we can explain the very powerful methods available we need to study in detail some simpler cases.

1.2. Uniform sampling

Let us start with a simple example. Consider the one–dimensional (N=1) integral:

$$I = \int_0^1 dx \cos(x) \tag{1.12}$$

this is a particular case of Eq.(1.1) in which the function $f_{\hat{\mathbf{x}}}(x)$ is equal to 1 in the interval $(0,1)$ and 0 otherwise. The name **uniform sampling** comes from the fact that the variable x is equally likely to take any value in the interval $(0,1)$ (or, generally speaking, in any finite interval (a,b)). Although one could, in principle, devise some physical process that would produce values of a variable x uniformly distributed in the interval $(0,1)$, it is preferable, for many reasons, to use simple algorithms that can be programmed on a computer and that produce "random–type" (or *pseudo–random*) numbers uniformly distributed in the interval $(0,1)$. We will not discuss in these lectures the interesting topic of pseudo–random number generation. An excellent exposition can be found in references (Knuth, 1981) and (James, 1990). Let us simply assume that there exists a fortran function, which we will call *ran_u*(), which returns an independent random number uniformly distributed in the interval $(0,1)$. By using such a function, we can implement very easily the algorithm of uniform sampling:

```
      m=1000
      r=0.0
      s=0.0
      do 99 k=1,m
      x=ran_u()
      gk=g(x)
      r=r+gk
      s=s+gk*gk
99    continue
      r=r/m
      s=s/m-r*r
      s=sqrt(s/m)
      write(6,*) r,'+/-',s
      end
      function g(x)
      g=cos(x)
      return
      end
```

The reader should find out the name given to the function *ran_u*() is his own computer and run the above program checking that the error in the output value scales as the inverse square root of the number of samples M.

1.3. Importance sampling for N=1

In the N=1 case, when the variable $X = x$ has only one component, a simple and important theorem allows to generate $x^{(k)}$ very easily. Recall that the probability distribution function $F_{\hat{\mathbf{x}}}(x)$ is defined by integration of the probability density function $f_{\hat{\mathbf{x}}}(x)$ as:

$$F_{\hat{\mathbf{x}}}(x) = \int_{-\infty}^{x} f_{\hat{\mathbf{x}}}(y)\,dy \tag{1.13}$$

Theorem 1 *If $f_{\hat{x}}(x)$ is a one-variable probability distribution function, then the variable $u = F_{\hat{x}}(x)$ is uniformly distributed in the interval* $(0,1)$.

As a consequence, we have the

Corollary 1 *If u is uniformly distributed in the interval $(0,1)$, then $x = F_{\hat{x}}^{-1}(u)$ is distributed according to $f_{\hat{x}}(x)$.*

(The proof of this theorem is left as an exercise for the reader). This theorem reduces the problem of generating random numbers according to a given distribution $f_{\hat{x}}(x)$ to the inversion of the corresponding probability distribution function. If, for instance, we want to use the method of importance sampling to compute the integral

$$I = \int_0^\infty dx e^{-x} x^2 \tag{1.14}$$

we can take $f_{\hat{x}}(x) = e^{-x}$ if $x \geq 0$, and $G(x) = x^2$. This choice for $f_{\hat{x}}(x)$ respects positivity and normalization. To generate values of $x^{(k)}$ according to $f_{\hat{x}}(x)$ we need to invert $F_{\hat{x}}(x)$, which in this case can be done analytically:

$$F_{\hat{x}}(x) = \int_0^x dy e^{-y} = 1 - e^{-x} = u \rightarrow x = F_{\hat{x}}^{-1}(u) = -\log(1-u) \tag{1.15}$$

Since $1-u$ is also uniformly distributed in the interval $(0,1)$ we can write simply:

$$x = -\log(u) \tag{1.16}$$

which is equivalent to $x = -\log(1-u)$ from the statistical point of view. A program to compute integral (1.14) can be:

```
      m=1000
      r=0.0
      s=0.0
      do 99 k=1,m
      x=ran_f()
      gk=g(x)
      r=r+gk
      s=s+gk*gk
99    continue
      r=r/m
      s=s/m-r*r
      s=sqrt(s/m)
      write(6,*) r,'+/-',s
      end
      function g(x)
      g=x*x
      return
      end
```

```
function ran_f()
ran_f=-log(ran_u())
return
end
```

A technical point is that, in some cases, the function $F_{\hat{x}}^{-1}(u)$ is not expressible in terms of elementary functions and some kind of approximative methods (such as numerical interpolation) might be needed to compute it. It is important to realize that the decomposition of a given integral in $f_{\hat{x}}(x)$ and $G(x)$ to put it in the form of Eq.(1.3) might not be unique. In the previous example, Eq.(1.14), we could have taken as well $f_{\hat{x}}(x) = xe^{-x}$ and $G(x) = x$. This choice, however, makes the probability distribution function $F_{\hat{x}}(x)$ difficult to invert.

An important case in which the function $F_{\hat{x}}^{-1}(u)$ is not easily calculable is that of Gaussian random numbers (of mean zero and variance 1) for which the probability density function is:

$$f_{\hat{x}}(x) = \frac{1}{\sqrt{2\pi}} e^{-x^2/2} \tag{1.17}$$

(random numbers y of mean μ and variance σ^2 are easily obtained by the linear transformation $y = \sigma x + \mu$). The inverse probability distribution function is

$$x = \sqrt{2}\, \mathrm{erf}^{-1}(2u - 1) \tag{1.18}$$

where $\mathrm{erf}^{-1}(z)$ is the inverse error function (Abramowitz and Stegun, 1972; Valls and Mazenko, 1986). The inverse error function does not usually belong to the set of predefined functions in a programming language, although some libraries (for example, the NAG library) do include it in their list of functions. An alternative to the generation of Gaussian distributed random numbers is the algorithm of Box-Muller-Wiener (Box and Muller, 1958; Ahrens and Dieter, 1972; 1988) which is based on a change of variables to polar coordinates. Namely,

$$\begin{aligned} x_1 &= \sqrt{\rho}\cos(\theta) \\ x_2 &= \sqrt{\rho}\sin(\theta) \end{aligned} \tag{1.19}$$

If x_1 and x_2 are independent Gaussian variables, it is easy to prove that ρ and θ are independent variables. ρ follows an exponential distribution and θ a uniform distribution, so that their inverse distribution functions can be expressed in terms of elementary functions. Namely,

$$\begin{aligned} f_\rho(\rho) = \frac{1}{2}e^{-\rho/2} &\to F_\rho(\rho) = 1 - e^{-\rho/2}, \quad \rho \ge 0 \\ f_\theta(\theta) = \frac{1}{2\pi} &\to F_\theta(\theta) = \frac{\theta}{2\pi}, 0 \le \theta \le 2\pi \end{aligned} \tag{1.20}$$

The resulting algorithm is:

$$\begin{aligned} x_1 &= \sqrt{-2\log(u)}\cos(2\pi v) \\ x_2 &= \sqrt{-2\log(u)}\sin(2\pi v) \end{aligned} \tag{1.21}$$

Where u and v are independent random variables uniformly distributed in the interval $(0,1)$. The main advantage of this Box-Muller-Wiener algorithm is that it is exact, yielding two independent Gaussian variables, x_1, x_2 from two independent uniform variables, u, v. Its main disadvantage, though, is that it is extremely slow since it involves the calculation of trigonometric, logarithm and square root functions. In most of the applications a linear interpolation approximation to the inverse error function does produce sufficiently good quality Gaussian random numbers at a considerable gain in speed (Toral and Chakrabarti, 1993). A possible implementation of the Box-Muller-Wiener algorithm is the following:

```
function ran_g()
data is /-1/
is=-is
if (is.eq.1) then
        a=sqrt(-2.0*log(ran_u()))
        b=6.2831853072*ran_u()
        ran_g=a*cos(b)
        x2=a*sin(b)
        return
endif
ran_g=x2
return
end
```

(a practical note: it might be necessary to tell the compiler that the two functions *ran_u*() that appear in the previous program produce different values and need to be computed separately, sometimes compilers are too clever).

Another important case that deserves being considered explicitly, in despite of its simplicity, concerns the generation of events with a given probability. Imagine we want to simulate tossing a biased coin, with a $p = 0.6$ probability for heads (and 0.4 for tails). We need to generate a variable that takes some value (1, for example) 60% of the times and another value (0), 40% of the times. This can be done by comparing a random number u uniformly distributed in the interval $(0,1)$ with the given probability p. If $u \leq p$ then we take $x = 1$, otherwise we take $x = 0$. This is achieved by the program:

```
if (ran_u().lt.p) then
                    x=1
               else
                    x=0
endif
```

For N-dimensional variables $(x_1, \ldots, x_N)$, the situation is much more complex. The equivalent of Theorem 1 states that in order to generate values of the N–dimensional variable $(x_1, \ldots, x_N)$ we need (Rubinstein, 1981): (i) generate N-independent random numbers $(u_1, \ldots, u_N)$ uniformly distributed in the interval $(0,1)$ (this is the easy part) and (ii) solve the following set of equations:

$$
\begin{aligned}
F_{x_1}(x_1) &= u_1 \\
F_{x_2}(x_2|x_1) &= u_2 \\
&\cdots\cdots \\
F_{x_N}(x_N|x_1,\ldots,x_{N-1}) &= u_N
\end{aligned}
\tag{1.22}
$$

Where, for example, $F_{x_2}(x_2|x_1)$ is the conditional probability distribution function of the variable x_2 given that x_1 has taken a particular value, and so on. The calculation of the conditional probability distribution functions is generally at least as complicated as the calculation of the original integral we wanted to compute numerically and the above procedure is of little practical use. In order to develop alternative methods suitable for the generation of N-dimensional variables we need first to introduce the so–called rejection methods for 1 variable.

1.4. Rejection method for N=1

In those cases that the inverse probability distribution function is difficult to compute, the rejection method offers a very convenient alternative. Also, it is the basis of the N-dimensional methods which we will develop later. The method is based on the fact that the probability density function $f_{\hat{x}}(x)$ is proportional to the probability that the variable x takes a particular value. If, for example, $f_{\hat{x}}(x_1) = 2f_{\hat{x}}(x_2)$, we can affirm that the value x_1 is twice as probable as the value x_2. The rejection method (in its simplest version) proposes the values x_1 and x_2 with the same probability and then accepts the proposed value x with a probability $h(x)$ proportional to $f_{\hat{x}}(x)$, such that, in our example, x_1 will be accepted twice as many times as x_2. Consequently, x_1 will appear twice as many times as x_2, which is equivalent to saying that x_1 is twice as probable as x_2 as desired. We will illustrate the method by an example. Let us consider the probability density function:

$$
f_{\hat{x}}(x) = 6x(1-x), \qquad x \in (0,1) \tag{1.23}
$$

Which has the shape indicated in Fig.1.

We propose a value of x uniformly in the interval $(0,1)$. The proposed value has to be accepted with a probability $h(x)$ proportional to $f_{\hat{x}}(x)$, $h(x) = \alpha f_{\hat{x}}(x)$. The constant α is arbitrary but the resulting function $h(x)$ has to be interpreted as a probability, which means that we have to keep the bounds $0 \le h(x) \le 1$. This yields $\alpha \le \frac{1}{\max_x f_{\hat{x}}(x)}$. Obviously, the rejection method is more efficient when the acceptance probability is high, which means that one has to take the maximum possible value for α. In our example, the maximum for $f_{\hat{x}}(x)$ is at $x = 0.5$, $f_{\hat{x}}(0.5) = 1.5$, so that we take $\alpha = 2/3$ and, consequently, $h(x) = 4x(1-x)$, see Fig.1. The acceptance process is done as explained in the previous section by comparing the probability $h(x)$ with a random number uniformly distributed in $(0,1)$. The whole process is repeated until the proposed value is accepted. The final algorithm can be coded as:

```
function ran_f()
h(x)=4.0*x*(1.0-x)
```

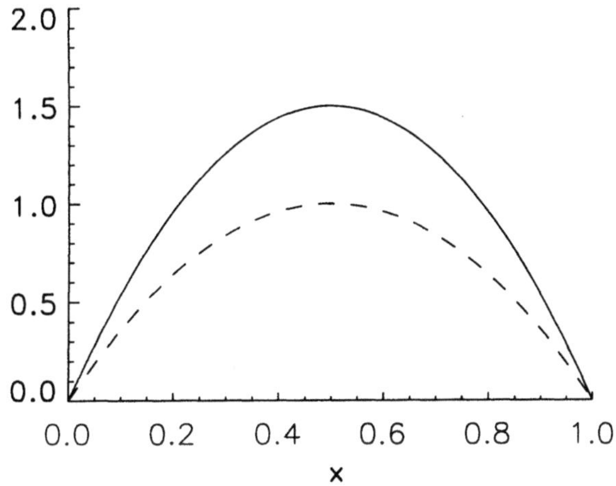

Fig. 1. Probability density function $f_{\hat{x}}(x) = 6x(1-x)$ (solid line) and acceptance probability $h(x) = 4x(1-x)$ (dashed line)

```
1     x=ran_u()
      if (ran_u().gt.h(x)) goto 1
      ran_f=x
      return
      end
```

We will now consider a more general version of the rejection method in which a value is proposed according to some probability distribution function $g_{\hat{x}}(x)$ and then accepted according to some probability $h(x)$. The probability distribution function of the joint process "propose the value x and accept it" is $g_{\hat{x}}(x)h(x)$. According to the Bayes theorem, the probability density function of the variable x *given that it has been accepted* is (Grimmett and Stirzaker, 1982):

$$f_{\hat{x}}(x) = \frac{g_{\hat{x}}(x)h(x)}{\int_{-\infty}^{\infty} dz g_{\hat{x}}(z)h(z)} \tag{1.24}$$

One interesting thing to notice is that the previous expression is properly normalized. This is very important in those cases where the normalization constant for $f_{\hat{x}}(x)$ is not known explicitly. Looking at Eq.(1.24) one can see that there is some freedom in choosing the functions $g_{\hat{x}}(x)$ and $h(x)$. The only requirement is that its product $g_{\hat{x}}(x)h(x)$ be proportional to the function $f_{\hat{x}}(x)$ we want to generate. An important constraint is that the proposal $g_{\hat{x}}(x)$ must be normalized and that the acceptance function $h(x)$ must be bound in the interval $(0,1)$. For instance, in the example of Eq.(1.23), we could have considered $g_{\hat{x}}(x) = 2x$. The function $h(x)$ should then be proportional to $3(1-x)$. Since $0 \le h(x) \le 1$ one takes simply $h(x) = 1 - x$. The generation of $g_{\hat{x}}(x) = 2x$ is straightforward since the inverse probability distribution function is $G_{\hat{x}}^{-1}(u) = \sqrt{u}$. The proposed value is accepted with probability $1 - x$, i.e. rejected with probability x. This can be coded as:

```
      function ran_f()
1     x=sqrt(ran_u())
      if(ran_u().lt.x) goto 1
      ran_f=x
      return
      end
```

An efficient method is one that requires, on average, a small average number of proposals before acceptance. This average number is given by the inverse of the overall acceptance probability, p_a which is nothing but the denominator of expression (1.24):

$$p_a \equiv p(accept) = \int_{-\infty}^{\infty} dz g_{\hat{x}}(z)h(z) \tag{1.25}$$

It is easy to show that of the two methods developed for the function (1.23) the first algorithm is more efficient than the second.

Let us consider yet another example: to generate numbers distributed according to

$$f_{\hat{x}}(x) = C \exp\left[-\frac{x^2}{2} - x^4\right] \tag{1.26}$$

the obvious choice is (notice that the precise value of C is not needed):

$$\begin{aligned} g_{\hat{x}}(x) &= \frac{1}{\sqrt{2\pi}} e^{-\frac{x^2}{2}} \\ h(x) &= e^{-x^4} \end{aligned} \tag{1.27}$$

Here $g_{\hat{x}}(x)$ is nothing but a Gaussian distribution, and $h(x)$ is the optimal choice, given the choice for $g_{\hat{x}}(x)$, since $\max_x h(x) = 1$. The overall acceptance probability is, according to Eq.(1.25):

$$p_a = \int_{-\infty}^{\infty} dz \frac{1}{\sqrt{2\pi}} e^{-\frac{z^2}{2}} e^{-z^4} \approx 0.62028 \tag{1.28}$$

And the average number of trials needed to accept a value is $1/0.62028 \approx 1.6$.

1.5. Rejection with repetition

There is a modification of the rejection method consisting in the following: if the proposed value is not accepted then, instead of proposing a new one, the previous value is repeated. This method obviously produces correlated values (some of them will be equal to the previous ones) but, however, can be very efficient for vector computers in which the usual structure of the rejection method prevents vectorization. We still have to prove that the values generated with this method do follow the required distribution $f_{\hat{x}}(x)$. According to Bayes theorem, the probability density function $f_{\hat{\mathbf{x}}_n}(x)$ generated at the n–proposal step will have two different contributions corresponding to the acceptance or rejection, respectively, of the proposed value:

$$f_{\hat{\mathbf{x}}_n}(x) = f_{\hat{\mathbf{x}}_n}(x|accept)p(accept) + f_{\hat{\mathbf{x}}_n}(x|reject)p(reject) \tag{1.29}$$

The first term of the right hand side can be written as the probability of acceptance given x (which is the function $h(x)$) times the probability density function of proposing x (which is $g_{\hat{\mathbf{x}}}(x)$):

$$f_{\hat{\mathbf{x}}_n}(x) = h(x)g_{\hat{\mathbf{x}}}(x) + f_{\hat{\mathbf{x}}_n}(x|reject)(1 - p_a) \tag{1.30}$$

If x has been rejected, the previous value is repeated which means that the probability density function at the n–step in the case of rejection is the same that the one at the $(n-1)$–step:

$$f_{\hat{\mathbf{x}}_n}(x) = h(x)g_{\hat{\mathbf{x}}}(x) + f_{\hat{\mathbf{x}}_{n-1}}(x)\left[1 - \int_{-\infty}^{\infty} dz h(z)g_{\hat{\mathbf{x}}}(z)\right] \tag{1.31}$$

where we have substituted p_a as given by Eq.(1.25). The solution of this linear recurrence relation is:

$$f_{\hat{\mathbf{x}}_n}(x) = (1 - p_a)^n \left[f_{\hat{\mathbf{x}}_0}(x) - \frac{h(x)g_{\hat{\mathbf{x}}}(x)}{\int_{-\infty}^{\infty} dz h(z)g_{\hat{\mathbf{x}}}(z)}\right] + \frac{h(x)g_{\hat{\mathbf{x}}}(x)}{\int_{-\infty}^{\infty} dz h(z)g_{\hat{\mathbf{x}}}(z)} \tag{1.32}$$

We can write this result in terms of the desired distribution $f_{\hat{\mathbf{x}}}(x)$, Eq. (1.24):

$$f_{\hat{\mathbf{x}}_n}(x) = (1 - p_a)^n \left[f_{\hat{\mathbf{x}}_0}(x) - f_{\hat{\mathbf{x}}}(x)\right] + f_{\hat{\mathbf{x}}}(x) \tag{1.33}$$

Given that $0 < p_a \leq 1$, one concludes that the solution tends to the desired distribution in the limit $n \to \infty$ independently of the initial distribution $f_{\hat{\mathbf{x}}_0}$:

$$lim_{n\to\infty} f_{\hat{\mathbf{x}}_n}(x) = f_{\hat{\mathbf{x}}}(x) \tag{1.34}$$

If the initial proposed numbers x_0 are distributed according to $f_{\hat{\mathbf{x}}}(x)$, i.e. if $f_{\hat{\mathbf{x}}_0}(x) = f_{\hat{\mathbf{x}}}(x)$, then we would have $f_{\hat{\mathbf{x}}_n}(x) = f_{\hat{\mathbf{x}}}(x)$, $\forall n$ and the numbers we obtain are distributed according to the desired distribution from the beginning. However, if the initial numbers are not distributed according to $f_{\hat{\mathbf{x}}}(x)$ (as it is usually the case) we can still affirm that, thanks to the factor $(1-p)^n$, the initial condition will be lost after a number sufficiently large of proposals. This process of rejecting the, say, M_0 initial values produced because they are not distributed yet according to the required distribution is called *thermalization*. The initial time needed for thermalization can be computed from the non-linear correlation function (Binder and Heermann, 1988):

$$\rho_x^{NL}(k) = \frac{\langle x^{(k)}\rangle - \langle x\rangle}{\langle x^{(0)}\rangle - \langle x\rangle} \tag{1.35}$$

The time M_0 can be measured as a characteristic time for the decay of the above function towards the stationary value 0. For the method of rejection with repetition the non-linear correlation function coincides with the auto-correlation function (see later):

$$\rho_x^{NL}(k) = (1-p_a)^k \tag{1.36}$$

The thermalization time M_0 can be defined as the time it takes the above function to reach a small value ϵ.

$$M_0 = \frac{\ln \epsilon}{\ln(1-p_a)} \tag{1.37}$$

If, for instance, $\epsilon = 10^{-8}$, $p_a = 0.5$, we have $M_0 \approx 27$.

We can illustrate now the enhancement of the errors due to the correlations. In this method of rejection with repetition, one can compute the correlation function of the variable x as (see Eq.(1.10)):

$$\rho_x(k) = \lim_{k' \to \infty} \frac{\langle x^{(k)} x^{(k+k')} \rangle - \langle x \rangle^2}{\langle x^2 \rangle - \langle x \rangle^2} = (1-p_a)^k \tag{1.38}$$

From Eq.(1.9) we deduce that the autocorrelation time is:

$$\tau_x = \sum_{k=1}^{\infty} \rho_x(k) = \frac{1-p}{p} \tag{1.39}$$

Remember that, according to Eq.(1.8) the statistical error gets enhanced by the factor $(2\tau_x + 1)^{1/2}$.

We finally mention that it is possible to interpret this algorithm of rejection with repetition as a Markov succession (Grimmett and Stirzaker, 1982). Indeed, Eq.(1.31) can be cast in the form:

$$f_{\hat{\mathbf{x}}_n}(x) = \int_{-\infty}^{\infty} f(x|y) f_{\hat{\mathbf{x}}_{n-1}}(y)\, dy \tag{1.40}$$

where the transition probability is:

$$f(x|y) = h(x) g_{\hat{\mathbf{x}}}(x) + \left[1 - \int_{-\infty}^{\infty} dz\, h(z) g_{\hat{\mathbf{x}}}(z)\right] \delta(x-y) \tag{1.41}$$

1.6. The algorithm of Metropolis et al.

Although it is very powerful, the rejection method can fail if the distribution is very peaked around an (unknown) value x_0 (see Fig.2). If our proposal does not take into account that only a very small part of the variable space (that around x_0) is important, we are bound to have a extremely high rejection probability yielding the rejection method completely useless. The solution comes by making a proposal that does not differ much of the present value of the random variable we are sampling, i.e. by having the transition probabilities of the Markov succession depend explicitly on the present value, say y, of the variable. This implies that the functions $h(x)$ and $g_{\hat{\mathbf{x}}}(x)$ depend on y and become two-variable functions, $g_{\hat{\mathbf{x}}}(x|y)$, $h(x|y)$. The transition probability Eq.(1.41) then becomes also a two variable function:

$$f(x|y) = h(x|y)g_{\hat{\mathbf{x}}}(x|y) + \left[1 - \int_{-\infty}^{\infty} h(z|y)g_{\hat{\mathbf{x}}}(z|y)\,dz\right]\delta(x-y) \tag{1.42}$$

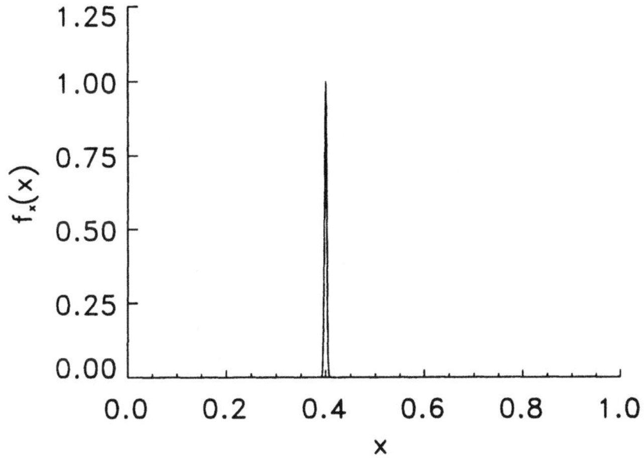

Fig. 2. Function $f_{\hat{\mathbf{x}}}(x)$ for which the rejection method can fail

The equation giving the evolution of the probability density function at the n-step is still Eq. (1.40), but now the recursion equation for $f_{\hat{\mathbf{x}}_n}(x)$ does not tell us in a straight way which is the stationary ($n \to \infty$) distribution. We would like to have $f_{\hat{\mathbf{x}}}(x)$ as stationary solution of the recursion equation, i.e:

$$f_{\hat{\mathbf{x}}}(x) = \int_{-\infty}^{\infty} f(x|y) f_{\hat{\mathbf{x}}}(y)\,dy \tag{1.43}$$

By using the normalization condition $\int_{-\infty}^{\infty} f(y|x)\,dy = 1$ and the definition (1.42) for the transition probability $f(x|y)$, one gets easily:

$$\int dy\,[g_{\hat{\mathbf{x}}}(x|y)h(x|y)f_{\hat{\mathbf{x}}}(y) - g_{\hat{\mathbf{x}}}(y|x)h(y|x)f_{\hat{\mathbf{x}}}(x)] = 0 \tag{1.44}$$

A sufficient condition to fulfill this relation is given by the *detailed balance condition*:

$$g_{\hat{\mathbf{x}}}(x|y)h(x|y)f_{\hat{\mathbf{x}}}(y) = g_{\hat{\mathbf{x}}}(y|x)h(y|x)f_{\hat{\mathbf{x}}}(x) \tag{1.45}$$

to be satisfied by the proposal $g_{\hat{\mathbf{x}}}(x|y)$ and the acceptance probability $h(x|y)$. Let us remark that if the detailed balance condition is satisfied and if $f_{\hat{\mathbf{x}}_0}(x) = f_{\hat{\mathbf{x}}}(x)$ then $f_{\hat{\mathbf{x}}_n}(x) = f_{\hat{\mathbf{x}}}(x)$, $\forall n$. Also, if the transition probabilities satisfy the condition of *ergodicity*, which, roughly speaking, means that any value for the random variable can be achieved starting from any other value after a sufficient number of steps, then we can assure that $f_{\hat{\mathbf{x}}}(x)$ is the only stationary solution of the

Markov succession and that $\lim_{n\to\infty} f_{\hat{\mathbf{x}}_n}(x) = f_{\hat{\mathbf{x}}}(x)$, independently of the value of $f_{\hat{\mathbf{x}}_0}(x)$ (see Kalos and Whitlock, 1986).

It is important to stress that any functions $g_{\hat{\mathbf{x}}}(x|y)$ and $h(x|y)$ that satisfy ergodicity and the detailed balance condition are suitable to use in this rejection method. The most widely used solution is the one given by the **the Metropolis algorithm** (Metropolis et al., 1953) in which $g_{\hat{\mathbf{x}}}(x|y)$ is a given function usually of the form:

$$g_{\hat{\mathbf{x}}}(x|y) = \frac{1}{2\Delta}, \qquad \text{if} \quad |x-y| \leq \Delta \tag{1.46}$$

i.e. x is sample uniformly from the interval $(y-\Delta, y+\Delta)$. Once the value of x has been proposed one looks for acceptance probabilities $h(x|y)$ verifying the detailed balance condition:

$$\frac{h(x|y)}{h(y|x)} = \frac{g_{\hat{\mathbf{x}}}(y|x) f_{\hat{\mathbf{x}}}(x)}{g_{\hat{\mathbf{x}}}(x|y) f_{\hat{\mathbf{x}}}(y)} \equiv q(x|y) \tag{1.47}$$

When searching for solutions of these equations one has to remember that $h(x|y)$ is a probability and must also satisfy the condition $0 \leq h(x|y) \leq 1$. A possible solution is the *Metropolis solution*:

$$h(x|y) = \min\left(1, q(x|y)\right) \tag{1.48}$$

Another widely used solution is the *Glauber solution* (Glauber, 1963):

$$h(x|y) = \frac{q(x|y)}{1+q(x|y)} \tag{1.49}$$

A family of solutions is obtained by:

$$h(x|y) = \sqrt{q(x|y)}\,\omega\left(q(x|y)\right) \tag{1.50}$$

where $\omega(z)$ is any function satisfying

$$\omega(z) = \omega(z^{-1}) \tag{1.51}$$

In particular, the Metropolis solution is recovered taking:

$$\omega(z) = \min(z, z^{-1}) \tag{1.52}$$

and the Glauber solution taking

$$\omega(z) = \frac{1}{z^{1/2}+z^{-1/2}} \tag{1.53}$$

Before we use the Metropolis algorithm to study some models of field theory, we want to finish this section by showing the use of the Metropolis algorithm with a simple example in a one variable case. We will consider the Gaussian distribution:

$$f_{\hat{\mathbf{x}}}(x) = Ae^{-x^2/2} \tag{1.54}$$

the normalization constant A is irrelevant for the Metropolis method. The proposal will be a value belonging to the interval $(y-\Delta, y+\Delta)$. Notice that this choice (Eq.(1.46)) satisfies the symmetry relation $g_{\hat{x}}(x|y) = g_{\hat{x}}(y|x)$. This proposal will be accepted with a probability given, for example, by the Metropolis solution Eq.(1.48):

$$h(x|y) = \min\left(1, q(x|y)\right) = \min(1, e^{(y^2-x^2)/2}) \tag{1.55}$$

i.e if $|x| \leq |y|$ accept with probability 1, otherwise accept with probability $e^{(y^2-x^2)/2}$. The Metropolis algorithm can be coded as:

```
function ran_g(y,delta)
x=y+delta*(2*ran_u()-1)
if(abs(x).gt.abs(y)) then
    if (exp(0.5*(y-x)*(y+x)).lt.ran_u()) then
                    ran_g=y
                    return
    endif
endif
y=x
ran_g=y
return
end
```

Intuitively, the effect of this algorithm is to accept with a larger probability those proposals which tend towards the origin, where the probability has a maximum.

An important warning concerns the fact that we must keep the old value of the variable if the new one has been rejected. We can not keep proposing until a value is accepted (at variance with what happened in the ordinary rejection method in which the two procedures were correct). If one does not keep the old value, a reasoning similar to the one used in the rejection method, section 1.4, leads to the following transition probabilities, see Eq.(1.24):

$$f(x|y) = \frac{h(x|y)g(x|y)}{\int_{-\infty}^{\infty} dz h(z|y)g(z|y)} \tag{1.56}$$

and:

$$f(y|x) = \frac{h(y|x)g(y|x)}{\int_{-\infty}^{\infty} dz h(z|x)g(z|x)} \tag{1.57}$$

Now it is very difficult to satisfy the detailed balance conditions, Eq.(1.45), namely:

$$\frac{h(x|y)g(x|y)}{\int_{-\infty}^{\infty} dz h(z|y)g(z|y)} f_{\hat{x}}(y) = \frac{h(y|x)g(y|x)}{\int_{-\infty}^{\infty} dz h(z|x)g(z|x)} f_{\hat{x}}(x) \tag{1.58}$$

since the integrals are generally difficult to compute (in particular, the Metropolis solution does not satisfy this condition). The reader can test the following modification of the Metropolis algorithm for the Gaussian distribution:

```
      function ran_g(y,delta)
1     x=y+delta*(2*ran_u()-1)
      if(abs(x).gt.abs(y)) then
         if (exp(0.5*(y-x)*(y+x)).lt.ran_u()) goto 1
      endif
      y=x
      ran_g=y
      return
      end
```

and check that it does not produce Gaussian random numbers (it is worse for large Δ).

For efficiency of the Metropolis algorithm Δ must be chosen such that the autocorrelation time of the algorithm, τ, takes its minimum value. It is easy to understand why there must be a minimum of τ as a function of Δ. If Δ is very small, the new proposal will be very close to the old one (hence highly correlated), τ will be large and $h(x|y)$ will be very close to 1 such that the acceptance will be also close to 1. On the other hand, if Δ is large, the acceptance probability will be small and the proposed value will be often rejected, such that the new value will be equal to the old one (again highly correlated). A rule of thumb tells us that Δ must be chosen such that the acceptance probability is neither too high, neither too small, i.e. of the order of 50%.

1.7. Rejection method in the N–dimensional case

We really do not need to develop any new concepts. The nice thing about the Metropolis algorithm is that is easily generalizable to the N-dimensional case. Simply replace x and y in the above expressions by the N-dimensional variable vectors $(x_1, \dots, x_N)$ and $(y_1, \dots, y_N)$, respectively, and all of the above formulae and concepts will still apply. In order to keep the acceptance probability within reasonable limits, the proposal can not be too different from the old value. This is achieved by changing only one of the N variables at one time. The variable to be updated can be chosen randomly amongst the set of the N variables or sequentially. If chosen randomly the function $g_{\hat{\mathbf{x}}}(x|y)$ is explicitly given by:

$$g_{\hat{\mathbf{x}}}(x|y) = \frac{1}{N} \sum_{i=1}^{N} \frac{1}{2\Delta} \prod_{j \neq i} \delta(x_j - y_j) \tag{1.59}$$

Since it verifies the symmetry condition $g_{\hat{\mathbf{x}}}(x|y) = g_{\hat{\mathbf{x}}}(y|x)$, the function $q(x|y)$ is simply $q(x|y) = f_{\hat{\mathbf{x}}}(x)/f_{\hat{\mathbf{x}}}(y)$. The acceptance probability can be chosen as the Metropolis solution $h(x|y) = \min(1, q(x|y))$ or other solution.

1.8. Heat bath

In the Metropolis algorithm the acceptance probability $h(x|y)$ was determined (with some freedom) once the proposal $g_{\hat{\mathbf{x}}}(x|y)$ had been specified. In the so called **heat bath** algorithm, which is useful for N-dimensional variables, the proposed value for the variable is always accepted, i.e. one chooses

$$h(x|y) = 1 \tag{1.60}$$

The detailed balance condition becomes

$$g_{\hat{\mathbf{x}}}(x|y)f_{\hat{\mathbf{x}}}(y) = g_{\hat{\mathbf{x}}}(y|x)f_{\hat{\mathbf{x}}}(x) \tag{1.61}$$

In the 1-variable case, a trivial solution (remember that $g_{\hat{\mathbf{x}}}(x|y)$ must be a probability density function) is: $g_{\hat{\mathbf{x}}}(x|y) = f_{\hat{\mathbf{x}}}(x)$ which is independent of the old value y. In the N-variable case, a solution is found by changing only one of the N variables at a time:

$$g_{\hat{\mathbf{x}}}(x|y) = \frac{1}{N}\sum_{i=1}^{N} g_{\hat{\mathbf{x}}}(x_i|y)\prod_{j\neq i}\delta(x_j - y_j) \tag{1.62}$$

where each of the functions $g_{\hat{\mathbf{x}}}(x_i|y)$ satisfies detailed balance:

$$\begin{aligned} &g_{\hat{\mathbf{x}}}(x_i|y)f_{\hat{\mathbf{x}}}(x_1,\ldots,x_{i-1},y_i,x_{i+1},\ldots,x_N) = \\ &g_{\hat{\mathbf{x}}}(y_i|x)f_{\hat{\mathbf{x}}}(x_1,\ldots,x_{i-1},x_i,x_{i+1},\ldots,x_N) \end{aligned} \tag{1.63}$$

A solution of this functional equation is obtained by taking $g_{\hat{\mathbf{x}}}(x_i|y) = g_{\hat{\mathbf{x}}}(x_i)$, independent of y, as the following conditional probability:

$$g_{\hat{\mathbf{x}}}(x_i) = f_{\hat{\mathbf{x}}}(x_i|x_1,\ldots,x_{i-1},x_{i+1},\ldots,x_N) \tag{1.64}$$

It is trivial to verify detailed balance, Eq.(1.63), if one remembers the definition of the conditional probability:

$$f_{\hat{\mathbf{x}}}(x_i|x_1,\ldots,x_{i-1},x_{i+1},\ldots,x_N) = \frac{f_{\hat{\mathbf{x}}}(x_1,\ldots,x_{i-1},x_i,x_{i+1},\ldots,x_N)}{f_{\hat{\mathbf{x}}}(x_1,\ldots,x_{i-1},x_{i+1},\ldots,x_N)} \tag{1.65}$$

Intuitively, what the heat bath method does when updating variable x_i, is to select the value of x_i according to the conditional probability given that the rest of the variables is fixed. These fixed variables act a a *heat bath* for variable x_i. In many cases, it turns out that the conditional probability $g_{\hat{\mathbf{x}}}(x_i) = f_{\hat{\mathbf{x}}}(x_i|x_1,\ldots,x_{i-1},x_{i+1},\ldots,x_N)$ takes a simple form and can be sampled by any of the 1-variable methods explained in this chapter.

2. Φ^4 Model

2.1. Introduction and basic definitions

We will illustrate the use of Monte Carlo techniques in field theory with the scalar Φ^4 model. This model has been used in many different contexts, in such different topics as quantum field theory (Zinn Justin, 1989) or in the study of structural phase transitions (Cowley, 1980; Bruce, 1980; Bruce and Cowley, 1980).

Let us consider a d-dimensional (hyper-cubic) regular lattice Λ, consisting of $N = L^d$ points. Every point $i = 1, \ldots, N$ of this lattice has d coordinates: $i = (j_1, \ldots, j_d)$. Periodic boundary conditions are assumed on this lattice. On every site of the lattice there is a scalar variable Φ_i. The set of all variables is $[\Phi] \equiv (\Phi_1, \ldots, \Phi_N)$. We also introduce a Hamiltonian function $\mathcal{H}$ given by:

$$\beta\mathcal{H}([\Phi]) = \sum_{i=1}^{N} a_0^d \left[\frac{-b}{2}\Phi_i^2 + \frac{u}{4}\Phi_i^4 + \frac{K}{2}\sum_{\mu=1}^{d}\left(\frac{\Phi_{i_\mu} - \Phi_i}{a_0}\right)^2 \right] \tag{2.1}$$

The sum over μ runs over d-nearest neighbours of site i, i.e. if the coordinates of i are $(j_1, \ldots, j_d)$ then the coordinates of i_μ are $(j_1, \ldots, j_\mu + 1, \ldots, j_d)$. The continuum limit of the system can be obtained by letting the lattice spacing a_0 tend to 0 and the system size L to ∞. In this limit, the sums are replaced by integrals and the sum over μ tends to a gradient. The continuum Hamiltonian is then:

$$\beta\mathcal{H}([\Phi]) = \int d\boldsymbol{r} \left[\frac{-b}{2}\Phi(\boldsymbol{r})^2 + \frac{u}{4}\Phi(\boldsymbol{r})^4 + \frac{K}{2}|\nabla\Phi(\boldsymbol{r})|^2 \right] \tag{2.2}$$

This expression for the Hamiltonian is the preferred one for analytical treatment, such as series expansions or renormalization–group treatments in momentum space (Amit, 1984). The lattice version is the preferred one for numerical studies, besides the fact that it yields a regularization of the Hamiltonian (Parisi, 1988). We will consider from now on only the lattice version of the Φ^4 model. We can set the lattice spacing $a_0 = 1$, since it can be rescaled in the parameters b, u and K.

The first two terms of the sum appearing in the Hamiltonian (2.1) are local terms (depending only on the field at location i) and can be thought of as local potential terms $V(\Phi_i)$:

$$V(\Phi_i) = \frac{-b}{2}\Phi_i^2 + \frac{u}{4}\Phi_i^4 \tag{2.3}$$

For stability of this local potential, the parameter u must be greater than 0. The local potential changes qualitative when the parameter b changes sign (see Fig.3). If $b < 0$ the local potential has only one minimum at $\Phi_i = 0$. On the other hand, when $b > 0$ there are two minima of equal depth located at $\Phi_i = \pm\sqrt{b/u}$.

The third term in (2.1), the one multiplied by K, is called the interaction term. When $K > 0$, the contribution of this term is always non negative and the ground state of the Hamiltonian $\mathcal{H}$ is reached when the contribution of this term is zero, i.e. when all the fields take the same values $\Phi_i = \Phi_0, \forall i$. This is the situation considered by the simple **mean field** theory (Landau and Lifshitz,

1980). Since all the fields take the same value, the mean field Hamiltonian reduces to:

$$N^{-1}\beta\mathcal{H}^{MF}(\Phi_0) = \frac{-b}{2}\Phi_0^2 + \frac{u}{4}\Phi_0^4 \tag{2.4}$$

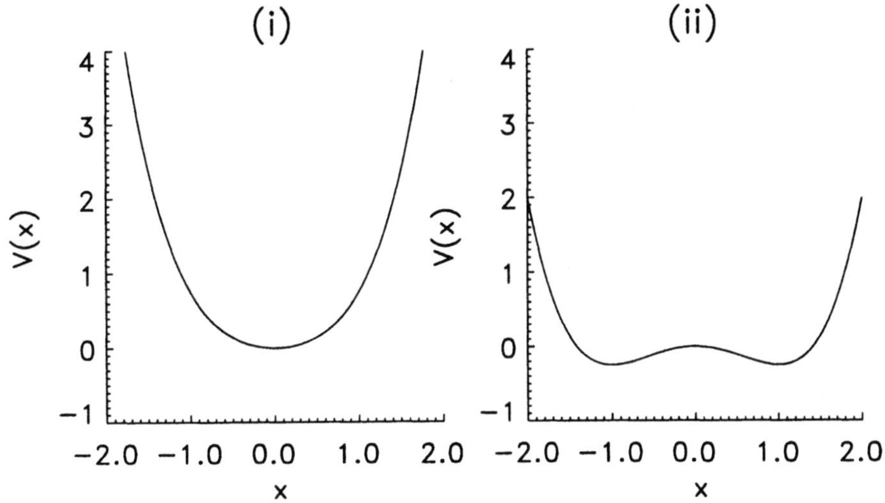

Fig. 3. Local Potential $V(\Phi)$, Eq.(2.3), in the cases (i) $b > 0$ and (ii) $b < 0$

The value of the mean field value Φ_0 depends on the constants b, u and is given by:

$$\Phi_0 = \begin{cases} 0 & \text{if} \quad b < 0 \\ \sqrt{\frac{b}{u}} & \text{if} \quad b > 0 \end{cases} \tag{2.5}$$

(see Fig.4). To go beyond mean field theory one needs to consider fluctuations: the fields Φ_i do not take a constant value, but fluctuate from one lattice site to another. In this case, we need a statistical description which, according to the principles of Statistical Mechanics (Pathria, 1972), is given by the canonical distribution. The probability density function $f([\Phi])$ governing the statistical properties of the fields $[\Phi]$ at inverse temperature β is:

$$f([\Phi]) = \frac{e^{-\beta\mathcal{H}[\Phi]}}{\mathcal{Z}(b,u,K)} \tag{2.6}$$

The denominator of this expression is called the **partition function** and it is the multiple integral of the numerator for all the field values:

$$\mathcal{Z}(b,u,K) = \int_{-\infty}^{\infty} d\Phi_1 \ldots \int_{-\infty}^{\infty} d\Phi_N e^{-\beta\mathcal{H}([\Phi])} \equiv \int d\Phi e^{-\beta\mathcal{H}([\Phi])} \tag{2.7}$$

The magnitudes of interest are computed as averages of field functions, $G([\Phi])$, with the probability density function $f([\Phi])$:

$$\langle G([\Phi])\rangle = \mathcal{Z}^{-1}\int d\Phi G([\Phi])e^{-\beta\mathcal{H}([\Phi])} \tag{2.8}$$

Example of quantities of interest are the *magnetization*, m defined as:

$$m = \langle\frac{1}{N}\sum_{i=1}^{N}\Phi_i\rangle \tag{2.9}$$

the energy, e:

$$e = \left\langle\frac{\mathcal{H}[\Phi]}{N}\right\rangle \tag{2.10}$$

the magnetic susceptibility, κ_T:

$$\kappa_T = N\left[\langle\left(\frac{1}{N}\sum_{i=1}^{N}\Phi_i\right)^2\rangle - \langle m\rangle^2\right] \tag{2.11}$$

the specific heat, C_V:

$$C_V = N\left[\langle\left(\frac{\mathcal{H}[\Phi]}{N}\right)^2\rangle - \langle e\rangle^2\right] \tag{2.12}$$

and many others.

In general, these quantities differ considerably from the mean field values. In Figure 4 we can see the difference between the magnetization computed from a numerical calculation in two dimensions and the mean field result (which is independent of dimension).

It is known that the Φ^4 model belongs to the universality class of the Ising model (Amit, 1984). This means that both models share the same critical exponents and scaling functions. Here, however, we want to point out a different relation between the two models, namely, that the Ising model can be obtained as a suitable limit of the Φ^4 model. The limit is obtained as follows: expanding the square in the "gradient" term and using the periodic boundary conditions, one can rewrite the Hamiltonian in the form:

$$\beta\mathcal{H}([\Phi]) = \sum_{i=1}^{N}\left[\left(dK - \frac{b}{2}\right)\Phi_i^2 + \frac{u}{4}\Phi_i^4\right] - K\sum_{\langle i,j\rangle}\Phi_i\Phi_j \tag{2.13}$$

where the sum over $\langle i,j\rangle$ means sum over all the possible pairs of i–j nearest neighbours in the lattice. Introducing a rescaled field variable

$$S_i = \frac{\Phi_i}{\sqrt{\frac{b-2dK}{u}}} \tag{2.14}$$

one can write:

$$\beta\mathcal{H}([S]) = \alpha\sum_{i=1}^{N}\left[-\frac{1}{2}S_i^2 + \frac{1}{4}S_i^4\right] - \beta J\sum_{<i,j>}S_iS_j \tag{2.15}$$

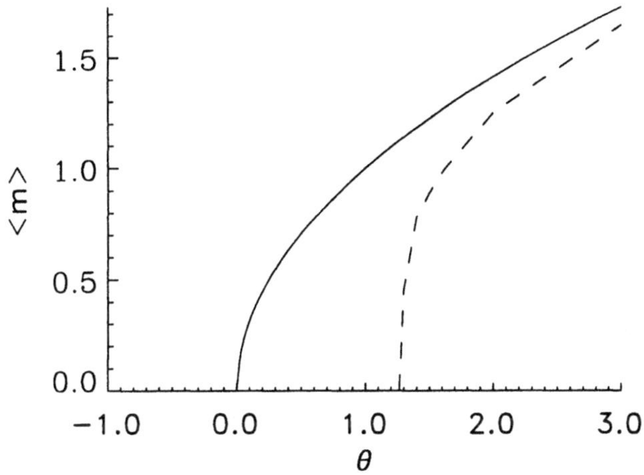

Fig. 4. Mean field magnetization, Eq.(2.5) (solid line) and the result of a numerical calculation using Monte Carlo techniques in the two-dimensional system (dashed line). In this plot we use the value $\chi = 1$ (see Eq.(2.21) for a definition of the parameters).

with

$$\alpha = \frac{(-b + 2dK)^2}{u}, \qquad \beta J = \frac{K(b - 2dK)}{u} \tag{2.16}$$

The Gibbs factor is

$$e^{-\beta \mathcal{H}([S])} = \prod_{i=1}^{N} \left\{ \exp\left[-\alpha(-\frac{1}{2}S_i^2 + \frac{1}{4}S_i^4)\right] \exp\left[\beta J \sum_{<i,j>} S_i S_j\right] \right\} \tag{2.17}$$

The local factor of this distribution is always double peaked centered around the values ± 1 (see Fig.5). Now we can think of the parameter α as controlling the width of the local field distribution around these two maxima. A large value for α implies that the field is bound to take essentially the values ± 1, and S_i becomes what is called a *spin variable*. In the limit $\alpha \to \infty$ with J kept constant we recover the Ising model. This limit is obtained in the original Φ^4 model by taking

$$\begin{aligned} &b \to \infty \\ &u \to \infty \\ &\frac{b}{u} \to \text{constant} \qquad \beta J = K\frac{b}{u} \end{aligned} \tag{2.18}$$

In this limit, one obtains for the Gibbs factor:

$$e^{-\beta \mathcal{H}([S])} = \prod_{i=1}^{N} \left[\delta(S_i^2 - 1)\right] \exp\left[\beta J \sum_{<i,j>} S_i S_j\right] \tag{2.19}$$

This is equivalent to writing:

$$\mathcal{H}([S]) = -J \sum_{<i,j>} S_i S_j \qquad\qquad S_i = \pm 1 \tag{2.20}$$

Which is nothing but the Hamiltonian of the Ising model. For smaller values of α, the field can fluctuate around the two maxima, hence the name of *soft spins model* that sometimes is given to the Φ^4 model in this situation (Cowley, 1980; Bruce, 1980; Bruce and Cowley, 1980).

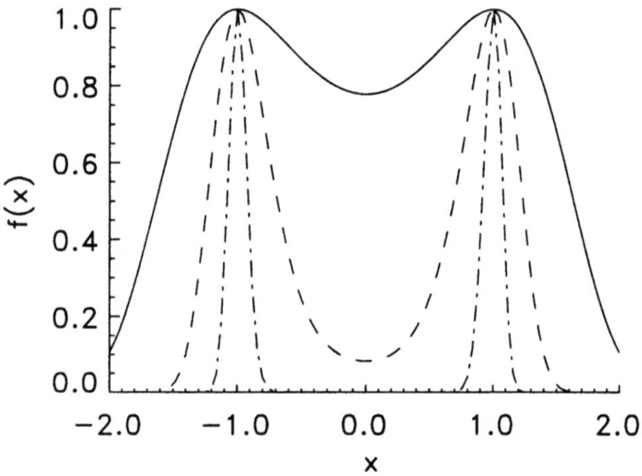

Fig. 5. Function $f(x) = \exp\left[-\alpha(-\frac{1}{2}S_i^2 + \frac{1}{4}S_i^4)\right]$ appearing in Eq.(2.17), in the cases $\alpha = 1$ (solid line), $\alpha = 10$ (dashed line), $\alpha = 100$ (dotted–dashed line). When $\alpha \to \infty$, the function $f_{\hat{x}}(x)$ tends to the sum of two delta functions located at $x = \pm 1$.

2.2. Monte Carlo methods

Before we proceed further and show how one can obtain numerically the quantities of interest in the Φ^4 model, we notice that one of the three parameters of the model, u, b, K, is redundant since it can be absorbed in the field scale. Many ways of redefining parameters have been used in the literature (see (Toral and Chakrabarti, 1990) for a review). We will use here a simple reparametrization in which the field is rescaled by a factor $K^{1/2}$ hence yielding a parameter independent interaction term (this rescaling is obviously valid only in the ferromagnetic case $K > 0$). Specifically, we introduce a new field ϕ, and two new parameters θ and χ by the definitions:

$$\begin{aligned} \phi &= K^{1/2}\Phi \\ \theta &= \frac{b}{K} \\ \chi &= \frac{u}{K^2} \end{aligned} \tag{2.21}$$

The Hamiltonian, in terms of these new variables, can be written as:

$$
\begin{aligned}
\beta\mathcal{H}([\phi]) &= \sum_{i=1}^{N}\left[\frac{-\theta}{2}\phi_i^2 + \frac{\chi}{4}\phi_i^4 + \frac{1}{2}\sum_{\mu=1}^{d}\left(\phi_{i_\mu} - \phi_i\right)^2\right] \\
&= \sum_{i=1}^{N}\left[\frac{\tilde{\theta}}{2}\phi_i^2 + \frac{\chi}{4}\phi_i^4 - \sum_{\mu=1}^{2d}\phi_i\phi_{i_\mu}\right]
\end{aligned}
\tag{2.22}
$$

(the sum over μ runs over the $2d$ nearest neighbours of site i). Here we have introduced

$$
\tilde{\theta} = 2d - \theta \tag{2.23}
$$

This Hamiltonian can be explicitly separated into the local part and the interaction part, $\beta\mathcal{H} = \beta\mathcal{H}_0 + \beta\mathcal{H}_I$ with

$$
\beta\mathcal{H}_0 = \sum_{i=1}^{N}\left[\frac{\tilde{\theta}}{2}\phi_i^2 + \frac{\chi}{4}\phi_i^4\right] \tag{2.24}
$$

$$
\beta\mathcal{H}_I = -\sum_{<i,j>}\phi_i\phi_j \tag{2.25}
$$

Now it is about time we start applying what we learnt in the previous chapter about Monte Carlo techniques! To implement the Metropolis algorithm we need to take the following steps:

(i) Select (randomly or sequentially) one of the N field variables, ϕ_i.
(ii) Propose a change to another close value, ϕ'_i, chosen randomly in the interval $\phi_i - \Delta, \phi_i + \Delta$, with Δ a suitable value selected such that the acceptance probability is around 0.5.
(iii) Compute the change in energy, $\Delta\mathcal{H}$ that this proposed change produces.
(iv) Accept the proposed value ϕ'_i with probability $\min(1, e^{-\beta\Delta\mathcal{H}})$.

In computing the change of energy one does not have to use the full expression (2.22), but rather notice that most of the terms disappear when subtracting the old and the new energy, such that the change in energy is simply:

$$
\beta\Delta\mathcal{H} = \frac{\tilde{\theta}}{2}({\phi'}_i^2 - \phi_i^2) + \frac{\chi}{4}({\phi'}_i^4 - \phi_i^4) + ({\phi'}_i - \phi_i)\sum_{\mu=1}^{2d}\phi_{i_\mu} \tag{2.26}
$$

This simple Metropolis algorithm has some convergence problems due to the fact that most of the trials belong to a small probability region. This is particularly true in the vicinities of the Ising limit in which the values around 0 have a very small probability (see again Fig.5). Several methods have been proposed as an alternative to the Metropolis algorithm and all of these methods fit into the proposal/acceptance scheme developed in the previous chapter: use a proposal probability density function $g_{\hat{\mathbf{x}}}(x|y)$ (where $x \equiv (\phi'_1, \ldots, \phi'_N)$ and $y \equiv (\phi_1, \ldots, \phi_N)$ stand for the complete field configuration after and before the proposal, respectively) and an acceptance probability $h(x|y)$, satisfying detailed balance:

$$
g_{\hat{\mathbf{x}}}(x|y)h(x|y)\frac{e^{-\beta\mathcal{H}(y)}}{\mathcal{Z}} = g_{\hat{\mathbf{x}}}(y|x)h(y|x)\frac{e^{-\beta\mathcal{H}(x)}}{\mathcal{Z}} \tag{2.27}
$$

As usual, only one variable will be updated at a time, what makes the proposal $g_{\hat{\mathbf{x}}}(x|y)$ become a function $g_{\hat{\mathbf{x}}}(\phi'_i|y)$. In the approach of reference (Milchev et al., 1986) the proposal $g_{\hat{\mathbf{x}}}({\phi'}_i|y)$ is chosen independent of y and proportional to the Gibbs factor of the local term of the Hamiltonian:

$$g_{\hat{\mathbf{x}}}({\phi'}_i) = C\exp(-\frac{\tilde{\theta}}{2}{\phi'}_i^2 - \frac{\chi}{4}{\phi'}_i^4) \tag{2.28}$$

this is a one-variable probability density function that can be sampled by any of the methods explained in chapter 1. The authors in reference (Milchev et al., 1986) chose a numerical inversion method. Once this proposal has been taken, in the detailed balance condition only the interaction part of the Hamiltonian appears explicitly,

$$h(x|y)e^{-\beta\mathcal{H}_I[y]} = h(y|x)e^{-\beta\mathcal{H}_I[x]} \tag{2.29}$$

For this equation we can use the Metropolis solution:

$$h(x|y) = \min(1, e^{-\beta\Delta\mathcal{H}_I}) \tag{2.30}$$

where the novel feature is that only the change of the interaction energy appears in the exponential. This adopts a very simple form:

$$\beta\Delta\mathcal{H}_I = ({\phi'}_i - \phi_i)\sum_{\mu=1}^{2d}\phi_{i\mu} \tag{2.31}$$

In general, this procedure can be used when the Hamiltonian can be splitted in a sum of local terms plus an interaction term. One can chose the new value of the variable, independently of the old value, according to the distribution dictated by the local term. This proposal is then accepted with the Metropolis probability using only the interaction term.

The heat-bath algorithm has also been used for this model (Toral and Chakrabarti, 1990). Let us remember than in the heat-bath algorithm, the acceptance probability is equal to 1 (i.e. the proposal is always accepted), one variable ${\phi'}_i$ is changed at a time, and the proposal probability density function $g_{\hat{\mathbf{x}}}({\phi'}_i)$ is obtained from a distribution in which all the other variables remain constant, i.e. we need to identify exactly where does the variable ${\phi'}_i$ appears, all the other terms will be considered constants. From Eq.(2.22) it is very easy to find out the expression for $g_{\hat{\mathbf{x}}}({\phi'}_i)$:

$$g_{\hat{\mathbf{x}}}({\phi'}_i) = A\exp\left(-\frac{\tilde{\theta}}{2}{\phi'}_i^2 - \frac{\chi}{4}{\phi'}_i^4 + {\phi'}_i\sum_{\mu=1}^{2d}\phi_{i_\mu}\right) \tag{2.32}$$

A is some normalization factor depending on the values of the other fields (which are considered to be constant in the heat–bath method). To sample this one-variable probability density function it will not be useful to use a numerical inversion method, because this function depends of the sum of the $2d$ neighbours of site i, which vary from trial to trial. To sample this distribution we can use instead a rejection technique. If we introduce $c = \sum_{\mu=1}^{2d} \phi_{i_\mu}$, we can write the above function as:

$$g_{\hat{\mathbf{x}}}(\phi'_i) = B\left[\exp\left(-\frac{\chi}{4}\phi'^4_i\right)\right]\left[\frac{1}{\tilde{\theta}^{-1/2}\sqrt{2\pi}}\exp\left(-\frac{(\phi'_i - c/\tilde{\theta})^2}{2\tilde{\theta}^{-1}}\right)\right] \tag{2.33}$$

B is another normalization constant. This is the product of the function $\exp(-\frac{\chi}{4}\phi'^4_i)$ and a Gaussian distribution of mean $c/\tilde{\theta}$ and variance $\tilde{\theta}^{-1}$ (for this we require $\tilde{\theta} > 0$). This Gaussian distribution can be generated by the relation $\phi'_i = c/\tilde{\theta} + \tilde{\theta}^{-1/2}$ran_g() where ran_g() is a Gaussian distributed random variable of mean 0 and variance 1. Finally, the value ϕ'_i is accepted with a probability $\exp(-\frac{\chi}{4}\phi'^4_i)$.

Another elegant way of implementing a Monte Carlo method for the Φ^4 model is that of Bruce (Bruce, 1985) in his study of the *border model* (this is nothing but the Φ^4 model in the case $\tilde{\theta} = 0$). In his approach, the proposal $g_{\hat{\mathbf{x}}}(x|y) \equiv g_{\hat{\mathbf{x}}}(\phi'_i)$ is also independent of the old configuration y (in the same vein than heat-bath), but $g_{\hat{\mathbf{x}}}(\phi'_i)$ is chosen to be the sum of two Gaussians which best approximate the actual local distribution of the field ϕ'_i:

$$\begin{aligned} g_{\hat{\mathbf{x}}}(\phi'_i) =& \frac{1}{2}\left[\frac{1}{\sigma_1\sqrt{2\pi}}\exp\left(-\frac{(\phi'_i - \mu_1)^2}{2\sigma_1^2}\right)\right. \\ &\left.+\frac{1}{\sigma_2\sqrt{2\pi}}\exp\left(-\frac{(\phi'_i - \mu_2)^2}{2\sigma_2^2}\right)\right] \end{aligned} \tag{2.34}$$

To sample this distribution one generates a Gaussian number, ran_g() of mean 0 and variance 1 and a uniform number, ξ, in $(0,1)$. If $\xi < 0.5$ then one chooses the field from the first Gaussian distribution, i.e. $\phi'_i = \mu_1 + \sigma_1$ran_g(), else, if $\xi \geq 0.5$, one chooses from the second, $\phi'_i = \mu_2 + \sigma_2$ran_g(). An initial guess of the parameters μ_1, μ_2, σ_1 and σ_2 is later refined with the information coming from the computed local distribution. The acceptance probability must satisfy detailed balance:

$$g_{\hat{\mathbf{x}}}(\phi'_i)h(x|y)\frac{e^{-\beta\mathcal{H}(y)}}{\mathcal{Z}} = g_{\hat{\mathbf{x}}}(\phi_i)h(y|x)\frac{e^{-\beta\mathcal{H}(x)}}{\mathcal{Z}} \tag{2.35}$$

introducing the *effective* interaction Hamiltonian, $\tilde{\mathcal{H}}([\phi])$:

$$\beta\tilde{\mathcal{H}}([\phi]) = \beta\mathcal{H}([\phi]) + \sum_{i=1}^{N} \ln g_{\hat{\mathbf{x}}}(\phi_i) \tag{2.36}$$

the detailed balance condition becomes:

$$h(x|y)e^{-\beta\tilde{\mathcal{H}}(y)} = h(y|x)e^{-\beta\tilde{\mathcal{H}}(x)} \tag{2.37}$$

from which it is clear that a possible solution for the acceptance probability is the Metropolis solution in terms of the function $\tilde{\mathcal{H}}$:

$$h(x|y) = \min\left(1, e^{-\beta\Delta\tilde{\mathcal{H}}}\right) \tag{2.38}$$

2.3. Histogram extrapolation

We have now developed some tools to study numerically the Φ^4 model. We shall next apply them to the computation of some magnitudes of interest. Let us focus first on the magnetization m (see Fig.4). This is a function of the parameters θ and χ. If we are interested, say, in the variation of m with θ for a fixed value of χ, we have to run our numerical algorithm for several values of θ. For each run, we have to make sure that the system has reached equilibrium (thermalization) and then produce a sufficient number of independent configurations such that the errors are kept under reasonable limits. Repeating this procedure for a large number of θ values in order to have smooth curves for the magnetization is a painful and slow procedure. There is a way of improving somehow this process by using information from simulations at a given value of the parameter θ to obtain results for another value θ'. The idea, which is extremely simple (and clever!), has been obtained and reobtained many times, but were Ferrenberg and Swendsen (see Ferrenberg and Swendsen (1989) and references therein) who dramatically demonstrated its utility in the case of the Ising model. We now describe this *histogram extrapolation technique* for the Φ^4 model.

The configurations obtained in a numerical simulation using the Monte Carlo method follow the Gibbs probability distribution which depend on the parameters θ and χ, that we now write out explicitly:

$$f([\Phi];\theta,\chi) = \frac{e^{-\beta\mathcal{H}([\Phi];\theta,\chi)}}{\mathcal{Z}(\theta,\chi)} \tag{2.39}$$

The type of averages one is interested on usually involve only the following functions:

$$\begin{aligned} m_0 &= \sum_{i=1}^{N} \phi_i \\ m_1 &= \sum_{i=1}^{N} \phi_i^2 \\ m_2 &= \sum_{i=1}^{N} \phi_i^4 \\ m_3 &= \sum_{i=1}^{N} \sum_{\mu=1}^{d} (\phi_{i_\mu} - \phi_i)^2 \end{aligned} \tag{2.40}$$

other functions m_i can be incorporated into what follows, but we restrict to these for simplicity. Since the Hamiltonian depends only on linear combinations of m_1, m_2 and m_3, the probability distribution function in terms of m_0, m_1, m_2 and m_3 is:

$$f(m_0,m_1,m_2,m_3;\theta,\chi) = \frac{N(m_0,m_1,m_2,m_3)\exp\left(\frac{\theta}{2}m_1 - \frac{\chi}{4}m_2 - \frac{1}{2}m_3\right)}{\mathcal{Z}(\theta,\chi)} \tag{2.41}$$

Here $N(m_0, m_1, m_2, m_3)dm_0dm_1dm_2dm_3$ is the number of configurations with values of m_i in the interval $(m_i, m_i + dm_i)$, $i = 0, 1, 2, 3$. These values of $N(m_0, m_1, m_2, m_3) \equiv N([m_i])$ are recorded during the simulation forming a histogram (hence the name of histogram extrapolation). The partition function can be expressed in terms of $N([m_i])$ as:

$$\mathcal{Z}(\theta, \chi) = \int \prod_i dm_i N([m_i]) \exp\left(\frac{\theta}{2}m_1 - \frac{\chi}{4}m_2 - \frac{1}{2}m_3\right) \tag{2.42}$$

The average value of any function $G([m_i])$ can be computed as:

$$\langle G([m_i])\rangle = \int \prod_i dm_i f([m_i]; \theta, \chi) G([m_i]) \tag{2.43}$$

The extrapolation scheme allows one to compute the probability density function $f([m_i]; \theta', \chi')$, for values θ', χ' of the model parameters, if $f([m_i]; \theta, \chi)$ is known. For our particular model, the method is based on the exact relation:

$$f([m_i]; \theta', \chi') = \frac{f([m_i]; \theta, \chi) \exp\left(\frac{\theta'-\theta}{2}m_1 - \frac{\chi'-\chi}{4}m_2\right)}{\int \prod_i dm_i f([m_i]; \theta, \chi) \exp\left(\frac{\theta'-\theta}{2}m_1 - \frac{\chi'-\chi}{4}m_2\right)} \tag{2.44}$$

Let us, in order to discuss the implication of this relation, consider the simpler case $\chi' = \chi$ for which the above formula reduces to:

$$f([m_i]; \theta', \chi) = \frac{f([m_i]; \theta, \chi) \exp\left(\frac{\theta'-\theta}{2}m_1\right)}{\int \prod_i dm_i f([m_i]; \theta, \chi) \exp\left(\frac{\theta'-\theta}{2}m_1\right)} \tag{2.45}$$

This identity is telling us that if we know the probability distribution function of the $m_i^{\prime}s$ at a given value of the parameter θ then we know it **for any other value** θ'. This is wonderful! Only one simulation at a selected value of θ is required! Once we have computed the probability density function $f([m_i]; \theta, \chi)$ we can obtain, by means of the previous formula, the probability density function $f([m_i]; \theta', \chi)$ and from the knowledge of $f([m_i]; \theta, \chi)$ we can compute any average we need using Eq.(2.43).

In practice, alas!, things are not so easy. Relation (2.45) is indeed exact, but what it is telling us is that the information contained in the tail of the distribution of $f([m_i]; \theta, \chi)$ is used to compute the distribution $f([m_i]; \theta', \chi)$, see Fig. 6. The only problem is that the errors in the tail of the distribution usually are very large, so making infeasible to extrapolate to a value of θ' very far from θ.

Of course, the width of the distribution $f([m_i]; \theta, \chi)$ will be very important to determine the extrapolation range $\Delta\theta$. If the distribution is very wide, one can extrapolate to farther away values than for a very narrow distribution. This is the reason why this histogram extrapolation technique is useful near the critical point where the distributions are wider.

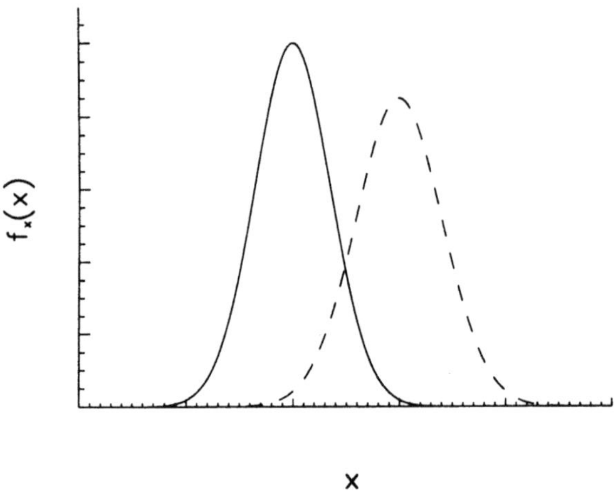

Fig. 6. Probability distribution functions $f([m_i];\theta,\chi)$ at two different values of θ using Eq.(2.43). According to Eq.(2.45) the region around the maximum of the function given by the dashed line is obtained from the tail of the function given by the solid line, where statistical errors are larger.

2.4. Finite size scaling

In Figures 7 to 10 we have plotted, for different system sizes and using the histogram extrapolation scheme, results for several quantities of interest: magnetization, energy, specific heat and susceptibility. Each of these figures has been produced by just two simulations whose location is marked with points in the figures. The rest of the curves have been obtained using the extrapolation scheme detailed above.

These figures show what is called *finite size effects* (Barber, 1983). Let us compare Fig. 7 for the magnetization with Fig.4 which depicts the mean field behavior and the expected true behavior. The theory predicts that there should be a value of $\theta = \theta_c$ below which the spontaneous magnetization is strictly zero. In fact, the magnetization behaviour near θ_c can be characterized by a critical exponent, β such that:

$$m(\theta) = \begin{cases} 0 & \text{if} \quad \theta < \theta_c \\ a\epsilon^{\beta} & \text{if} \quad \theta > \theta_c \end{cases} \tag{2.46}$$

Here $\epsilon = 1 - \frac{\theta}{\theta_c}$. This is a non-analytical behaviour which, strictly speaking, can only appear in a system of infinite size. For a finite system, the number of integrations appearing in the definition of the partition function and the calculation of averages is always finite, thus leading necessarily to an analytical result. We can understand intuitively why finite size effects will be more important near a second order phase transition. In this situation the correlation length which measures the linear range over which fields at different sites of the lattice are correlated, diverges (in an infinite system) with a power-law singularity:

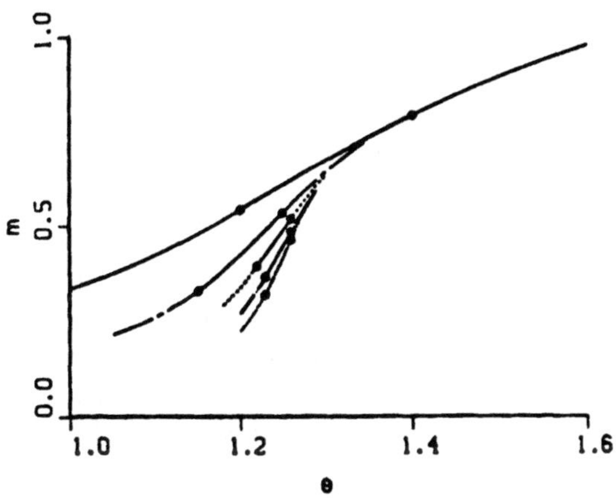

Fig. 7. Magnetization m for the two-dimensional ϕ^4 model for $\chi = 1$ and different values of the system side $L = 8, 16, 24, 32, 40$ (from bottom to top lines). For each value of L a continuous line has been drawn by running two simulations at the points marked with a dot and extrapolating to other values of θ using Eq.(2.45).

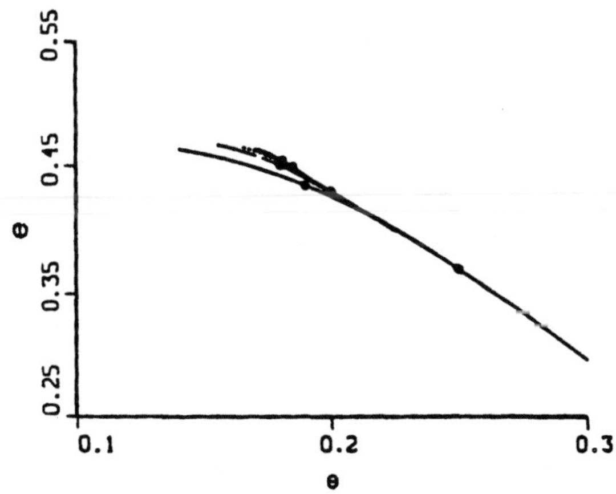

Fig. 8. Same than figure 7 for the energy e and $\chi = 0.1$.

$$\xi(\theta) \sim |\epsilon|^{-\nu} \tag{2.47}$$

For a finite system, the fields can not be correlated longer that the system side and we must have $\xi \sim L$. The theory of finite size tells us exactly how (and why!) the averages of interest behave (Cardy, 1988). The basic idea is that now the magnetization, say, becomes a *homogeneous* function of ξ and the system side L, $m(\xi, L) = \xi^x \tilde{m}(\xi/L)$. The unknown exponent x is obtained by demanding that in the infinite system, and close enough to the critical point, one recovers the known behaviour given by Eq.(2.46). This implies that the function $\tilde{m}(z)$ takes a finite limit when $z \to 0$ and then:

$$m(\theta) = \lim_{L\to\infty} m(\theta, L) = \xi^x \tilde{m}(0) \sim \left[\epsilon^{-\nu}\right]^x \sim \epsilon^{-x\nu} \tag{2.48}$$

compared to (2.46) one concludes $\beta = -x\nu$ and then the prediction for the magnetization near the critical point for a finite system is:

$$m(\theta, L) = \xi^{-\beta/\nu}\tilde{m}(\xi/L) = L^{-\beta/\nu}\tilde{\tilde{m}}[(1-\theta/\theta_c)L^{-1/\nu}] \tag{2.49}$$

The typical way of checking this scaling behavior is to plot $m(\theta, L)L^{\beta/\nu}$ vs the rescaled variable $(1-\theta/\theta_c)L^{-1/\nu}$. If, as it is usually the case, the critical value θ_c and the critical exponents ν, β are not known, this procedure implies a three-parameter fit which is very difficult to do in practice. One can use the equivalent scaling relations for the specific heat and the susceptibility:

$$\begin{aligned} C(\theta, L) &= L^{\alpha/\nu}\tilde{\tilde{C}}[(1-\theta/\theta_c)L^{-1/\nu}] \\ \kappa(\theta, L) &= L^{\gamma/\nu}\tilde{\tilde{\kappa}}[(1-\theta/\theta_c)L^{-1/\nu}] \end{aligned} \tag{2.50}$$

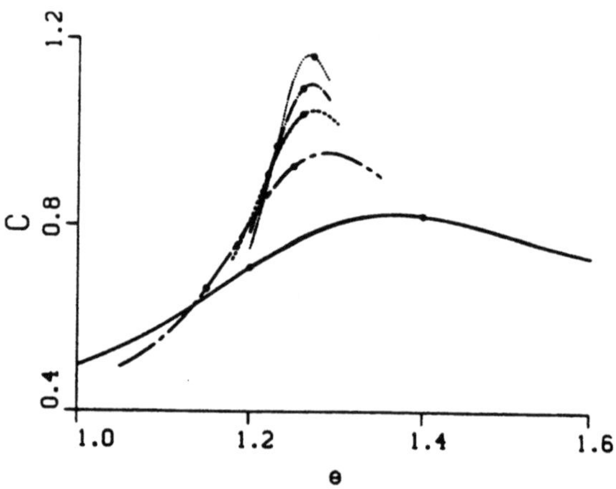

Fig. 9. Specific heat for the two–dimensional ϕ^4 model for $\chi = 1$. Same symbols and meanings than in figure 7.

From Figures 9 and 10 we see that both $C(\theta, L)$ and $\kappa(\theta, L)$ develop maxima at locations $\theta_c^1(L)$ and $\theta_c^2(L)$, respectively. By using the above expressions for the specific heat and the susceptibility one can see that these values behave as:

$$\begin{aligned} \theta_c^1(L) &= \theta_c + a_1 L^{-1/\nu} \\ \theta_c^2(L) &= \theta_c + a_2 L^{-1/\nu} \end{aligned} \tag{2.51}$$

From where we deduce that a plot of $\theta_c^1(L)$ and $\theta_c^2(L)$ vs $L^{-1/\nu}$ must yield a straight line whose interception at the origin is precisely θ_c. This can be seen in Fig.11 for which the critical value $\theta_c = 1.265 \pm 0.010$ for $\chi = 1$ is deduced.

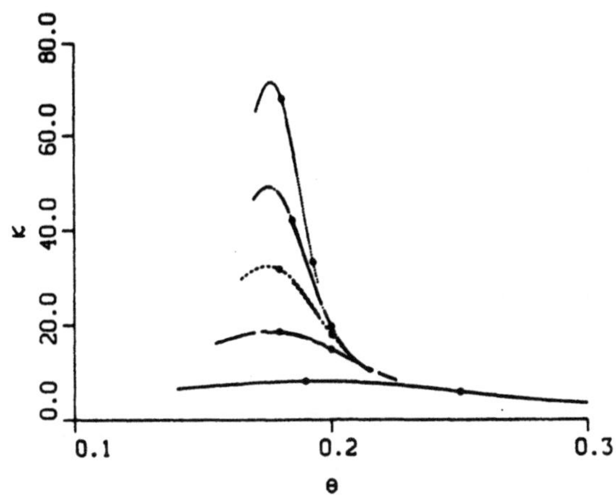

Fig. 10. Magnetic susceptibility for the two-dimensional ϕ^4 model for $\chi = 0.1$. Same symbols and meanings than in figure 7.

In this figure we have used the known value $\nu = 1$ for the $2-d$ ϕ^4 model. When the value of ν is not known one could use directly the three parameter fit $\theta_c^i(L) = \theta_c + a_i L^{-1/\nu}$ to obtain θ_c although the quality of the fit usually is not good enough to allow also for a very accurate measurement of ν. Once the value of θ_c is known one can use the remaining finite size scaling relations to obtain the critical exponents.

3. Field Dynamics

3.1. Introduction

Up to now we have focused only on the equilibrium properties of the scalar Φ^4 model. If one wants to write dynamical equations for the fields one finds the difficulty that the Φ^4 Hamiltonian, so successful for equilibrium properties, does not contain enough information (e.g. kinetic energy terms) to determine the dynamics of the model. Instead of adding more terms to the Hamiltonian, the usual approach to study dynamical properties is to consider that the field variables follow some phenomenological equations of motion. These equations of motion will be stochastic in nature to reflect our ignorance of the detailed effect of the microscopic variables in the field variables (Gunton et al., 1983b). In addition, they must respect the basic symmetries of the system and lead, in the limit of infinite time, to the equilibrium canonical distribution given by the Gibbs factor.

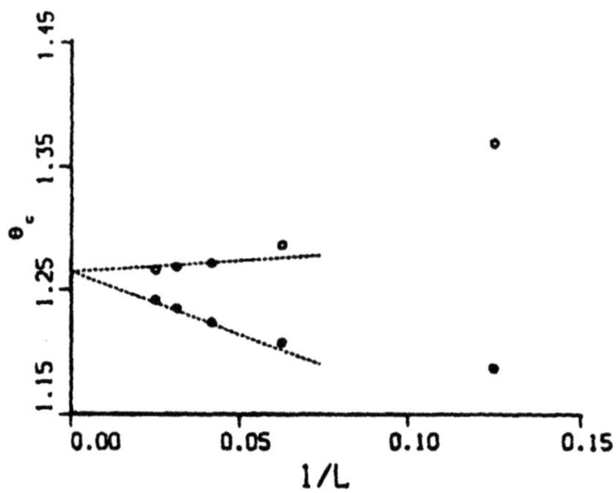

Fig. 11. Pseudo critical parameters $\theta_c^1(L)$ (open dots) and $\theta_c^2(L)$ (black dots) defined as the location of the maxima of the specific heat and susceptibility, respectively, plotted vs. L^{-1}. The extrapolation to the origin, according to Eq.(2.51) yields the critical value $\theta_c = 1.265 \pm 0.010$.

3.2. The Langevin equation

As discussed before, one writes down semi-phenomenological stochastic equations (Langevin equations) to model the time evolution of the field. For the simplest case of the Φ^4 model without any further symmetry (remember than the main symmetry of the model is the $\Phi \to -\Phi$), one is guided by the fact that at sufficiently low temperatures the final configuration must be a minimum of the Hamiltonian. Then, the time evolution is defined by a purely relaxational model such that the speed of approaching the ground state is proportional to the gradient of the Hamiltonian. One adds a stochastic contribution (usually called the *noise* term) that ensures the correct thermal fluctuations in the stationary (equilibrium) state. The resulting Langevin equation is:

$$\frac{\partial \Phi(\boldsymbol{r},t)}{\partial t} = -\Gamma \frac{\delta \mathcal{H}}{\delta \Phi(\boldsymbol{r},t)} + \eta(\boldsymbol{r},t) \tag{3.1}$$

Here Γ is a positive coefficient usually assumed to be constant, although in some cases one also studies the effect of the dependence of Γ on the field Φ (Langer et al., 1975; Lacasta et al., 1992, 1993). As in the equilibrium case, one assumes a lattice discretization in which, instead of a real variable in every point of the space, only points on a d-dimensional lattice Λ are considered. In the lattice case, the functional derivative $\delta\mathcal{H}/\delta\Phi(\boldsymbol{r},t)$ becomes a partial derivative:

$$\frac{d\Phi_i(t)}{dt} = -\Gamma a_0^{-d} \frac{\partial \mathcal{H}}{\partial \Phi_i(t)} + \eta_i(t) \tag{3.2}$$

In the absence of the noise term, it is easy to show that the system evolves towards a minimum of the Hamiltonian. We have:

$$\frac{d\mathcal{H}}{dt} = \sum_{i=1}^{N} \frac{\partial \mathcal{H}}{\partial \Phi_i} \frac{d\Phi_i}{dt} = -\Gamma a_0^{-d} \sum_{i=1}^{N} \left(\frac{\partial \mathcal{H}}{\partial \Phi_i} \right)^2 \leq 0 \tag{3.3}$$

The statistical properties of the noise terms, $\eta_i(t)$, have to be determined in order to obtain the correct equilibrium distribution. They are supposed to be Gaussian distributed of mean zero and correlations given by:

$$\langle \eta_i(t) \eta_j(t') \rangle = D a_0^{-d} \delta_{i,j} \delta(t - t') \tag{3.4}$$

characteristic of the so called *white noise* (Gardiner, 1985). D is a parameter, to be determined, representing the intensity of the noise and, consequently, D must increase with temperature. In the limit of the lattice spacing a_0 going to 0, the Kronecker-delta becomes a Dirac-delta function:

$$\langle \eta(\boldsymbol{r}, t) \eta(\boldsymbol{r}', t') \rangle = D \delta(\boldsymbol{r} - \boldsymbol{r}') \delta(t - t') \tag{3.5}$$

From now on, we will take the lattice spacing as unit of length $a_0 = 1$. Since η_i is Gaussian, these correlation functions completely determine its statistical properties. We now relate the noise intensity D to temperature T. For later reference, we will be more general and consider the set of stochastic differential equations:

$$\frac{d\Phi_i(t)}{dt} = F_i(t) + \sum_{j=1}^{N} G_{ij} \eta_i(t) \tag{3.6}$$

With the same statistical properties for the noise term than before, Eq.(3.4). G_{ij} is a constant matrix. The complete statistical description of the fields $[\Phi]$ is given by the joint probability density function $P([\Phi]; t)$, which satisfies the Fokker-Planck equation (Gardiner, 1985). The description given by the Langevin equations is equivalent to the one given by the Fokker-Planck equation:

$$\frac{\partial P([\phi]; t)}{\partial t} = \sum_{i=1}^{N} \frac{\partial}{\partial \Phi_i} \left[-F_i P + \frac{D}{2} \sum_{j=1}^{N} H_{ij} \frac{\partial P}{\partial \Phi_j} \right] \tag{3.7}$$

The matrix H is $H = GG^{\top}$. In the case we are concerned with, Eq.(3.2), $H = \mathrm{I}$, unity matrix, and F_i is related to the partial derivative of the Hamiltonian:

$$\frac{\partial P([\phi]; t)}{\partial t} = \sum_{i=1}^{N} \frac{\partial}{\partial \Phi_i} \left[\Gamma \frac{\partial \mathcal{H}}{\partial \Phi_i} P + \frac{D}{2} \frac{\partial P}{\partial \Phi_i} \right] \tag{3.8}$$

It is easy to verify that the stationary solution of this equation is:

$$P_{st}([\Phi]) \propto \exp \left[-\frac{2\Gamma}{D} \mathcal{H}([\Phi]) \right] \tag{3.9}$$

Since we want the statistical stationary properties of $[\Phi]$ to be determined by the Gibbs distribution, $\exp\left[-\beta \mathcal{H}([\Phi])\right]$ we are led to identify:

$$\frac{2\Gamma}{D} = \beta \Longrightarrow D = 2k_B T\Gamma \tag{3.10}$$

D is then proportional to temperature, as we had anticipated. By using the expression for the lattice Hamiltonian equation (2.1) we arrive at the following equation:

$$\frac{d\Phi_i(t)}{dt} = \Gamma k_B T \left(b\Phi_i - u\Phi_i^3 + K\nabla_L^2\Phi_i\right) + \eta_i(t) \tag{3.11}$$

∇_L^2 is the lattice laplacian operator defined by:

$$\nabla_L^2\Phi_i = \sum_{\mu=1}^{2d}(\Phi_{i_\mu} - \Phi_i) \tag{3.12}$$

here the sum over μ runs over the $2d$ nearest neighbours of site i. By redefining time $t \rightarrow t\Gamma K k_B T$ and by introducing the same field rescaling and parameters than in the equilibrium case, Eq. (2.21), we arrive at the somewhat simpler equation:

$$\frac{d\phi_i(t)}{dt} = \theta\phi_i - \chi\phi_i^3 + \nabla_L^2\phi_i + \xi_i(t) \tag{3.13}$$

where the new noise variables ξ_i satisfy:

$$\langle\eta_i(t)\eta_j(t')\rangle = 2\delta_{i,j}\delta(t-t') \tag{3.14}$$

The Langevin equation we have considered so far is the simplest dynamical model than can be written for a scalar field and in the famous taxonomy of Hohenberg and Halperin (Hohenberg and Halperin, 1977) is simply called *model A*. Langevin-type equations can describe more complex systems, such as *model B* which is obtained formally by replacing the constant coefficient Γ in Eq.(3.1) defining model A by $-\Gamma\nabla^2$. The resulting equation of motion is:

$$\frac{\partial\Phi(\boldsymbol{r},t)}{\partial t} = \nabla^2\left(\Gamma\frac{\delta\mathcal{H}([\Phi])}{\delta\Phi(\boldsymbol{r},t)}\right) + \eta(\boldsymbol{r},t) \tag{3.15}$$

where $\eta(\boldsymbol{r},t)$ are Gaussian distributed random variables of mean zero and correlations:

$$\langle\eta(\boldsymbol{r},t)\eta(\boldsymbol{r}',t')\rangle = -D\delta(t-t')\nabla^2\delta(t-t') \tag{3.16}$$

We let the physical interpretation of this model to the next chapter. Here we will mention that model B is suitable to describe the relaxational evolution of a *conserved* field, i.e. one for which the spatial integral of the field is time independent:

$$m \equiv \int d\boldsymbol{r}\Phi(\boldsymbol{r},t) \longrightarrow \frac{dm}{dt} = 0 \tag{3.17}$$

The lattice discretization of model B equation leads to:

$$\frac{d\Phi_i(t)}{dt} = \nabla_L^2\left(\Gamma\frac{\partial\mathcal{H}([\Phi])}{\partial\Phi(\boldsymbol{r},t)}\right) + \eta_i(t) \tag{3.18}$$

Again, in order to ensure that the correct equilibrium distribution is obtained, one needs to tune the correct value for D. The Fokker-Planck equation is now:

$$\begin{aligned}\frac{\partial P([\phi];t)}{\partial t} &= \sum_{i=1}^{N} \frac{\partial}{\partial \Phi_i}\left[-\Gamma \nabla_L^2 \left(\frac{\partial \mathcal{H}}{\partial \Phi_i}\right) P + \frac{D}{2}\sum_{j=1}^{N} \frac{\partial}{\partial \Phi_j}\left(-\nabla_L^2 \delta_{i,j} P\right)\right] \\ &= -\sum_{i=1}^{N} \frac{\partial}{\partial \Phi_i} \nabla_L^2 \left[\Gamma \frac{\partial \mathcal{H}}{\partial \Phi_i} P + \frac{D}{2}\frac{\partial P}{\partial \Phi_i}\right]\end{aligned} \tag{3.19}$$

The equilibrium distribution is given by $\exp\left[-\frac{2\Gamma}{D}\mathcal{H}([\Phi])\right]$ and we are led to identify $D = 2\Gamma k_B T$ as in model A.

3.3. Numerical solution of stochastic differential equations

The Langevin equations for models A and B that we considered in the previous section can not be solved exactly. A great deal of what we know about the behavior of the solution of these equations comes from numerical simulations. In this section we want to study how to handle numerically Langevin equations of the general form:

$$\frac{dx_i(t)}{dt} = F_i(t) + \sum_{j=1}^{N} G_{ij}\eta_i(t) \qquad i = 1,\dots,N \tag{3.20}$$

F_i are functions that depend on the set of variables $x_1,\dots,x_N$. If $G_{i,j}$ also depend on $x_1,\dots,x_N$, one talks of *multiplicative* noise and the resulting stochastic differential equations will be considered in the Stratonovich sense (van Kampen, 1981). If, on the other hand, $G_{i,j}$ are constant functions, one talks of *additive* noise. The noise variables $\eta_i(t)$ are Gaussian distributed, independent random variables of mean zero and correlations given by:

$$\langle \eta_i(t)\eta_j(t')\rangle = D\delta_{i,j}\delta(t-t') \tag{3.21}$$

Of course, the appearance of the Dirac-delta functions tells us that we might have some problems of mathematical rigor in we insist in considering the noise variables η_i as simple functions. They should be considered as distributions, but most of the results we can get with more mathematical rigor are also obtained with the usual function formalism ... and some care. In particular, since the noise variables are not differentiable, we can not use the standard Taylor expansions that help so much in developing algorithms for numerically solving ordinary (not stochastic) differential equations (Press et al., 1986). Instead, we must use integration expansions (Kloeden and Platen, 1992). We can express these ideas more clearly with a simple example.

Let us consider a one variable stochastic differential equation:

$$\frac{dx(t)}{dt} = F(t) + \eta(t) \tag{3.22}$$

where $F(t)$ is a sufficiently smooth (differentiable) function and $\eta(t)$ is a Gaussian distributed variable of mean zero and correlations:

$$\langle \eta(t)\eta(t')\rangle = D\delta(t-t') \tag{3.23}$$

In fact, this equation is so simple that we can solve it exactly. However, let us develop a numerical algorithm to solve it. The algorithm generates a recursion relation that will allow us to compute $x(t+h)$ given $x(t)$, h is the integration step. This is the same structure than in ordinary differential equations in which numerical methods do not give the solution for every value of time t but only at regular intervals $t_n = t_0 + nh$, $n = 0, 1, \ldots$ separated by a integration step h. In order to find the recursion relation we integrate (3.22) between t_n and $t_n + h$ to obtain:

$$\int_{t_n}^{t_n+h} dt \frac{dx(t)}{dt} = x(t_n + h) - x(t_n) = \int_{t_n}^{t_n+h} ds F(s) + \int_{t_n}^{t_n+h} ds \eta(s) \tag{3.24}$$

the first term of the right hand side of this equation can be approximated by

$$\int_{t_n}^{t_n+h} ds F(s) = hF(t_n) + o(h^2) \tag{3.25}$$

The second term is a Gaussian variable since it is the integral of a Gaussian variable. Hence, its statistical properties are completely determined by the mean value and the correlations. Let us define the random variable:

$$\omega_h(t) = \int_t^{t+h} ds \eta(s) \tag{3.26}$$

whose mean value is:

$$\langle \omega_h(t)\rangle = \int_t^{t+h} ds \langle \eta(s)\rangle = 0 \tag{3.27}$$

and whose correlations are given by:

$$\langle \omega_h(t)\omega_h(t')\rangle = \int_t^{t+h} ds \int_{t'}^{t'+h} du \langle \eta(s)\eta(u)\rangle = \int_t^{t+h} ds \int_{t'}^{t'+h} du D\delta(s-u) \tag{3.28}$$

This integral is an easy exercise on delta-function integration. The result is:

$$\langle \omega_h(t)\omega_h(t')\rangle = \begin{cases} D(h - |t-t'|) & |t-t'| \le h \\ 0 & |t-t'| \ge h \end{cases} \tag{3.29}$$

The important thing to notice is that for the times t_n that appear in the recursion relation (3.24), one has $\langle \omega_h(t_n)\omega_h(t'_n)\rangle = Dh\delta_{n,n'}$, so that the variables $\omega_h(t_n)$ are independent, Gaussian, random variables of zero mean and variance $\langle \omega_h(t_n)^2\rangle = Dh$. Hence they can be written as

$$\omega_h(t_n) = \sqrt{Dh}u^{(n)} \tag{3.30}$$

where $u^{(n)}$ are independent Gaussian variables of mean zero and variance unity which can be generated, for instance, by the Box-Muller-Wiener method. The final recursion relation reads:

$$\begin{aligned} x^{(0)} &= x_0 \\ x^{(n+1)} &= x^{(n)} + hF^{(n)} + \sqrt{Dh}u^{(n)} \end{aligned} \tag{3.31}$$

with the obvious notation $x^{(n)} = x(t_0 + nh)$, $F^{(n)} = F(x(t_0 + nh))$, etc. This simple example has taught us that the stochastic contribution is of order $h^{1/2}$ whereas the deterministic contribution is of order h. This singular expansion is the main reason for the failure of algorithms based on a naïve Taylor series on integer powers of h. Let us now consider another example:

$$\frac{dx(t)}{dt} = F(x) + G(x)\eta(t) \tag{3.32}$$

Integration of this equation leads to (we have taken $t_0 = 0$ to simplify notation):

$$x(h) = x(0) + \int_0^h dsF(x(s)) + \int_0^h dsG(x(s))\eta(s) \tag{3.33}$$

To proceed, we Taylor expand the functions $F(x)$ and $G(x)$:

$$\begin{aligned} F(x(s)) &= F(x(0)) + \left.\frac{dF}{dx}\right|_{x=x(0)} (x(s) - x(0)) + o(x(s) - x(0))^2 \\ G(x(s)) &= G(x(0)) + \left.\frac{dG}{dx}\right|_{x=x(0)} (x(s) - x(0)) + o(x(s) - x(0))^2 \end{aligned} \tag{3.34}$$

By substituting these expansions in (3.33) we notice is that at lowest order, $h^{1/2}$, we have

$$x(s) - x(0) = G(x(0)) \int_0^s du\eta(u) \tag{3.35}$$

and hence, to have an algorithm correct to $o(h)$ it suffices to consider the following terms in the expansion:

$$\begin{aligned} x(h) =& x(0) + F(x(0))h + G(x(0))\omega_h(0) + \\ & G(x(0)) \left.\frac{dG}{dx}\right|_{x=x(0)} \int_0^h ds \int_0^s du\eta(s)\eta(u) + o(h^{3/2}) \end{aligned} \tag{3.36}$$

The double integral gives $\omega_h(0)^2/2$. Replacing ω_h by $\sqrt{Dh}u$, see Eq.(3.30), we arrive at the following recursion relation, known as the Milshtein algorithm (Milshtein, 1974):

$$\begin{aligned} x^{(0)} &= x_0 \\ x^{(n+1)} &= x^{(n)} + hF^{(n)} + \sqrt{Dh}u^{(n)} + \frac{1}{2}G^{(n)} \left.\frac{dG}{dx}\right|^{(n)} Dh[u^{(n)}]^2 \end{aligned} \tag{3.37}$$

The Milshtein method is an algorithm correct to order h and can then be considered as the stochastic equivalent of the Euler method to solve numerically ordinary differential equations. Sometimes in the literature (Greiner et al., 1988) the name Euler method for stochastic differential equations (in the Stratonovich sense) is given to the above expression in which the term $[u^{(n)}]^2$ is replaced by its mean value $\langle [u^{(n)}]^2 \rangle = 1$:

$$x^{(n+1)} = x^{(n)} + h \left[F^{(n)} + \frac{D}{2} G^{(n)} \left. \frac{dG}{dx} \right|^{(n)} \right] + \sqrt{Dh} u^{(n)} \tag{3.38}$$

This is an algorithm with a worst convergence that the Milshtein algorithm. However, if one is interested only on computing the moments $\langle x(t)^k \rangle$ of the variable $x(t)$ one can prove that this Euler algorithm has the same accuracy than the Milshtein algorithm. Given that both algorithms have approximately the same computational complexity, it does not seem justified to use the poorer Euler algorithm instead of the Milshtein algorithm.

It is possible but very tedious to develop higher order algorithms. The good news is that very rarely one really needs them. This is so because, in general, one has to solve numerically the Langevin equations and average the results for different realizations of the noise (and maybe initial conditions). This generates a source of statistical errors coming from the averages which are, in many occasions, greater than the systematic errors due to the order of convergence of the numerical method. So it is usually better to spend the computer time in reducing the statistical errors by increasing the number of samples in the average rather than using a more complicated, higher order, algorithm. However, we would like to mention briefly the stochastic Runge-Kutta type methods because, although they do not yield better results than the Milshtein method for the convergence of the stochastic terms, they do treat better the deterministic terms, increasing in many cases the numerical stability. The simplest stochastic Runge-Kutta type method is the Heun method which, for Eq.(3.32) reads:

$$\begin{aligned} k &= hF(x(0)) \\ l &= \sqrt{Dh} u^{(0)} G(x(0)) \\ y &= x(0) + k + l \\ x(h) &= x(0) + \frac{h}{2}\left[F(x(0)) + F(y)\right] + \frac{\sqrt{Dh} u^{(0)}}{2}\left[G(x(0)) + G(y)\right] \end{aligned} \tag{3.39}$$

One can prove by Taylor expanding the functions $F(x)$ and $G(x)$ that the Heun method reproduces, to order h, the stochastic part of the Milshtein method (Gard, 1987). For the purely deterministic case, $D = 0$, the Heun algorithm reduces to the standard 2nd order Runge–Kutta method.

So far, we have only considered one-variable equations. It is very complicated to develop algorithms for the general N-variable case of Eq.(3.20). The interested reader can find in (Kloeden and Platen, 1992) an extensive account of these general methods. One can always use the following extension of the Euler method:

$$
\begin{aligned}
x_i^{(n+1)} =& x_i^{(n)} + h\left[F_i^{(n)} + \frac{D}{2}\sum_{j=1}^{N}\sum_{k=1}^{N} G_{j,k}^{(n)} \left.\frac{\partial G_{i,k}}{\partial x_j}\right|^{(n)}\right] \\
&+\sqrt{Dh}\sum_{j=1}^{N} G_{i,j}^{(n)} u_j^{(n)}
\end{aligned}
\tag{3.40}
$$

where the $u_i^{(n)}$ are independent Gaussian variables of mean 0 and variance 1:

$$
\langle u_i^{(n)} u_{i'}^{(n')}\rangle = \delta_{n,n'}\delta_{i,i'} \tag{3.41}
$$

In the case of *diagonal* noise, (Kloeden and Platen, 1992), i.e. one in which the noise terms do not couple variable at different lattice locations:

$$
G_{i,j}(x_1,\ldots,x_N) = G_i(x_i)\delta_{i,j} \tag{3.42}
$$

one can generalize the Milshtein algorithm to:

$$
x_i^{(n+1)} = x_i^{(n)} + hF_i^{(n)} + \sqrt{Dh}u_i^{(n)} + \frac{1}{2}G_i^{(n)} \left.\frac{dG_i}{dx_i}\right|^{(n)} Dh[u_i^{(n)}]^2 \tag{3.43}
$$

In next chapters we will show some applications of the numerical integration of Langevin field equations. In Chapter 4 the Langevin equation will be combined with Monte Carlo techniques to yield a very useful tool to study equilibrium properties: the *hybrid Monte Carlo algorithm*. In Chapter 5 we will be concerned with the dynamics of growth in two very different situations: phase separation and growth of random surfaces.

4. Hybrid Monte Carlo

4.1. Introduction

The numerical integration of Langevin equations described in the previous chapter has been used extensively as an alternative to Monte Carlo methods to sample the canonical distribution. The idea is to run the Langevin equation up to the stationary state (a process similar to thermalization for Monte Carlo) and then average the different field configurations produced over time. An advantage of this procedure as compared to Monte Carlo is that it is not necessary to compute the change in energy every time a single variable is updated. This is important when the computation of the Hamiltonian is extremely time consuming. Such situation arises in lattice gauge theories that include dynamical fermions (Zinn Justin, 1989). The effective Hamiltonian is so complicated for those systems that it is prohibitive having to compute $\mathcal{H}$ N times per Monte Carlo step.

An advantage of Monte Carlo methods over the Langevin integration is that the only errors in Monte Carlo are of statistical origin due to the limited number of configurations in the sample (and also to the correlations amongst them) and are known to behave as the inverse square root of the number of configurations. On top of the statistical errors, the numerical integration of the Langevin equation produces systematic errors due to the finiteness of the time step used in the integration. These systematic errors are much more cumbersome, since their numerical importance is not known a priori. A recent idea aims to combine the best of both worlds. In the so called *Hybrid Monte Carlo* algorithm (Duane et al., 1987) the numerical integration of the stochastic differential equation is used as the proposal to which an acceptance/rejection step is applied. Before we develop this beautiful idea, we need to develop some concepts related to the microcanonical ensemble.

4.2. Molecular dynamics

Molecular dynamics (Allen and Tildesley, 1987) offers a direct numerical approach to the behavior of a system with many degrees of freedom. Let $[X] = (x_1, x_2, \ldots, x_N)$ be microscopic variables and $[P] \equiv (p_1, p_2, \ldots, p_N)$ their conjugate momenta. The Hamiltonian function is:

$$\mathcal{H}(X,P) = \mathcal{V}(X) + \mathcal{T}(P) = \mathcal{V}(X) + \frac{P^2}{2} = \mathcal{V}(X) + \sum_{i=1}^{N} \frac{p_i^2}{2} \tag{4.1}$$

The representative point of the system $[x, p]$ evolves in phase space according to Hamilton equations

$$\begin{aligned} \frac{dx_i}{dt} &= p_i \\ \frac{dp_i}{dt} &= F_i = -\frac{\partial \mathcal{H}}{\partial x_i} \end{aligned} \tag{4.2}$$

and the movement is confined to the hypersurface of constant energy, $\mathcal{H}(X,P) = E$. This property can be used to perform averages on the microcanonical ensemble according to the ergodic principle that allows substitution of ensemble averages $\langle G \rangle$ by time averages $\bar{G}$:

$$\bar{G} = \lim_{T\to\infty} \frac{1}{T - t_0} \int_{t_0}^{T} dt G\left(x(t), p(t)\right) \tag{4.3}$$

To perform this time average we can integrate numerically Hamilton equations. In this way we can obtain equilibrium properties as well as dynamical (transitory) properties. However, similarly to what happened when using the numerical integration of Langevin equations to sample the canonical distribution, two kinds of errors are present: (i) statistical errors due to the finite number of samplings and (ii) systematic errors due to the time discretization used in the numerical solution.

The simplest integration scheme is the Euler method in which the derivative of a function $b(t)$ is approximated by:

$$\frac{db(t)}{dt} = \frac{b(t+\delta t) - b(t)}{\delta t} + O(\delta t) \tag{4.4}$$

Using this approximation, Hamilton equations become:

$$\left.\begin{aligned} \frac{x_i(t+\delta t)-x_i(t)}{\delta t}+O(\delta t) &= p_i(t) \\ \frac{p_i(t+\delta t)-p_i(t)}{\delta t}+O(\delta t) &= F_i(t) \end{aligned}\right\} \tag{4.5}$$

From where one gets the following recursive relations:

$$\left.\begin{aligned} x_i(t+\delta t) &= x_i(t)+\delta t p_i(t)+O(\delta t)^2 \\ p_i(t+\delta t) &= p_i(t)+\delta t F_i(t)+O(\delta t)^2 \end{aligned}\right\} \tag{4.6}$$

The local integration error is of order $o(\delta t)^2$. After n integration steps, $n\delta t = t$ the integration error is $\epsilon = O(n(\delta t)^2) = O(t\delta t) = O(\delta t)$. This error is usually too large and demands choosing a very small time step for accuracy and numerical stability. Of course, there are many other higher-order integration methods. Amongst all of them the *leap frog* algorithm will be particularly suited to our needs. The algorithm is:

$$\begin{aligned} x_i(t+\delta t) &= x_i(t)+\delta t\left(p_i(t)+\frac{\delta t}{2}F_i(t)\right) \\ p_i(t+\delta t) &= p_i(t)+\frac{\delta t}{2}\left(F_i(t)+F_i(t+\delta t)\right), \qquad i=1,\ldots,N \end{aligned} \tag{4.7}$$

and has a local error of order $o(\delta t)^3$. Although this is an improvement over the Euler method, still discretization errors show up, for instance, in the fact that the energy is not exactly conserved. These energy fluctuations (of numerical origin) imply that the microcanonical ensemble is not sampled exactly.

Other properties of the Hamiltonian dynamics are, however, preserved by this algorithm. In particular leap frog satisfies:

(i) Time reversal. If:

$$\begin{pmatrix} X(t) \\ P(t) \end{pmatrix} \xrightarrow{\delta t} \begin{pmatrix} X(t+\delta t) \\ P(t+\delta t) \end{pmatrix} \tag{4.8}$$

then reversal of the momenta will take us back to the original value for $[X]$ and the reversed value for the momenta $[P]$:

$$\begin{pmatrix} X(t+\delta t) \\ -P(t+\delta t) \end{pmatrix} \xrightarrow{\delta t} \begin{pmatrix} X(t+2\delta t) \\ P(t+2\delta t) \end{pmatrix} = \begin{pmatrix} X(t) \\ -P(t) \end{pmatrix} \tag{4.9}$$

(ii) Area preserving. The Jacobian of the change of variables induced during the time evolution is equal to one:

$$J\left(\frac{X(t+\delta t),P(t+\delta t)}{X(t),P(t)}\right) = 1 \tag{4.10}$$

But, let us repeat it once again, it does not conserve energy:

$$\delta\mathcal{H} = \mathcal{H}(t+\delta t)-\mathcal{H}(t) = O[(\delta t)^m] \neq 0 \tag{4.11}$$

Summarizing, the numerical integration by the leap-frog method induces a mapping given by Eq.(4.7) that evolves the whole system from one point of phase space to another point with some error in the conservation of energy or, in other words, leap-frog defines a global change of variables that almost conserves energy. The idea naturally appears to combine this mapping with the acceptance/rejection technique typical of the Monte Carlo method. This is the main idea behind the hybrid Monte Carlo method introduced by Duane et al. (1987) that we now develop in more detail.

4.3. Hybrid Monte Carlo

Let us consider a scalar field $\Phi = (\Phi_1, \ldots, \Phi_N)$ on the regular hypercubic lattice Λ. The statistical properties of the system are given by the Gibbs factor $\exp[-\mathcal{H}([\Phi])]$ (we take $\beta = 1$ to simplify notation), with a Hamiltonian $\mathcal{H}([\Phi])$. In order to define a Hamiltonian dynamics we introduce some **fictitious** momenta fields $[Q] = (Q_1, \ldots, Q_N)$ and a new Hamiltonian function

$$\hat{\mathcal{H}}([\Phi, Q]) = \mathcal{H}([\Phi]) + \frac{Q^2}{2} = \mathcal{H}(\Phi) + \sum_{i=1}^{N} \frac{Q_i^2}{2} \tag{4.12}$$

Since

$$\exp\left[-\hat{\mathcal{H}}([\Phi, Q])\right] = \exp\left[-\mathcal{H}([\Phi])\right] \exp\left[-\frac{Q^2}{2}\right] \tag{4.13}$$

from the statistical point of view, the momenta variables $(Q_1, \ldots, Q_N)$ are simply Gaussian distributed independent variables. By independent we imply not just independent of each other but also of the fields $[\Phi]$.

According to our plan, we want to make a proposal $g_{\hat{\mathbf{x}}}(\Phi'|\Phi)$ by using the numerical solution of Hamilton equations. We specify some initial values for the momenta $[Q]$ according to the Gaussian distribution $\exp\left[-\frac{Q^2}{2}\right]$ and then integrate numerically, obtaining new values for the field:

$$\begin{pmatrix} \Phi \\ Q \end{pmatrix} \overset{\delta t}{\rightarrow} \begin{pmatrix} \Phi' \\ Q' \end{pmatrix} \tag{4.14}$$

This mapping can be done, for instance, by using the leap-frog algorithm, Eqs.(4.7):

$$\begin{aligned} \Phi_i' &= \Phi_i + \frac{\delta t^2}{2} F_i + \delta t Q_i \\ Q_i' &= Q_i + \frac{\delta t}{2}(F_i + F_i') \end{aligned} \tag{4.15}$$

We use this mapping to construct the proposal probability $g_{\hat{\mathbf{x}}}(\Phi'|\Phi)$. More specifically, since the probability distribution function of the Q variables is $\exp(-Q^2/2)$ we can write out the probability density function $g_{\hat{\mathbf{x}}}(\Phi'|\Phi)$ as:

$$g_{\hat{\mathbf{x}}}(\Phi'|\Phi) d\Phi' = dQ e^{-\frac{Q^2}{2}} \tag{4.16}$$

being Q the value of the momenta necessary to go from Φ to Φ'. The key point now is to choose an acceptance probability such that detailed balance is satisfied with the hamiltonian $\mathcal{H}([\Phi])$:

$$e^{-\mathcal{H}([\Phi])} g_{\hat{x}}(\Phi'|\Phi) h(\Phi'|\Phi) = e^{-\mathcal{H}([\Phi'])} g_{\hat{x}}(\Phi|\Phi') h(\Phi|\Phi') \tag{4.17}$$

The proposal $g_{\hat{x}}(\Phi|\Phi')$ is given again by integration of Hamilton equations with a suitable value for the momenta, $[Q'']$:

$$\begin{pmatrix} \Phi' \\ Q'' \end{pmatrix} \overset{\delta t}{\rightarrow} \begin{pmatrix} \Phi \\ Q''' \end{pmatrix} \tag{4.18}$$

from where one gets for $g_{\hat{x}}(\Phi|\Phi')$:

$$g_{\hat{x}}(\Phi|\Phi') d\Phi = dQ'' e^{-\beta \frac{Q''^2}{2}} \tag{4.19}$$

being Q'' the value of the momenta necessary to go from Φ' to Φ. The detailed balance condition becomes:

$$e^{-\frac{Q^2}{2}} e^{-\mathcal{H}(\Phi)} h(\Phi'|\Phi) d\Phi dQ = e^{-\frac{Q''^2}{2}} e^{-\mathcal{H}(\Phi')} h(\Phi|\Phi') d\Phi' dQ'' \tag{4.20}$$

Now the following properties apply:

(i) If the numerical integration satisfies time reversal, $Q'' = -Q'$ hence the Jacobian dQ''/dQ' is equal to 1.

$$e^{-\frac{Q^2}{2}} e^{-\mathcal{H}(\Phi)} h(\Phi'|\Phi) d\Phi dQ = e^{-\frac{Q'^2}{2}} e^{-\mathcal{H}(\Phi')} h(\Phi|\Phi') d\Phi' dQ' \tag{4.21}$$

(ii) If it satisfies the area conserving property $dQd\Phi = dQ'd\Phi'$

$$e^{-\frac{Q^2}{2}} e^{-\mathcal{H}(\Phi)} h(\Phi'|\Phi) = e^{-\frac{Q'^2}{2}} e^{-\mathcal{H}(\Phi')} h(\Phi|\Phi') \tag{4.22}$$

Which, using Eq.(4.13), can be written as:

$$e^{-\hat{\mathcal{H}}(\Phi,Q)} h(\Phi'|\Phi) = e^{-\hat{\mathcal{H}}(\Phi',Q')} h(\Phi|\Phi') \tag{4.23}$$

We already know a possible solution to this equation, the Metropolis solution:

$$h(\Phi'|\Phi) = \min(1, e^{-\delta\hat{\mathcal{H}}}) \tag{4.24}$$

where $\delta\hat{\mathcal{H}} = \hat{\mathcal{H}}(t+\delta t) - \hat{\mathcal{H}}(t))$ is the change of total energy produced in the evolution by time step δt. Notice that, although we have made a global change by updating all the variables Φ_i at once, $\delta\hat{\mathcal{H}}$ is controllable and the acceptance probability can be made to stay within reasonable limits (close to 50%). It is not necessary to make the acceptance/rejection decision every time step δt. Instead, usually one integrates during n time steps before acceptance/rejection.

We give now an explicit implementation of the *Hybrid Monte Carlo method*:

(i) Generate independent Gaussian momenta $[Q]$.

(ii) Update the system $[\Phi, Q]$ using leap–frog (or any other time reversible, area preserving method) during a time $\Delta t = n\delta t$

$$\begin{pmatrix} \Phi(t) \\ Q(t) \end{pmatrix} \overset{\delta t}{\rightarrow} \begin{pmatrix} \Phi(t+\delta t) \\ Q(t+\delta t) \end{pmatrix} \overset{\delta t}{\rightarrow} \cdots \overset{\delta t}{\rightarrow} \begin{pmatrix} \Phi(t+\Delta t) \\ Q(t+\Delta t) \end{pmatrix}$$

(iii) Compute $\Delta\hat{\mathcal{H}} = \hat{\mathcal{H}}(t+\Delta t) - \hat{\mathcal{H}}(t)$

(iv) Accept configuration $\Phi' = \Phi(t+\Delta t)$ with probability $\min(1, e^{-\Delta\hat{\mathcal{H}}})$

Hybrid Monte Carlo constitutes an important tool for the simulation of field theories and it has been applied successfully to $SU(n)$ field theory (Sexton and Weingarten, 1992), Lennard-Jones systems (Mehlig et al., 1992a), Φ^4 scalar field (Mehlig et al., 1992b), $X-Y$ model (Gupta, 1992), polymer systems (Forrest and Toral, 1994; Irbeck, 1994), etc. and it is today a routine tool of great applicability.

There is another important aspect of the Hybrid Monte Carlo algorithm, namely its relation with Langevin dynamics. Remember the Langevin equation defining model A. Eq.(3.1):

$$\frac{\partial\Phi_i}{\partial\tau} = -\frac{\delta\mathcal{H}}{\delta\Phi_i} + \eta_i(\tau) = F_i(\tau) + \eta_i(\tau) \tag{4.25}$$

where the noise terms satisfies:

$$< \eta_i(\tau)\eta_j(\tau') >= 2\delta_{ij}\delta(\tau-\tau') \tag{4.26}$$

The numerical solution using the Euler algorithm proceeds via the following recursion relation:

$$\Phi_i(\tau+\delta\tau) = \Phi_i(\tau) + \delta\tau F_i(\tau) + \sqrt{2\delta\tau}u_i(\tau) \tag{4.27}$$

where $u_i(\tau)$ are Gaussian variables of mean 0 and correlations $\langle u_i(\tau)u_j(\tau)\rangle = \delta_{ij}$. This is exactly the same evolution scheme of the leap-frog algorithm, Eq.(4.15), if we make the following relation between the time steps of the two methods: $\delta\tau = \delta t^2/2$. We conclude that Hybrid Monte Carlo uses the same updating scheme than Langevin (with a different unit of time), but the presence of the acceptance/rejection step makes the method independent of the time step used in the Langevin integration, so avoiding the systematic errors so difficult to deal with in the Langevin method.

4.4. Generalized hybrid Monte Carlo

It is clear that Hybrid Monte Carlo relies upon a numerical integration of the differential equations for a dynamical system in which the energy is conserved. We will now introduce a general class of dynamical equations that also conserve energy. This will give rise to a generalization of the hybrid Monte Carlo method (Ferreira and Toral, 1993; Toral and Ferreira, 1994).

The generalization goes as follow: to each field Φ_i we assign a vector momenta variable $\boldsymbol{Q}_i = (q_i^1, \ldots, q_i^D)$ of arbitrary dimension D. We introduce the following equations of motion:

$$\begin{aligned} \frac{d\Phi_i}{dt} &= \sum_{s=1}^{D}\sum_{j=1}^{N}(\mathcal{A}^s)_{ij}Q_j^s \\ \frac{dq_i^s}{dt} &= \sum_{j=1}^{N}(\mathcal{A}^s)_{ji}F_j \qquad\qquad s = 1,\ldots,D \end{aligned} \tag{4.28}$$

or, written in more compact vector notation:

$$\begin{aligned} \frac{d\Phi}{dt} &= \sum_{s=1}^{D} \mathcal{A}^s Q^s \\ \frac{dQ^s}{dt} &= (\mathcal{A}^s)^\top F \qquad\qquad s = 1, \ldots, D \end{aligned} \tag{4.29}$$

where the $\mathcal{A}^s$ are some linear operators which can be represented as a matrix, and F_j represents the force as computed from the Hamiltonian $-\frac{\partial}{\partial \Phi_j}\mathcal{H}$. We also introduce a total Hamiltonian $\hat{\mathcal{H}}$, including the kinetic energy terms:

$$\hat{\mathcal{H}}([\Phi, Q]) = \mathcal{H}(\Phi) + \sum_{i=1}^{N}\sum_{s=1}^{D} \frac{(q_i^s)^2}{2} \tag{4.30}$$

The reader can easily verify that the proposed dynamics in Eqs.(4.28) exactly conserves energy, i.e., $d\hat{\mathcal{H}}/dt = 0$. The standard hybrid Monte Carlo can be obtained from the above dynamics setting $D = 1$ and $\mathcal{A}$ equal to the identity operator.

For the approximate integration of the previous equations of motion the leap–frog scheme can be used, introducing a discrete mapping $[\Phi(t), Q(t)] \rightarrow [\Phi(t + \delta t), Q(t + \delta t)] = G^{\delta t}([\Phi(t), Q(t)])$, dependent on the time step δt chosen. The leap-frog approximation reads:

$$\begin{aligned} \Phi' &= \Phi + \delta t \sum_{s=1}^{D} \mathcal{A}^s Q^s + \frac{(\delta t)^2}{2} \sum_{s=1}^{D} \mathcal{A}^s (\mathcal{A}^s)^\top F([\Phi]) \\ Q'^s &= Q^s + \frac{\delta t}{2} (\mathcal{A}^s)^\top \left(F([\Phi]) + F([\Phi'])\right) \end{aligned} \tag{4.31}$$

It can be shown that this leap-frog approximation for arbitrary matrices $\mathcal{A}^s$ satisfies the properties of time reversibility and area preserving. However, again as a result of the time discretization used in the leap–frog scheme, the total energy is no longer conserved although its variation can be controlled by varying δt. We define yet another mapping obtained iterating n times the previous mapping, i.e. $\mathcal{G} = (G^{\delta t})^n$. To satisfy detailed balance the configuration obtained when one applies $\mathcal{G}$ is accepted with probability $\min[1, \exp(-\Delta\hat{\mathcal{H}})]$, where $\Delta\hat{\mathcal{H}} = \hat{\mathcal{H}}(\mathcal{G}([\Phi, Q])) - \hat{\mathcal{H}}([\Phi, Q])$. As in the standard hybrid Monte Carlo, the momenta variables are refreshed after every acceptance/rejection step according to a Gaussian distribution of independent variables.

We have defined a general class of hybrid Monte Carlo-type methods characterized by a particular choice of matrices $\mathcal{A}^s$. One can choose the matrices $\mathcal{A}^s$ that better suit a particular problem. This generalized hybrid Monte Carlo method turns out to be related to the method introduced in reference (Batrouni et al., 1985) using the numerical integration of a Langevin equation with a matrix time–step. This method is based upon the observation that the stationary solution of the Langevin equation (4.25) can be obtained approximately by the recursion relation:

$$\Phi_i(\tau+\delta\tau) = \Phi_i(\tau) + \sum_{j=1}^{N}\left[-\delta\tau\epsilon_{ij}\frac{\delta\mathcal{H}}{\delta\Phi_j} + \sqrt{2\delta\tau}\sqrt{\epsilon}_{ij}u_j\right] \tag{4.32}$$

Where ϵ_{ij} is an arbitrary matrix and u_j is a Gaussian variable of mean zero and correlations $\langle u_i u_j\rangle = \delta_{ij}$. Comparing with (4.31) we can see that it corresponds exactly to the one step leap-frog approximation of the generalized hybrid Monte Carlo method introduced above taking $D = 1$ and if we identify: $(\delta t)^2/2 = \delta\tau$ and $\mathcal{A}\mathcal{A}^\top = \epsilon$. The main difference between the two methods is the presence of an acceptance/rejection step in the generalized hybrid Monte Carlo. In this sense, we can say that the generalized hybrid Monte Carlo method introduced in this section makes exact (in the sense that averages are not biased by the choice of the time step) the numerical integration of the Langevin equation using a matrix time step introduced in reference (Batrouni et al., 1985).

4.5. Fourier acceleration

Near a second order phase transition simulations become very time consuming due to the large correlations which appear in the neighbourhood of the critical point. This enhancement of the time correlations, known as **critical slowing down**, greatly increases the statistical errors in a Monte Carlo simulation making it necessary to average over a large number of configurations in order to obtain values with an acceptable error. The physical origin of critical slowing down is related to the fact that, near the critical point, fluctuations of all wavelengths are present, each with a characteristic time scale. In a Langevin integration, for example, the integration step has to be small enough to treat correctly the short–wavelength components but, on the other hand, the integration has to proceed for a sufficient number of steps to have the long–wavelengths mode evolve significantly. In the **Fourier acceleration** method (Parisi, 1983; Batrouni et al., 1985), an integration technique is developed such that each mode evolves with its own effective time step, large for long wavelengths and small for small wavelengths. Whether Fourier acceleration is successful in completely overcoming critical slowing down for a particular system depends on to which extent the evolution of the different Fourier modes can be considered indeed independent of each other.

Fourier acceleration has been traditionally framed in the language of Langevin equations. But we know already that numerical solutions of Langevin equations introduce systematic errors due to the finiteness of the time step used in the numerical integration. It is possible to develop exact Fourier acceleration methods in the context of the generalized hybrid Monte Carlo algorithm developed in the previous section. The basic ingredient is the following:

- Use matrices $\mathcal{A}^s$ whose elements are diagonal in Fourier space.

The optimal choice for these matrices depends on the system we are considering. Let us now discuss the example of the Gaussian model, for which the technique is most successful in reducing critical slowing down. Of course, the example is purely academic, since the Gaussian model can be solved analytically without having to resource to numerical methods.

The Gaussian model is defined by the following Hamiltonian for the scalar fields $[\Phi_i]$ (Parisi, 1988):

$$\mathcal{H} = \sum_{i=1}^{N} \left[\frac{\mu}{2} \Phi_i^2 + \frac{1}{2} \mid \nabla_L \Phi_i \mid^2 \right] \tag{4.33}$$

index i runs over the $N = L^d$ sites of a d-dimensional regular lattice Λ, with periodic boundary conditions; ∇_L is the usual lattice discretized version of the gradient operator; μ is a parameter called the *field mass*. Although this model does not show a real phase transition, the critical slowing down occurs for the critical value" $\mu = 0$.

To implement the generalized hybrid Monte Carlo method we choose the number of momenta variables associated to a given field equal to 1 (i.e. $D = 1$ in Eq.(4.28)). The total Hamiltonian $\hat{\mathcal{H}}$, Eq.(4.30), including the kinetic energy terms can be written in terms of the Fourier transform of fields and momenta:

$$\hat{\mathcal{H}} = \sum_{k=1}^{N} \left[\frac{\omega_k^2}{2} |\hat{\Phi}_k|^2 + \frac{1}{2} |\hat{Q}_k|^2 \right] \tag{4.34}$$

the sum runs over the N points in the reciprocal lattice, $k = (k_1, \ldots, k_d)$. Here $\omega_k^2 = \mu + 4\sum_{j=1}^{d} \sin^2(k_j/2)$ and $\hat{\Phi}_k$ and $\hat{Q}_k$ stand for the fields and momenta variables in Fourier space. Notice that the total Hamiltonian is a sum of terms each one depending on a single mode k, such that the modes evolve independently of each other.

According to the discussion above, we implement Fourier acceleration by choosing the matrix $\mathcal{A}$, generating the dynamics, diagonal in Fourier space. Let us denote by $\hat{A}_k$ the diagonal elements of the matrix $\mathcal{A}$ in Fourier space. After n leap-frog steps, the evolution equations (4.31) imply:

$$\begin{bmatrix} \omega_k \hat{\Phi}_k(n\delta t) \\ \hat{Q}_k(n\delta t) \end{bmatrix} = M_k^n \begin{bmatrix} \omega_k \hat{\Phi}_k(0) \\ \hat{Q}_k(0) \end{bmatrix} \qquad k = 1, \ldots, N \tag{4.35}$$

The dynamical properties of the algorithm are determined by matrices M_k^n, given by:

$$M_k^n = \begin{pmatrix} \cos(n\theta_k) & \sin(n\theta_k)/\cos(\theta_k/2) \\ -\cos(\theta_k/2)\sin(n\theta_k) & \cos(n\theta_k) \end{pmatrix} \tag{4.36}$$

where we have introduced $\theta_k = \cos^{-1}(1 - c_k^2/2)$ and $c_k = \hat{A}_k \omega_k \delta t$. We see explicitly how in this model different modes evolve independently of each other. The fact that the evolution equations are linear greatly simplifies the analysis of this problem (this is similar to the standard hybrid Monte Carlo where $\mathcal{A} = 1$, Kennedy and Pendleton, 1991).

We can now answer the question of which is the optimal choice for matrix $\mathcal{A}$. If we choose the following values for the diagonal elements of $\mathcal{A}$ in Fourier space:

$$\hat{A}_k = 1/\omega_k = \left[\mu + 4\sum_{j=1}^{d} \sin^2(k_j/2)\right]^{-1/2} \tag{4.37}$$

the evolution matrices M_k^n become independent of μ and k such all the modes evolve with the same effective time step $A_k \omega_k \delta t = \delta t$. This choice for $\hat{A}_k$ reduces completely critical slowing down because correlation times (which are related to the eigenvalues of M_k^n) become independent of the mass μ even when μ tends to zero and the model becomes critical.

5. Applications in Domain Growth

5.1. Dynamics of first order phase transitions

5.1.1. The Cahn-Hilliard-Cook equation

Many binary (A-B) mixtures which are homogeneous at high temperature T_0, phase separate when quenched below a certain critical value (T_c) of the temperature. In the final equilibrium configuration two phases coexist: the A and B phases, each one rich in the A and B material respectively (see Fig.12). We assume that the equilibrium properties of such mixture are given by the Gibbs distribution at the final temperature T_f and we are concerned here with the way the mixture reaches this thermal equilibrium, i.e. with the dynamical process of phase separation (Gunton et al., 1983; Binder, 1990). In many occasions the relevant equations of motion are those of model B defined in a previous chapter, equations that we now justify from a more physical point of view.

The relevant field Φ in this case is a scalar field representing the local concentration difference of the two components of the mixture $\Phi(\boldsymbol{r},t) = \rho_A(\boldsymbol{r},t) - \rho_B(\boldsymbol{r},t)$. This field obviously follows a conservation law:

$$m \equiv \frac{1}{V}\int_V d\boldsymbol{r}\Phi(\boldsymbol{r},t) \longrightarrow \frac{dm}{dt} = 0 \tag{5.1}$$

(V is the system volume). This global conservation law can be expressed in local terms by means of a continuity equation:

$$\frac{\partial \Phi(\boldsymbol{r},t)}{\partial t} + \nabla \cdot \boldsymbol{J}(\boldsymbol{r},t) = 0 \tag{5.2}$$

The conservation current $\boldsymbol{J}(\boldsymbol{r},t)$ provides the driving force for field diffusion. Equilibrium is reached when the chemical potentials of the two substances are equal and uniform. Thus, we write the current $\boldsymbol{J}(\boldsymbol{r},t)$ as the sum of the gradient of the difference of chemical potentials of each component $\mu = \mu_A - \mu_B$ (this is the well known Fick's law) plus a random, noise type term to take properly into account thermal fluctuations:

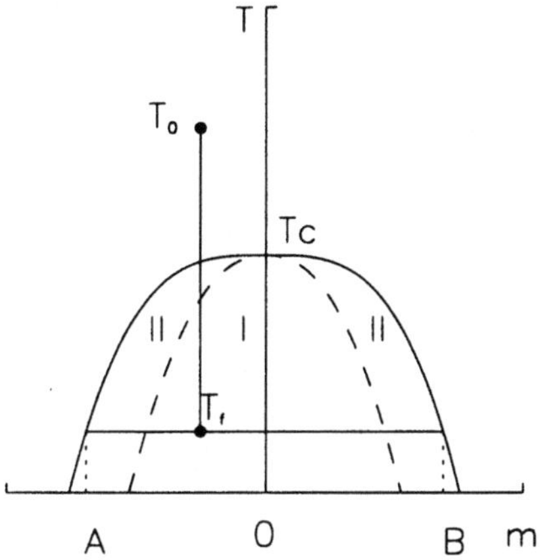

Fig. 12. Schematic phase diagram of a binary mixture $A-B$. The system is homogeneous at temperature T_0. After quenching at temperature T_f below T_c the system phase separates. At sufficiently late times, phase A and B coexist. The evolution proceeds differently if the quench is in region I or II, separated by the spinodal line (dashed line).

$$\boldsymbol{J}(\boldsymbol{r},t) = -\Gamma \nabla \mu + \boldsymbol{\eta}(\boldsymbol{r},t) \tag{5.3}$$

Γ is a positive coefficient called *mobility*. The noise variables $\boldsymbol{\eta} = (\eta_1, \ldots, \eta_d)$ are Gaussian distributed, of mean zero and correlations:

$$\langle \eta_k(\boldsymbol{r},t)\eta_{k'}(\boldsymbol{r}',t')\rangle = 2k_B T \delta_{k,k'}\delta(r-r')\delta(t-t') \tag{5.4}$$

The chemical potential is obtained as the derivative of $\mathcal{H}$ with respect to the field:

$$\mu(\boldsymbol{r},t) = \frac{\delta \mathcal{H}([\Phi])}{\delta \Phi(\boldsymbol{r},t)} \tag{5.5}$$

Combining Eqs.(5.2) to (5.5), we reobtain model B equation, which in the context of phase separation is called the Cahn-Hilliard-Cook (CHC) equation (Cahn and Hilliard, 1958; Cook, 1970):

$$\frac{\partial \Phi(\boldsymbol{r},t)}{\partial t} = \nabla^2 \left(\Gamma \frac{\delta \mathcal{H}([\Phi])}{\delta \Phi(\boldsymbol{r},t)} \right) + \xi(\boldsymbol{r},t) \tag{5.6}$$

where $\xi(\boldsymbol{r},t) = \nabla \cdot \eta(\boldsymbol{r},t)$. If we adopt expression (2.1) for the Hamiltonian we arrive at the following lattice form for the CHC equation:

$$\frac{d\Phi_i(t)}{dt} = \Gamma k_B T \nabla_L^2 \left(-b\Phi_i + u\Phi_i^3 - K\nabla_L^2 \Phi_i\right) + \xi_i(t) \tag{5.7}$$

If we perform the same field and time rescaling than in the section 3.2 for model A, we obtain the simpler equation:

$$\frac{d\phi_i(t)}{dt} = \nabla_L^2 \left(-\theta b \phi_i + \chi \phi_i^3 - \nabla_L^2 \phi_i\right) + \xi_i(t) \tag{5.8}$$

with correlations:

$$\langle \xi_i(t)\xi_j(t')\rangle = -2\delta(t-t')\nabla_L^2 \delta_{i,j} \tag{5.9}$$

The Cahn–Hilliard–Cook equation is a very complicated stochastic, nonlinear, partial differential equation that has so far defied an exact analytical treatment. Very successful systematic approximations valid for early times have been developed (Grant et al., 1985). For the late stages of the dynamics the most successful theory is that of Langer, Bar-on and Miller (Langer et al., 1975), based on some approximative mode decoupling. This theory, however, is very difficult to improve upon due to its non-perturbative nature. It seems at this stage that it is necessary to resource to numerical studies to analyze the long–time behaviour of the equation. One can integrate the CHC equation using any of the stochastic numerical methods developed in section 3.3. In Figure 13 we have plotted configurations resulting of a numerical integration using the simple Euler method. In this figure we observe that the evolution depends on the location of the quench point in phase diagram. For critical ($m = 0$) or near critical concentrations, (region I of the phase diagram in Fig.12), the system is unstable against long wavelength, small amplitude fluctuations. This generates initially a highly interconnected pattern that coarsens with time. The system is said to evolve by *spinodal decomposition*. If, on the other hand, the system is in region II, between the so–called spinodal line and the coexistence curve, the system is unstable against the formation of nuclei of the minority phase. These nuclei evolve in time in such a way that large droplets grow at the expense of small droplets. This is the evolution for *nucleation and growth*. Although this simple picture is thought to be of general validity, it is not possible to sharply separate the domains in the phase diagram where the two processes dominate (Binder, 1990; Chakrabarti, 1992).

5.1.2 The Kawasaki model

A more microscopic approach to the dynamics of phase separation is the one given by the Kawasaki dynamical version of the Ising model (Kawasaki, 1972). In each site of a regular lattice Λ we define two occupation variables, C_i^A and C_i^B, such that C_i^A takes the value 1 (0) if in site i there is (there is not) an A-atom. C_i^B is defined in a similar way. If we consider the situation in which every site of the lattice is occupied by either A-atoms or B-atoms, i.e. that there are no vacancies, C_i^A and C_i^B verify $C_i^A + C_i^B = 1$. We define a new variable S_i as:

$$S_i = C_i^A - C_i^B \tag{5.10}$$

from where it is easy to deduce:

$$C_i^A = \frac{1+S_i}{2} \qquad C_i^B = \frac{1-S_i}{2} \tag{5.11}$$

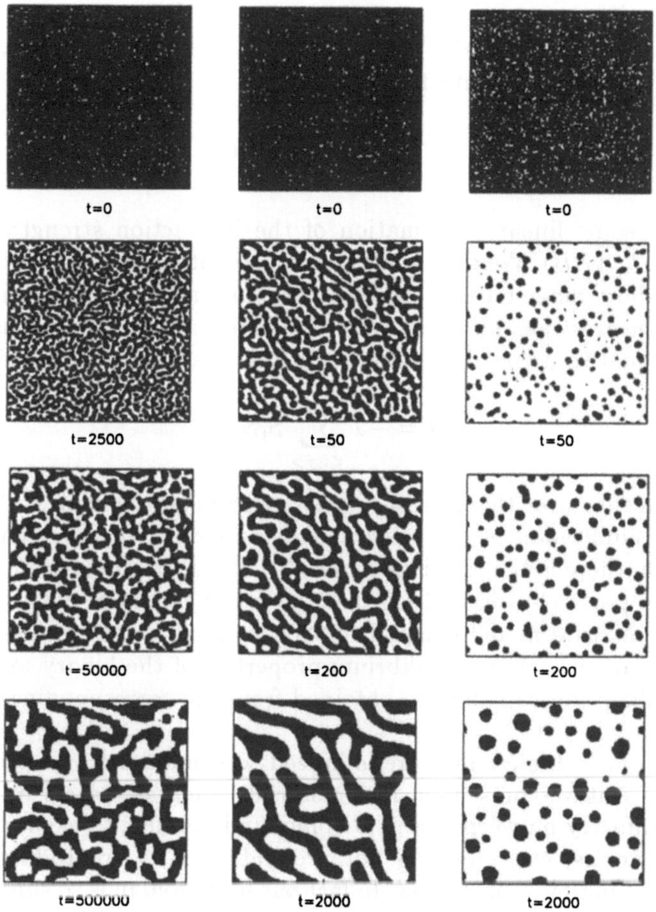

Fig. 13. Time evolution of the Kawasaki model and the Cahn–Hilliard–Cook (CHC) model. Positive (negative) values for the fields or the spins are indicated by white (black) regions. Notice the similarity between the Kawasaki model (left column) and the CHC (center column) both at critical concentration ($m = 0$). In these two cases, the system evolves by spinodal decomposition. In the right column, we plot a solution of the CHC model for an off–critical quench (in this case $m = 0.2$). Notice that evolution proceeds via the formation of nuclei of the minority phase that coarsen and grow.

S_i is an Ising type variable taking the value $+1$ (-1) if in site i there is a A-atom (B-atom). Let us denote by $J_{i,j}(AA)$, $J_{i,j}(BB)$, $J_{i,j}(AB)$ the strength of the interaction between AA, BB and AB atoms, respectively, at locations i and j in the lattice. The Hamiltonian is:

$$\mathcal{H} = \sum_{i,j} \left[J_{i,j}(AA) C_i^A C_j^A + J_{i,j}(BB) C_i^B C_j^B + J_{i,j}(AB) C_i^A C_j^B \right] \quad (5.12)$$

Writing the occupation variable C_i^A and C_i^B in terms of the spin variables S_i one gets, apart from a constant factor:

$$\mathcal{H} = -\sum_{i,j} J_{i,j} S_i S_j \quad (5.13)$$

Where $J_{i,j}$ is some linear combination of the interaction strengths $J_{i,j}(AA)$, $J_{i,j}(BB)$ and $J_{i,j}(AB)$. We take now the approximation that the forces between atoms are of short range nature, such that the only relevant interactions are the ones between nearest neighbour sites on the lattice and we further assume that they are all equal to a constant value J. In this case, the Hamiltonian reduces to:

$$\mathcal{H} = -J \sum_{<i,j>} S_i S_j \quad (5.14)$$

where the sum runs over the nearest neighbour pairs on the lattice. This is nothing but the Hamiltonian of the celebrated Ising model. In our binary mixture model, though, there is the restriction that $N^{-1} \sum_i S_i = m$ must be a fixed constant. One can prove (Pathria, 1972) that the grand canonical partition function of the binary mixture model is equal to the canonical partition function of the standard Ising model. Many equilibrium properties of the binary mixture, such as the phase diagram, can thus be obtained from the corresponding ones of the Ising model.

Our model for a binary mixture is indeed too simplified and it is absent of many realistic features present in real materials, such as defects in the lattice, no perfect symmetry between A and B compounds leading to asymmetric phase diagrams, vacancies, etc. One can not pretend, then, to reproduce all of the features present in these substances. In fact our hamiltonian is so simple that it is absent of any "natural" dynamics such as the one given by Hamilton equations. Kawasaki introduced an *stochastic* dynamics for the binary mixture model in the same spirit that Glauber dynamics for the Ising model (Glauber, 1963). The thermal fluctuations, mediated by the lattice phonons, induce random transitions in which atoms exchange positions in the lattice, see Fig.14 (in the ordinary Glauber dynamics, spins flip randomly their value, this is forbidden here due to the conservation law). This exchange process aims to mimic, in a very crude way, the diffusion process occurring in real materials.

In order to ensure that the system will asymptotically reach the correct equilibrium state given by the Gibbs distribution, Kawasaki assigns a probability to the microscopic exchange process. The equation governing the evolution of the system is a master equation for the probability density of configurations $P([S_1, \ldots, S_N]; t)$:

$$\frac{d}{dt} P([S];t) = -P([S];t) \sum_{[S']} P([S] \to [S']) + \sum_{[S']} P([S'];t) P([S'] \to [S]) \quad (5.15)$$

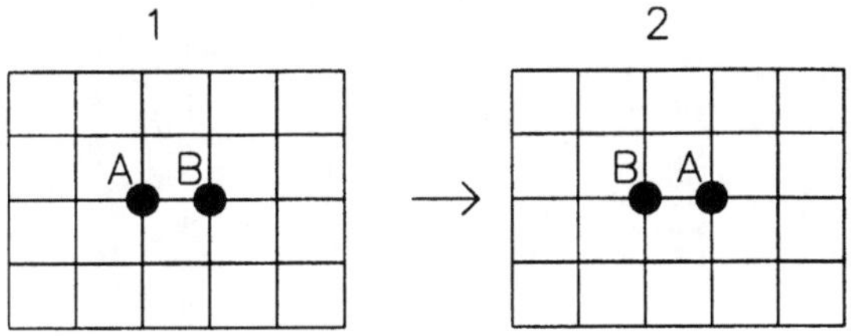

Fig. 14. Kawasaki exchange model basic process

The transition rate for the process $1 \to 2$ (see Fig. 14) is taken in Kawasaki formulation as:

$$P(1 \to 2) = \frac{1}{2\tau}\left[1 - \tanh\left(\frac{\Delta\mathcal{H}}{2k_B T}\right)\right] \tag{5.16}$$

Where $\Delta\mathcal{H}$ is the change in energy required to go from state 1 to state 2. τ simply fixes the unit of time. The previous expression is chosen by mathematical simplicity, but any function of the form $P(1 \to 2) = \exp\left(-\frac{\Delta\mathcal{H}}{k_B T}\right)\omega\left(\frac{\Delta\mathcal{H}}{k_B T}\right)$ where $\omega(z) = \omega(z^{-1})$ will satisfy detailed balance for the master equation and hence will ensure the correct equilibrium state (see discussion in section 1.6).

The Kawasaki exchange model and the Cahn–Hilliard–Cook model exhibit many common features. In fact, it is possible to obtain, using some approximations, the Cahn–Hilliard–Cook model by means of a coarse–graining of the Kawasaki exchange model (Langer, 1971; Gunton et al., 1983b). In Fig.13 we show the evolution of the CHC and Kawasaki models for the case of a quench of critical concentration. Observe the great similarity between the evolution of the two models.

5.1.3 Dynamical scaling

An important property of the late stages of the dynamics of phase separation process is that of *dynamical scaling* (Binder and Stauffer, 1974; Lebowitz et al., 1982). This scaling description is suggested by the fact that systems at different times look similar provided a space rescaling is performed. More precisely: if we take a snapshot of a configuration of the system at a given time t_1 and make it larger by a factor λ it turns out that the resulting figure is *statistically indistinguishable* of a snapshot of the same system at a later time $t_2 = \lambda^{1/a} t_1$, a being a dynamical scaling exponent (see Fig.15). This dynamical scaling description is based on the physical assumption that, at the late stages of the evolution, only one length, related to the typical domain size, is relevant. Mathematically, the scaling description implies that the pair correlation function $G(\boldsymbol{r}, t)$:

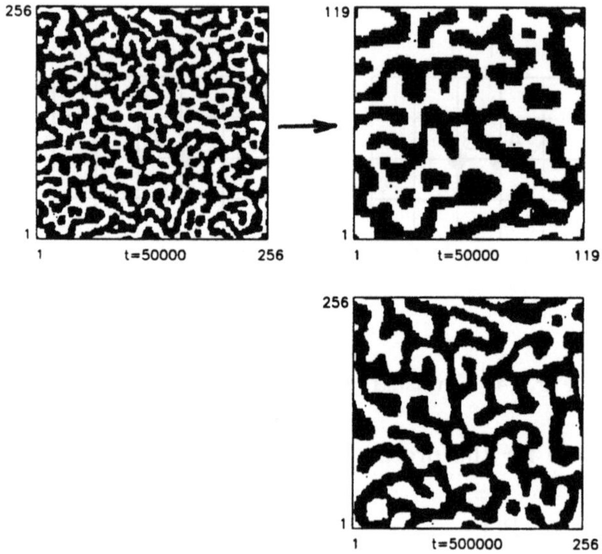

Fig. 15. Self similarity after rescaling of configurations for the two-dimensional CHC model in the case of a critical quench $m = 0$. A snapshot of the configuration at time $t_1 = 50,000$ (upper row, left) is magnified by a factor $256/119\ approx 2.15$. The resulting figure is very similar (from the statistical point of view) to a snapshot of the same system at time $t_2 = t_1 \times 2.15^3 \approx 500,000$, showing that the dynamical exponent in Eq.(5.20) is close to 1/3.

$$G(\boldsymbol{r},t) = \langle \Phi(\boldsymbol{r},t)\Phi(\boldsymbol{r}+\boldsymbol{r}',t)\rangle, \tag{5.17}$$

or, rather, its circular average:

$$G(r,t) = \frac{\int d\Omega G(\boldsymbol{r},t)}{\int d\Omega} \tag{5.18}$$

(Ω denotes the angular variables), which, in principle, is a two variable function, depends only on a combination of them, namely:

$$G(r,t) = g(r/R(t)) \tag{5.19}$$

where the *scaling function* $g(x)$ is a one variable function of the scaling variable $x = r/R(t)$. $R(t)$ is a measure of the typical domain size and can be defined, for instance, as the first zero of the pair correlation function. The time behaviour of $R(t)$ is described by a power-law:

$$R(t) \sim t^a \tag{5.20}$$

This expresses again the idea that the system at two different times t_1, t_2 will look similar if a space rescaling of magnitude λ for the system at time t_1 is performed such that $\lambda R(t_1) = R(t_2)$ or, according to the previous expression, $\lambda = (t_2/t_1)^a$. Another important feature of the dynamical scaling description is that of the universality of the scaling function $g(x)$ and the scaling exponent a. The value $a = 1/3$ seems to hold for the CHC and Kawasaki models independently of dimension and the quench location. The function $g(x)$ does depend on dimension and quench location (Chakrabarti et al., 1991; 1993), although it seems that its variation is very small for quenches close to the critical concentration (Toral et al., 1989; Fratzl and Lebowitz, 1989).

Dynamical scaling is usually checked in terms of the experimentally accessible structure function $S(\boldsymbol{k}, t)$, defined as the Fourier transform of the pair correlation function $G(\boldsymbol{r}, t)$. The scaling relation for $S(k,t)$ is obtained from the one holding for $G(r,t)$ as:

$$S(k,t) = R(t)^d \mathcal{F}(kR(t)) \tag{5.21}$$

$\mathcal{F}(x)$ is the scaling function, d is the system dimensionality. In Figure 16 we plot the time evolution of the angularly averaged structure factor $S(k,t)$ from a numerical solution of the CHC equation in 3 dimensions in the case of a critical quench. We can see that the structure factor starts from a rather flat shape and, as time goes on, develops a maximum whose location k_m moves towards smaller values of k and whose intensity increases. The presence of this maximum signals the phase separation process and the location of the peak is related to the inverse typical linear length of the domains. In Figure 16 we also plot $S(k,t)R(t)^{-d}$ versus $kR(t)$ to check the scaling relation (5.21). One concludes that scaling is well satisfied at the times studied. Similar results hold for other dimensions and models, as well as for experiments (Binder, 1990).

5.2. Growth of rough surfaces

Many surfaces in nature grow roughly. Examples include such diverse systems as biological tumors, mountain ranges, sand piles, a fluid flowing through a porous medium and many others (Lagally, 1990; Thomas, 1982). To study this widespread phenomenon, scientists have developed microscopic models and field models [1].

The simplest of the microscopic models is one due to Eden (Eden, 1958). The Eden model can be used to study the spread of epidemics. In this model one considers a regular lattice in which a site is initially marked as "infected". At a given time a randomly selected infected site infects one of its nearest neighbours, again chosen at random. After some time, the infected area is rather compact with a rough surface, see Fig.17. From a careful analysis of this figure one can deduce that the surface has fractal properties and that the dependence of the

[1] Recent reviews on kinetic roughening in surface growth models can be found in: Family and Vicsek, (1991); Krug and Spohn, (1991); Sander, (1991); Wolf, (1990); Family, (1990) and Vicsek, (1989)

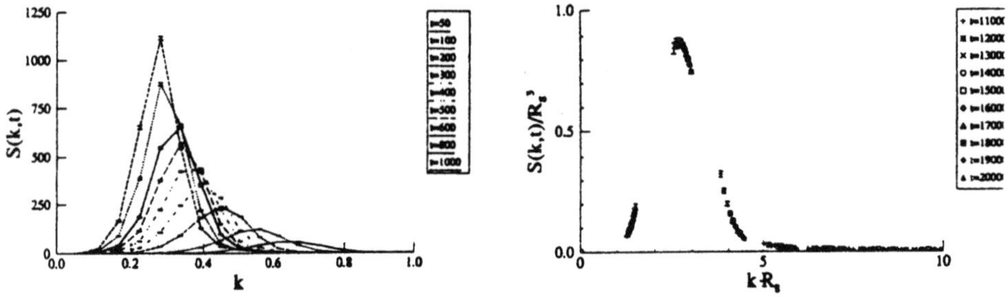

Fig. 16. Raw data (left figure) for the structure factor for the time evolution of the three-dimensional CHC in the case of a critical quench ($m = 0$). In the right figure we have rescaled the date according to Eq.(5.21) to show that dynamical scaling holds reasonably well for this system at the times studied.

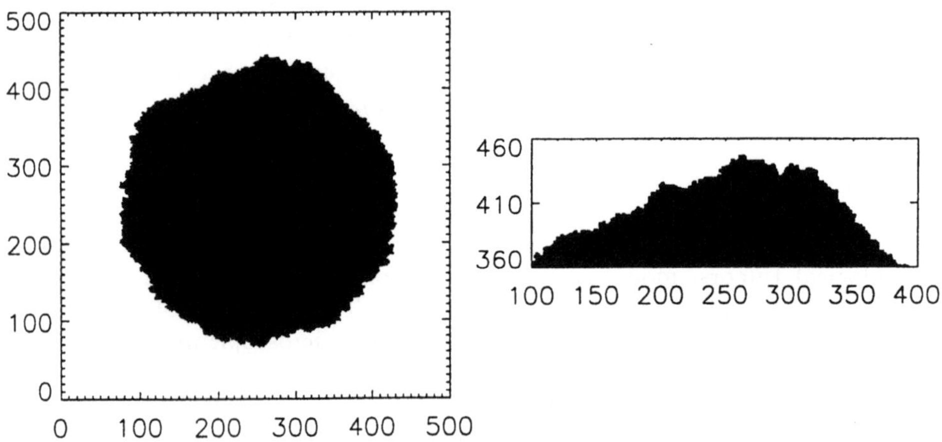

Fig. 17. A cluster grown by the Eden model rules (see text). The right figure is a detail of the cluster boundary to show that it is a rough surface.

roughness with time and with the linear dimensions of the Eden cluster can be described by a power-law with some characteristic exponents.

Another kind of models aim to mimic the evolution of a surface that grows by external flux of particles (Meakin et al., 1986). In the simplest *solid on solid* model particles are added to a regular lattice. The incident particles add to the height h_i in lattice site i. The presence of a surface tension is simulated by rejecting the addition of particles that produce a difference in the height between two nearest neighbours in the lattice larger that some fixed amount (Kim and Kosterlitz, 1989).

One of the most successful field theory models to study surface growth is that of Kardar, Parisi and Zhang (KPZ) (Kardar et al., 1986). In the KPZ model one defines a field $h(\boldsymbol{r},t)$ giving the surface height at location $\boldsymbol{r}$ on a $(d-1)$–dimensional substrate at time t. Several physical processes contribute to the time evolution of this field.

(i) Tendency to grow normal to the surface. A simple geometrical argument shows that in this case the evolution is given by:

$$\frac{\partial h(\boldsymbol{r},t)}{\partial t} = \lambda\sqrt{1+(\nabla h)^2} \approx \frac{\lambda}{2}(\nabla h)^2$$

to the lowest order in the gradient of the surface field. Here λ is the surface mean growth velocity.

(ii) Particle diffusion on the surface.

$$\nu \nabla^2 h$$

ν is a constant related to the surface tension.

(iii) Random variations of the incident flux. These are modeled by a stochastic, noise type term:

$$\eta(\boldsymbol{r},t)$$

Adding the contribution of the three terms we get the celebrated KPZ equation:

$$\frac{\partial h(\boldsymbol{r},t)}{\partial t} = \nu\nabla^2 h(\boldsymbol{r},t) + \frac{\lambda}{2}(\nabla h)^2 + \eta(\boldsymbol{r},t) \tag{5.22}$$

The $\eta(\boldsymbol{r},t)$ noise is assumed to be Gaussian distributed of mean zero and correlations:

$$\langle \eta(\boldsymbol{r},t)\eta(\boldsymbol{r}',t')\rangle = 2D\delta(\boldsymbol{r}-\boldsymbol{r}')\delta(t-t') \tag{5.23}$$

Some of the parameters of this equation, λ, ν, D are redundant. One can reparametrize the field $h \to (\nu/2D)^{1/2}h$ and the time $t \to \nu t$ to obtain a simpler equation:

$$\frac{\partial h(\boldsymbol{r},t)}{\partial t} = \nabla^2 h(\boldsymbol{r},t) + \frac{\lambda}{2}(\nabla h)^2 + \eta(\boldsymbol{r},t) \tag{5.24}$$

where

$$\langle \eta(\boldsymbol{r},t)\eta(\boldsymbol{r}',t')\rangle = \delta(\boldsymbol{r}-\boldsymbol{r}')\delta(t-t') \tag{5.25}$$

Although many exact results are known of this stochastic non–linear partial differential equation, an explicit solution is still missing. Numerical studies have played an important role in characterizing the solutions of the KPZ equation (Moser et al., 1991). In the numerical studies in a one–dimensional substrate one typically uses the following discretization of the KPZ equation:

$$\frac{dh_j(t)}{dt} = \frac{h_{j+1}+h_{j-1}-2h_j}{a_0^2} + \frac{\lambda}{2}\left(\frac{h_{j+1}-h_{j-1}}{2a_0}\right)^2 + \eta_j(t), \tag{5.26}$$

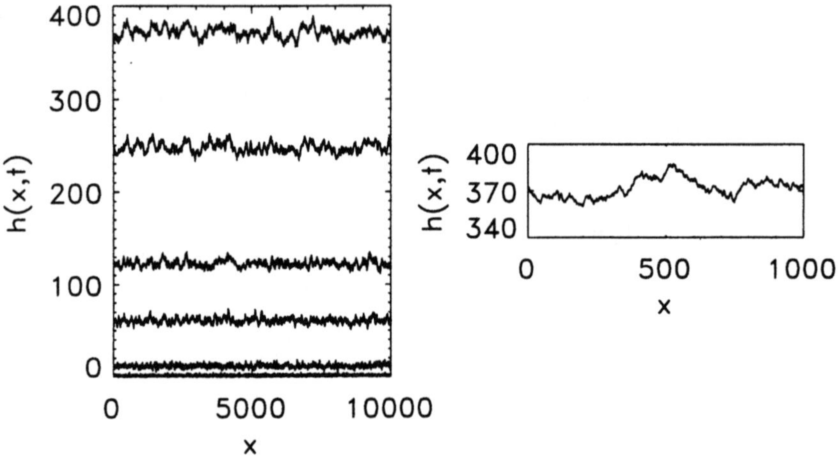

Fig. 18. Plot of the height field $h(x,t)$ from a numerical simulation of the KPZ equation in one–dimensional substrate for increasing times. We can see in this figure that the surface grows rough and that the roughness increases with time. The right figure shows a detail of the surface for the latest time.

(periodic boundary conditions are considered). The lattice spacing is usually taken equal to 1, $a_0 = 1$. The numerical integration of the previous set of stochastic differential equations can proceed by any of the methods explained in Chapter 3, although the Heun method is particularly well suited in this case since it allows to use much larger values for the integration step Δt than the one allowed in the Euler method (Forrest and Toral, 1993). If we write the previous equations in the general vector form:

$$\frac{\partial \boldsymbol{h}(t)}{\partial t} = \boldsymbol{F}(\boldsymbol{h}(t)) + \boldsymbol{\eta}(t), \tag{5.27}$$

The Heun method uses the following recurrence relation (see Eqs.(3.39)):

$$\begin{aligned} \boldsymbol{g}_1 &= \boldsymbol{F}(\boldsymbol{h}(t)) \\ \boldsymbol{g}_2 &= \boldsymbol{F}(\boldsymbol{h}(t) + \Delta t \boldsymbol{g_1} + \sqrt{\Delta t}\boldsymbol{u}) \\ \boldsymbol{h}(t+\Delta t) &= \boldsymbol{h}(t) + \frac{\Delta t}{2}(\boldsymbol{g}_1 + \boldsymbol{g}_2) + \sqrt{\Delta t}\boldsymbol{u} \end{aligned} \tag{5.28}$$

Where $\boldsymbol{u} = (u_1, \ldots, u_N)$ are independent Gaussian variables of mean zero and variance unity.

In Figure 18 we can see the evolution of the field for a 1-d substrate of length L. It is clear from this figure that the surface grows rough, similarly to the case of the microscopic Eden model. A quantitative measure of the surface roughness is given by the quantity $w(t, L)$

$$w(t,L) = \sqrt{\langle \overline{h^2} - \overline{h}^2 \rangle} \tag{5.29}$$

Where the bar denotes a spatial average $\overline{h} = L^{-1}\sum_{i=1}^{L} h_i$ and the angular brackets denote an average over initial conditions and realizations of the noise. It is important to characterize the time evolution of $w(t,L)$. The self–similar character of the surface implies that the late–time behaviour for a system of infinite size must be given by a power–law relation of the form:

$$w(t,\infty) \sim t^{\beta} \tag{5.30}$$

This relation has been well verified in computer simulations and the exact value for β can be obtained theoretically for 1-d systems ($\beta = 1/3$, see Familly et al., 1991). The reader could think that the verification of this simple law should be an easy task. Not at all! In the rest of these notes, we just want to point out which are the main difficulties in trying to verify directly the, innocent looking, power–law given by Eq.(5.30).

The first and most obvious difficulty comes from the fact that solving numerically a set of stochastic differential equations is *always* difficult. One has to average over a large number of realizations to have small statistical errors and also make sure that results do not depend on the time step chosen for the numerical integration.

The second problem is that expression (5.30) is an **asymptotic** result, i.e. strictly valid only in the limit $t \to \infty$. However, earlier time regimes can also be described by a relation of the same functional form with an *effective* exponent $\beta_{eff} \neq \beta$. For example, at very early times, the noise term dominates (remember that it gives a contribution of order $\sqrt{\Delta t}$) and the roughness grows as the square root of time, $w(t,L) \sim t^{1/2}$. For intermediate times, the linear term dominates. It is not difficult to solve this linear case ($\lambda = 0$, the so–called Edwards-Wilkinson model (Edwards and Wilkinsons, 1982)) and show that the roughness grows as $w(t) \sim t^{1/4}$. It is only at the very late stages that the effect of the non–linearity fully develops and one obtains the true asymptotic regime $w(t) \sim t^{1/3}$. But it is very difficult to be sure that one is in the asymptotic regime and that the exponent obtained from a, say, log–log plot of $w(t)$ versus t is giving indeed the true asymptotic exponent instead of an asymptotic exponent corresponding to smaller times.

Another problem concerns finite size effects. The result (5.30) is valid only in the limit of an infinite system. If the system size L is finite (as it is bound to be in a numerical simulation), the surface roughness can saturate to an asymptotic value $w(t \to \infty, L) \sim L^{\zeta}$ ($\zeta = 3/2$ in 1-d) and one might never enter the time regime in which relation (5.30) holds. The dramatic effect of the finite size effects can be shown in detail for the linear case, $\lambda = 0$. In this case, the solution can be found as:

$$w_{(0)}^2(t,L) = \frac{1}{L}\sum_{k=1}^{L-1} \frac{1-\exp(-8\sin^2(\pi k/L)t)}{4\sin^2(\pi k/L)}. \tag{5.31}$$

The strong L dependence of this linear solution can be seen in Fig. [19]. Looking at the figure, we can say that only for $L \gtrsim 10^4$ there is a relative error $w_{(0)} < 1\%$

for $t \simeq 6000$. The moral of the story is that if one wants to avoid finite size effects one has to take systems at least at large as the ones needed in the linear case, as it is sustained by a perturbative solution of the KPZ equation (Forrest at al., 1993).

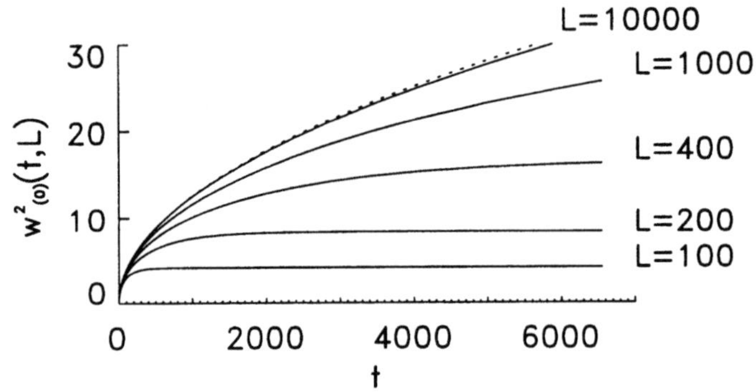

Fig. 19. Dependence with L of the square of the surface roughness, $w^2_{(0)}(t, L)$ for the linear solution of the KPZ equation, Eq.(5.31).

In fact, the finite size effects show up in a very determined way. It is now well established (Family and Vicsek, 1985; Jullien and Botet, 1985) that the roughness $w(L, t)$, for a finite substrate length L, follows an asymptotic scaling description in terms of both time t and L:

$$w(t, L) = t^{\beta} F(tL^{-\zeta/\beta}) \tag{5.32}$$

(ζ is the so-called *roughness exponent*). In principle, this relation can also be used to compute the exponent β. Again, the difficulty is that this relation is also valid for the linear case, $\lambda = 0$ *with different values for the exponents*, namely $w(t, L) = t^{1/4} F(tL^{-2})$ and one has to be careful enough to ensure that the non-linearity has fully developed. It is not strange, then, that the most powerful available computers have been used in this problem of computing precisely the exponents of the KPZ equation.

Financial support from the DGICYT, project number PB-92-0046 is acknowledged.

References

Abramowitz, M. and Stegun, I.A. (1972): *Handbook of Mathematical Functions*, Dover Publications.

Ahrens, J.H., Dieter, U. (1972): Comm. of the ACM, **15** 873; (1988): **31** 1331.

Allen, M.P., Tildesley, D.J. (1987): *Computer Simulation of Liquids*, Clarendon Press.

Amit, D.J. (1984):*Field Theory, the Renormalization Group and Critical Phenomena*, World Scientific.

Barber, M. (1983):*Phase Transitions and Critical Phenomena*, vol. 8, edited by C. Domb and J.Lebowitz, Academic Press.

Batrouni, G.G., Katz, G.R., Kronfled, A.S., Lepage, G.P., Svetitsky, B., Wilson, K.G. (1985): Phys. Rev. **D 32** 2736.

Binder, K., and Stauffer, D. (1974): Phys. Rev. Lett. **33**, 1006.

Binder, K. and Heermann, D.W. (1988): *Monte Carlo simulations in Statistical Physics*, Springer, Berlin, Heidelberg.

Binder, K. (1990):*Materials Science and Technology, Vol. 5 : Phase Transformation in Materials*, ed. P. Haasen, VCH (Weinheim, Germany), p-405.

Binder, K. (ed) (1992):*The Monte Carlo Method in Condensed Matter Physics*, Topics App. Phys. Vol 71 (Springer, Berlin, Heidelberg).

Box, G.E.P. and Muller, M.E. (1958): Ann. Math. Statist. **29**, 610.

Bruce, A.D. (1980): Adv. Phys. **29**, 111.

Bruce, A.D. and Cowley, R. (1980): Adv. Phys. **29**, 219.

Bruce, A.D. (1985): J.Phys. **A 18**, L873.

Cahn, J.W. and Hilliard, J. E., (1958): J. Chem. Phys. **28**, 258.

Cardy, J.L., editor (1988):*Finite Size Scaling*, North-Holland.

Chakrabarti, A., (1992): Phys. Rev. B **45**, 9620.

Chakrabarti, A., Toral, R. and Gunton, J.D., (1991): Phys. Rev. A **44**, 12133; (1993): Phys. Rev. E **47**, 3025.

Cook, H. (1970): Acta. Metall. **18** 297 (1970).

Cowley, R. (1980): Adv. Phys. **29**, 1.

Duane, S., Kennedy, A.D., Pendleton, B.J., and Roweth, D. (1987): Phys. Lett. **B 195** 216.

Eden, M. (1958):*Symposium on Information Theory in Biology*, edited by H. P. Yockey (Pergamon Press, New York), p.359.

Edwards, S.F. and Wilkinsons, D.R. (1982): Proc. R. Soc. Lon. A **381** 17.

Faires, J. and Burden, R. (1993): *Numerical Methods*, PWS Publishing Company, Boston.

Family, F. and Vicsek, T. (1985): J. Phys. A **18**, L75.

Family, F. and Vicsek, T. (eds.) (1991): *Dynamics of Fractal Surfaces*, World Scientific, Singapore.

Feller, W. (1971):*An introduction to probability theory and its applications*, vols. 1,2. John Wiley & Sons.

Ferreira, A.L. and Toral, R. (1993): Phys. Rev. **E 47** R3848.

Ferrenberg, A.M. and Swendsen, R.H. (1989): Comp. in Phys. **Sept/Oct**, 101.

Forrest, B.M. and Toral, R. (1993): J.Stat.Phys. **70**, 703.

Forrest, B.M. and Toral, R. (1994): J.Stat.Phys. **77**, 473.

Fratzl, P. and Lebowitz, J.L., (1989): Acta Metall. **37**, 3245; Fratzl, P. (1991): J. Appl. Cryst.**24**, 593.

Gard, T.C. (1987): "Introduction to Stochastic Differential Equations", Marcel Dekker inc, vol 114 of "Monographs and Textbooks in Pure and Applied Mathematics".

Gardiner, C.W. (1985): *Handbook of Stochastic Methods*, Springer-Verlag, Berlin.

Glauber, R.J. (1963): J. Math. Phys. **4**, 294.

Grant, M., San Miguel, M., Viñals, J. and Gunton, J.D. (1985): Phys. Rev. **B 31**, 3027.

Greiner, A., Strittmatter, W. and Honerkamp, J. (1988): J. Stat. Phys. **51**, 95.

Grimmett, G.R. and Stirzaker, D.R. (1982):*Probability and Random Processes*, Oxford U. Press.

Gunton, J.D., San Miguel, M. and Sahni, P.S. (1983): in *Phase Transitions and Critical Phenomena*, Vol. 8, edited by C. Domb and J. L. Lebowitz (Academic, London).

Gunton, J.D. and Droz, M. (1983): *Introduction ot the Theory of Metastable and Unstable States*, Springer-Verlag.

Gupta, S. (1992): Nuc. Phys. **B 370** 741.

Heermann 1986: *Computer Simulation Methods*, Springer–Verlag.

Hohenberg, P.C. and Halperin, B, (1977): Rev. Mod. Phys. **49**, 435.

Irback, A. (1994): J. Chem.Phys. **101** 1661.

James, F. (1990): Comp. Phis. Comm. **60**, 329.

Jullien, R. and Botet, R. (1985): J. Phys. A **18**, 2279.

Kalos, M.H. and Whitlock, P.A. (1986): *Monte Carlo Methods*, John Wiley and Sons.

Kardar, M., Parisi, G. and Zhang, Y.C. (1986): Phys. Rev. Lett. **56**, 889.

Kawasaki, K. (1972): in *Phase Transitions and Critical Phenomena*, vol. 2, edited by C. Domb and M.S.Green, Academic Press.

Kennedy A.D. and Pendleton B. (1991), Nucl.Phys. B (Proc. Suppl). **20**, 118.

Kim, J.M. and Kosterlitz, J.M. (1989): Phys. Rev. Lett. **62**, 2289.

Kloeden, P.E. and Platen, E. (1992):*Numerical Solution of Stochastic Differential Equations*, Springer-Verlag.

Knuth, D.E. (1981): "The Art of Computer Programming, vol.2 Seminumerical Algorithms", Addison-Wesley Pub. Co.

Krug, J. and Spohn, H. (1991): in *Solids Far From Equilibrium: Growth, Morphology and Defects*, edited by C. Godrèche, Cambridge Univ. Press, Cambridge, England.

Lacasta, A.M., Sancho, J.M., Hernández–Machado, A. and Toral, R. (1992): Phys. Rev. B **45**, 5276; (1993): Phys. Rev. B **48**, 6854.

Lagally M. (ed.) (1990):*Kinetics of Ordering and Growth at Surfaces*, (Plenum, New York), and references therein.

Landau, L.D. and Lifshitz, E.M. (1980):*Statistical Physics*, Pergamon Press.

Langer, J.S. (1971): Ann.Phys. **65**, 53.

Langer, J.S., Bar-on, B. and Miller, H.D. (1975): Phys. Rev. **A 11**, 1417.

Lebowitz, J.L., Marro, J. and Kalos, M.H. (1982): Act. Metall. **30**, 290 and references therein.

Meakin, P., Ramanlal, P., Sander, L.M., Ball, R.C. (1986): Phys. Rev. **A34**, 5091.

Mehlig, B., Ferreira, A.L. and Heermann, D.W. (1992a): Phys. Lett. **B291** 151.

Mehlig, B., Heermann, D.W. and Forrest, B.M. (1992b): Phys. Rev. **B 45** 679.

Metropolis, N., Rosenbluth, A.W., Rosenbluth, M.N., Teller, A.H., Teller, E. (1953): J. Chem. Phys. **21**, 917.

Milchev, A., Heermann, D. and Binder, K. (1986): J. Stat. Phys. **44**, 749.

Milshtein, G.N. (1974): Theory. Prob. Appl. **19**, 557; (1978): *ibid* **23**, 396.

Moser, K., Kertész, J., and Wolf, D.E. (1991): Physica A **178**, 215.

Parisi, G. (1984): in *Progress in Gauge Field Theory* (Cargese lectures), ed. G't Hooft et al., Plenum Press (reproduced in Parisi, G. (1992), *Field theory, disorder and simulations*, World Scientific.

Parisi, G. (1988):*Statistical Field Theory*, Addison Wesley.

Pathria, R.K. (1972):*Statistical Mechanics*, Pergamon Press.

Press, W.H., Flannery, B.P., Teulolsky, S.A. and Vetterling, W. (1986): "Numerical Recipes". Cambridge University Press.

Rubinstein, R.Y. (1981): *Simulation and the Monte Carlo method*, John Wiley & Sons.

Sander, L. (1991): in*Solids Far From Equilibrium: Growth, Morphology and Defects*, edited by C. Godrèche, Cambridge Univ. Press, Cambridge, England.

Sexton, J.C., Weingarten, D.W. (1992): Nuc. Phys. **B380** 665.

Sokal, A.D. (1991): Nuclear Physics B (Proc. Suppl.) **20** 55.

Thomas, T.R. (1982):*Rough Surfaces* (Longman, London).

Toral, R., Chakrabarti, A. (1990): Phys. Rev. **B 42**, 2445.

Toral, R., Chakrabarti, A. and Gunton, J.D. (1989): Phys. Rev. B **39**, 901; (1992): Phys. Rev. A **45**, R2147.

Toral, R., Chakrabarti, A. (1993): Comp. Phis. Comm. **74**, 327.

Toral, R. and Ferreira, A.L. (1994) *Proceeding fo the Conference Physics Computing, '94*, p. 265, R.Gruber, M. Tomasini, editors.

Valls, O.T. and Mazenko, G.F. (1986): Phys. Rev. **B 34**, 7941.

van Kampen (1981): *Stochastic Processes in Physics and Chemistry*, North-Holland.

Vicsek, T. (1989):*Fractal growth phenomena*, World Scientific, Singapore.

Zinn-Justin, J. (1989): *Quantum Field Theory and Critical phenomena*, Oxford. U. Press.

Simulating Granular Media on the Computer

H.J. Herrmann

P.M.M.H., (U.R.A. 857), E.S.P.C.I.,
10 rue Vauquelin, 75231 Paris, France

Abstract: Granular materials, like sand or powder, can present very intriguing effects. When shaken, sheared or poured they show segregation, convection and spontaneous fluctuations in densities and stresses. I will discuss the modeling of a granular medium on a computer by simulating a packing of elastic spheres via Molecular Dynamics. Dissipation of energy and shear friction at collisions are included. In the physical range the friction coefficient is found to be a linear function of the angle of repose. On a vibrating plate the formation of convection cells due to walls or amplitude modulations can be observed. The onset of fluidization can be determined and is in good agreement with experiments. Segregation of larger particles is found to be always accompanied by convection cells. There is also ample experimental evidence showing the existence of spontaneous density patterns in granular material flowing through pipes or hoppers. The Molecular Dynamics simulations show that these density fluctuations follow a $1/f^{\alpha}$ spectrum. I compare this behavior to deterministic one-dimensional traffic models. A model with continuous positions and velocities shows self-organized critical jamming behind a slower car. The experimentally observed effects are also reproduced by Lattice Gas and Boltzmann Lattice Models. Density waves are spontaneously generated when the viscosity has a nonlinear dependence on density which characterizes granular flow. We also briefly sketch a thermodynamic formalism for loose granular material. In a dense packing non-linear acoustic phenomena, like the pressure dependence of the sound velocity are studied. Finally the plastic shear bands occurring in large scale deformations of compactified granular media are investigated using an explicit Lagrangian technique.

1. Introduction

Many rather astonishing phenomena are known to occur when granular materials like sand or powders move [1 – 4] Examples are the so-called "Brazil nut" segregation [5 – 8], heap formation under vibration [9 – 11], density waves emitted from outlets [12] and 1/f noise in the power spectra of local forces [13]. All these effects originate in the ability of granular materials to form a hybrid state between a fluid and a solid: When the density exceeds a certain value, the critical dilatancy [14, 15], it is resistant to shear, like solids, while below this density it will "fluidify". This fluidified state can be rather complex, specially in the presence of density fluctuations and density gradients.

Particularly suited to study this fluidization is an experiment where sand is put on a loudspeaker or on a vibrating table [9 – 11, 16, 17]. Under gravity the sand jumps up and down and although kinetic energy is strongly dissipated, collisions among the grains reduce its density thereby allowing it to flow ("fluidization"). Under certain circumstances flow between top and bottom can occur in form of convection cells as has been observed experimentally in the case of inhomogeneities in the amplitude of the vibration [18]. More striking is that sand spontaneously can form heaps [9 – 11] as first described already in 1831 by Faraday. Also within these heaps convection occurs which might even be the motor for the heap formation: Inside the heap the sand rises, pops out at the top and then slides down on the surface. Usually these heaps have complicated shapes that change in time and sometimes one also observes ripples and other regular structures on their surface [19]. When particles of different sizes but equal density are put on the vibrating plate the larger particles tend to rise and after some time one observes a segregation into regions with larger particles and regions with smaller particles. When the vibration of the plate also has a horizontal component the material will flow in one direction, a technique often used in powder transport.

In order to formalize and quantify the complicated rheology of granular media various attempts have been made. Continuum equations of motion and a kinetic theory [3, 20], thermodynamic formulations [21 – 23], cellular automata [24 – 26] and a random walk approach [27] have been proposed. But many of the above mentioned effects have so far eluded a satisfactory explanation of phenomena, like size segregation or density fluctuations. This is because it is very difficult to incorporate into these theories static friction, local rotations and other relevant microscopic mechanisms.

To gain a better understanding of the rheological effects of granular media it is therefore very useful to do computer simulations [28, 29]. For over a decade discrete methods have been used where instead of a continuum one treats the granular material as an assemblage of particles interacting through their contacts. This technique was introduced by Cundall [30] to study the motion of rock masses. Since then it has been applied to statistical micromechanics [31, 32], constitutive behavior of granular soils [33], creep of soils [34], analysis of rock-support interaction [35] and other applications of soil mechanics [36]. These techniques have also been applied to model size segregation [8, 37], outflow from a hopper [38, 39], shear flow [40] and flow down an inclined chute [41].

In the following we will discuss these techniques. In particular we will present Molecular Dynamics (MD) [42, 43] simulations of inelastic particles with an additional shear friction. We present data in two dimensional systems for the onset of fluidization [44, 45] and give evidence for the occurrence of convection cells due to inhomogeneities in the vibration amplitude or due to walls [46, 47]. We also report on measurements of the velocity and density profiles of powder transported on a vibrating belt [48].

A series of experiments [12, 49 – 51] have given evidence for strong density fluctuations when granular material flows under the action of gravity. Baxter et al [12] used X-rays to visualize these wave-like patterns emanating from the outlet of a two dimensional wedge-shaped hopper. Similarly rather erratic shock-like density waves have been observed in flow through pipes [52]. Another experimentally observed ubiquitous phenomenon in granular media seems to be $1/f^\alpha$ noise. Baxter [53] observed power law decay in the frequency dependent forces that act on the wall of a hopper. For avalanches going down the slope of a sand pile theoretical considerations of self-organized criticality [54] led to the proposal that their size and life time distributions were power laws which was in fact only verified experimentally on very small piles [55].

In this course I will show using four models of different degree of theoretical abstraction [25, 52, 56 – 58] that the observed density patterns are composed of at least three different elements: (1) solitary waves of high density, similar to the kinematic waves of classical traffic jams, due to the concave density dependence of the flux, (2) low density waves due to the sharp increase of viscosity as function of density and (3) a background with a power-law spectrum due to self-organized criticality in the avalanches occurring inside the material because of instabilities intrinsic to inelastically colliding particles. All three phenomena are eventually consequence of dissipation and compressibility. Interestingly the fluctuations in stresses against the walls which also have a power-law spectrum don't seem to be directly related to these density patterns [59].

Loose granular media can in fact be described by a thermodynamic formalism [21] in which sand grains are treated in a similar way as molecules in a gas. This analogy reminds us that Molecular Dynamics was in fact originally developed to describe the motion of molecules.

Granular media at large densities have their own universe of funny effects. We all know that a packing of sand behaves like a solid when pushed but offers no resistance to pulling. Then there is the famous Reynold's experiment in which due to dilatancy a deformed elastic bag full of sand increases in volume when deformed. Another beach experiment [60] consists in making a sand tower having randomly spread inside some meters of fishing line. The tower withstands a considerable amount of pressure. Studying more in detail sound propagation in a granular packing shows a strong dependence of the wave velocity on pressure [61] and a subtle dependence of the local acceleration of a particle on the amplitude and the temperature due to the complex network that transmits the stresses. Large deformations of granular media occur on shear planes [62] due to a plastic instability [63].

In this course I will show results [64] of large scale MD (64.000 particles) simulations made on the CM-2 for the "mirage effect" of sound velocity and for standing waves in a rectangular box. In order to understand the formation of the fractal network of shear bands in a slowly deformed box I present results [65] from FLAC an explicit Lagrangian technique that makes use of the classical non-associate Mohr-Coulomb plastic yield criterium that is traditionally used as continuum theory for dense granular materials.

2. Molecular Dynamics Technique

As opposed to usual molecular gases the elementary units of granular materials are mesoscopic grains consisting of many atoms each ($10^{15} - 10^{25}$). When these objects interact (collide) the attractive potentials of the individual atoms are unimportant and completely different mechanisms must be considered. It is important that on a microscopic scale the surface of the grains is rough. Solid friction is the immediate consequence: When two touching grains are at rest with respect to each other a finite force F_s is needed to trigger relative motion (*static friction*), while moving against each other a finite force F_d is needed to maintain the motion (*dynamic friction*). $F_d < F_s$ and both only depend on the normal force and neither on the velocity nor on the area of contact (*Coulomb law*). No doubt, this picture is idealized and an entire discipline, called tribology, has evolved to study solid friction in depth [66]. For our purpose it is, however, more convenient to concentrate only on the basic mechanisms because we are interested in explaining the generic effects of granular rheology without entering into material-dependent details. Friction has a crucial consequence on the level of the grains, namely that the system does *not* conserve energy as opposed to what happens on the molecular level. Another source of dissipation can be plastic deformation of grains due to the normal force acting at collisions. Again it seems most important that dissipation does occur due to the normal momentum and the complications arising from the non-linearities of plasticity seems less relevant. We will therefore in the following assume very simple dissipation laws. In fact more complicated laws have been used increasing the number of parameters without giving qualitatively different answers.

Let us consider a system of N spherical particles of equal density and with diameters d chosen randomly from a homogeneous distribution of width w around d_0. When two particles i and j overlap (i.e. when their distance is smaller than the sum of their radia) three forces act on particle i:

1) an elastic restoration force

$$\boldsymbol{f}_{el}^{(i)} = k_n m_i \left[\mid \boldsymbol{r}_{ij} \mid - \frac{1}{2}(d_i + d_j) \right]^{\beta} \frac{\boldsymbol{r}_{ij}}{\mid \boldsymbol{r}_{ij} \mid} \quad , \tag{2.1a}$$

where k_n is the elastic modulus (normalized by the mass), $m_i \propto d_i^2$ the mass of particle i and $\boldsymbol{r}_{ij}$ points from particle i to j. For Hooke's law $\beta = 1$ but in the case of spheres in three dimensions one has to choose $\beta = 3/2$ [67] and for conical contacts one should take $\beta = 2$.

2) a dissipation due to the inelasticity of the collision

$$\boldsymbol{f}_{diss}^{(i)} = -\gamma_n m_i (\boldsymbol{v}_{ij} \cdot \boldsymbol{r}_{ij}) \frac{\boldsymbol{r}_{ij}}{\mid \boldsymbol{r}_{ij} \mid^2} = -\gamma m_i v_{ij}^n \quad , \tag{2.1b}$$

where γ_n is a phenomenological dissipation coefficient and $\boldsymbol{v}_{ij} = \boldsymbol{v}_i - \boldsymbol{v}_j$ the relative velocity;

3) a shear friction force which in its simplest from can be chosen as

$$f^{(i)}_{shear} = -\gamma_s m_i (\boldsymbol{v}_{ij} \cdot \boldsymbol{t}_{ij}) \frac{\boldsymbol{t}_{ij}}{|\boldsymbol{r}_{ij}|^2} = -\gamma_s m_i v^t_{ij} \quad , \tag{2.2a}$$

where γ_s is the shear friction coefficient and $\boldsymbol{t}_{ij}$ is the vector $\boldsymbol{r}_{ij}$ rotated by 90°. Eq. (2.2a) is a rather simplistic description of shear friction. In many applications (arching, heap formation) it is important to include real static friction [68] which can be done by a static friction force [31]: When two particles start to touch each other, one puts a "virtual" spring between the contact points of the two particles. If δs is the *total* shear displacement of this spring during the contact the restoring frictional force is $k_s \delta s$ (static friction). The maximum value of the restoring force is then according to Coulomb's criterion proportional to the normal force F_n and the proportionality constant is the friction coefficient μ. Cast into a formula this gives a friction force

$$f^{(i)}_{friction} = -\text{sign}(\delta s)\min(k_s \delta s, \mu F_n) \quad . \tag{2.2b}$$

where δs is the shear displacement integrated over the entire collision time. When particles are no longer in contact with each other the spring is removed. Main source of static friction is the geometrical roughness of the surfaces [69] and the same effects of particle stopping can be obtained also without (2.2b) by using particles of complicated shapes, like crosses or polygons [70, 71].

It is not straightforward to implement the above technique when the particles are allowed to rotate, i.e. able to roll on each other. In fact, when particles have strong deviations from the spherical shape rotations are suppressed. Often it is however useful to go an intermediate way and to include dynamic friction but not static friction and allow for the particles to have rotations [8, 38, 40]. In that case one uses a combination of Eqs. (2.2a) and (2.2b):

$$f^{(i)}_{dyn} = -\min(\gamma_s v^t_{ij}, \mu F_n) \tag{2.2c}$$

and also introduces equations of motion for the angular momentum of the particles.

When a particle collides with a wall the same forces act as if it would have encountered another particle of diameter d_0 at the collision point. One force that acts on all particles pulling them down is gravity, $g \approx -10\text{m/s}^2$. In most simulations presented here we use a fifth order predictor-corrector MD with $10^3 - 10^5$ iteration steps per cycle depending on the largest velocities appearing in the application. This algorithm vectorizes on the Cray-YMP, running at about 10 μsec per particle-update for $N = 200$. The programs were also run on 8 or 16 processors of an Intel iPSC/860 or on an IBM RS/6000.

3. Angle of Repose and Angle of Marginal Stability

As opposed to fluids the surface of granular media at rest is usually not flat but can show undulations (ripples, dunes) or form mountains (sand piles, heaps). This is due to the fact that up to a maximal angle, the so called angle of maximal stability θ_m, grains can arrange in a statically stable way. The angle of a heap of granular material can fluctuate between θ_m and a dynamic angle θ_r. Various experiments can be deviced to obtain these angles.

Numerical studies have been made to investigate the angles θ_r and θ_m in two dimensions [68]. They start by putting N particles randomly in a box and wait sufficiently long until under the action of gravity they have all settled on top of each other and come to rest. Next one side wall is removed. The particles violently fall out forming a big avalanche and after some time only a few remain inside the box. In Fig. 1 we see the pile that is left. It forms a certain angle which is what is called the angle of repose θ_r.

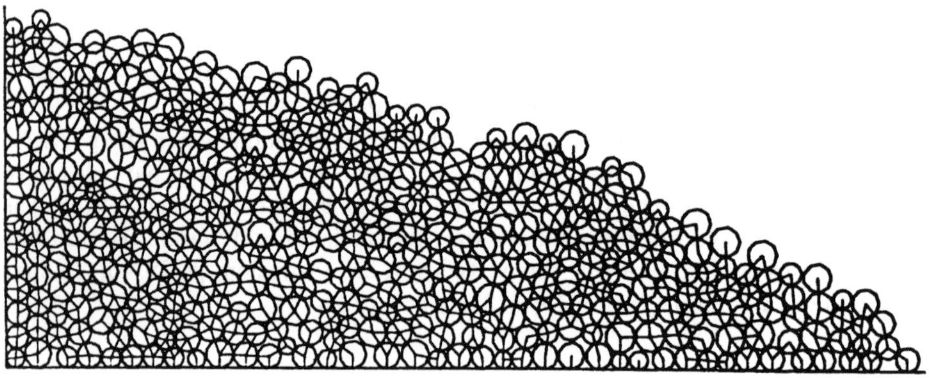

Fig. 1. Static pile obtained after the right wall from the box was removed. The parameters used are $\mu = 0.2$, $k_n = 10^6 g/sec^2$, $k_s = 10^4 g/sec^2$, $\gamma_n = 500$ and $\gamma_s = 5$ (from Ref. [68]).

In order to measure the angle θ_r from Fig. 1 one can divide the box into several vertical cells and for each cell find the highest position of the particles inside this cell. The line joining these positions can be considered as the surface of the pile. Next there are different ways to measure the slope: Either by joining the two end points, or by fitting a straight line through the points of the surface by linear regression, or by fitting a parabola through the cummulated values of the surface, etc. All definitions are slightly different but should give the same result for large enough systems.

In principle the static friction coefficient μ defined in (2.2c) should determine the slope of the pile. Since the angle of repose can easily be determined experimentally it is of interest to find the relation between θ_r and μ. In Fig. 2 we see that this relation is essentially linear for $\mu < 0.2$. Recently [72] in fact these calculations have been extended to larger values of μ and a curvature was observed which is close but not identical to the theoretical relation $\mu = tan\theta_r$. For the range of parameters which interest in practice, however, the linear relation is a very good fit.

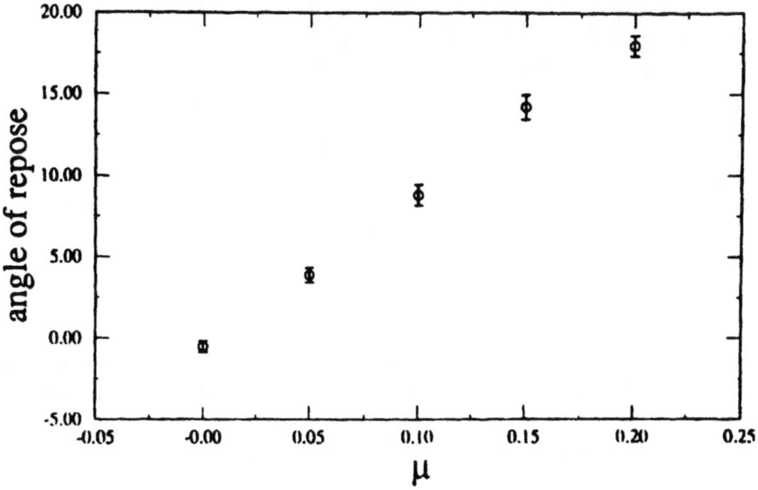

Fig. 2. Angle of repose θ_r as a function of μ for the same parameters as in Fig. 1 (from Ref. [68].

A method to obtain also the angle of maximal stability θ_m is the following [68]: Consider an empty box without a right wall and glue one layer of particles on the bottom. One monitors the largest velocity of the particles and each time this value falls below a sufficiently small value v_{max}, one inserts a new particle at the upper left corner of the box. This corresponds to experiments that have been used to measure avalanche statistics of small sand piles [55]. In Fig. 3 we show the angle as a function of time for very small piles. As predicted by the experiments [55] for such small piles the fluctuations of the angle are very large.

4. Simulating Granular Media on a Vibrating Plate

Let us now consider that the particles are placed into a container of length L that is open on the top and has either periodic boundary conditions or fixed walls in the horizontal direction. The bottom of the container is subjected to a vibrating motion described by:

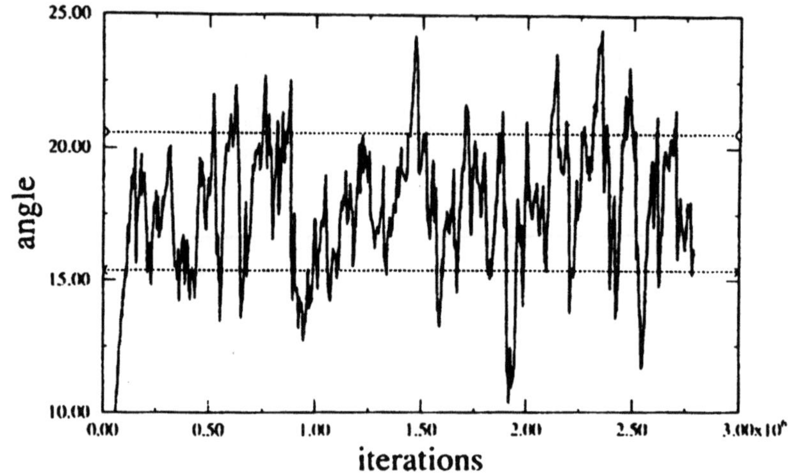

Fig. 3. Angle of the pile made by adding particles one by one as function of time (measured in MD-iterations). The dotted lines indicate the characteristic width of the fluctuations. The width of box is only four particle diameters, $v_{max} = 0.1$ and all the other parameters are the same as in Fig. 1 (from Ref. [68]).

$$z_0(t) = A(x)\sin(2\pi ft) \quad , \tag{3.1a}$$

f being the frequency and amplitude A. In some cases we will also consider an explicit spatial modulation of A of the form

$$A(x) = A_0(1 - B\cos(2\pi x/L)) \quad . \tag{3.1b}$$

For vibrating conveyor belts this plate undergoes harmonic oscillations in both horizontal (x) and vertical (z) directions according to

$$x(t) = A_x \sin(2\pi ft) \quad \text{and} \quad z(t) = A_z \sin(2\pi ft) \tag{3.2}$$

where f is the frequency and A_x and A_z are the amplitudes in x and z directions, respectively. The corresponding angle of the composed oscillation is $\alpha = \arctan(A_z/A_x)$.

Various initial positions of the particles can be considered: They can for instance be placed regularly on the bottom of the container or put at random positions inside a space several times as high as the dense packing. The initial velocities are either zero or randomly chosen. After that the particles are allowed to fall freely under gravity and relax for a time that corresponds to ten or twenty cycles of the vibration. The displacements, velocities and energies are then measured by averaging over up to 200 cycles.

Clément et al. [16] reported experimental observations of a "fluidized" state in a 2D vertical packing of steel spheres submitted to vertical vibrations. They periodically shake (at $f = 20Hz$) 300 steel beads inside a trapezoidal cell with side walls tilted by 30° with respect to the vertical axis. Positions and velocities of the particles were obtained by photographing the system periodically and then averaging over 15 snapshots taken at a given constant phase φ. Velocities were obtained from averages over a time interval τ around the phase φ. From a plot of the density of particles they argued fluidization to occur in the upper region of the packing. They found that the mean density does not depend on the phase of the vibration, implying the appearance of a steady state preserving the density profile at all times, independent of the up and down collective motion.

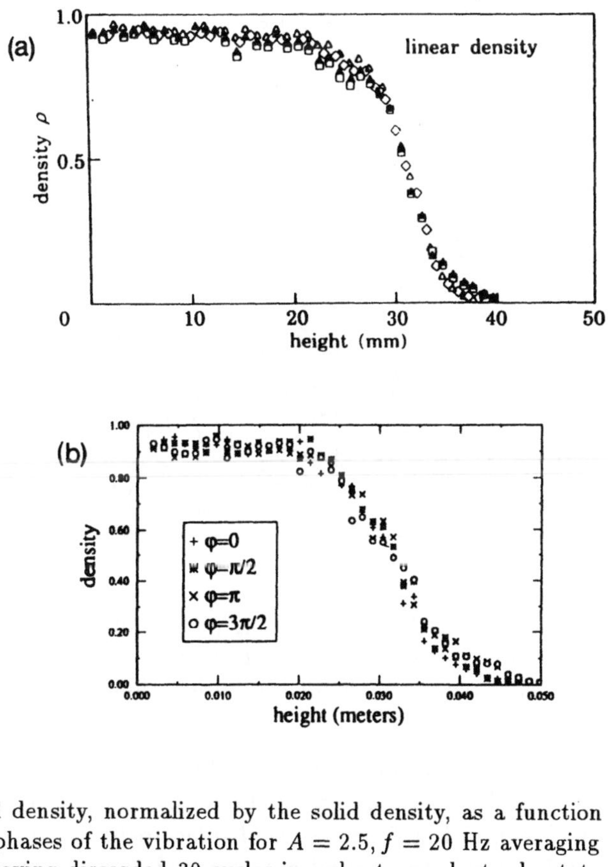

Fig. 4. Local density, normalized by the solid density, as a function of the height z for different phases of the vibration for $A = 2.5, f = 20$ Hz averaging over 30 shaking cycles after having discarded 30 cycles in order to reach steady state; (a) experiment of Ref. [16] and (b) simulation of Ref. [44].

Simulations of precisely the same geometry and number of particles as in the experiment have been performed [44] using the MD technique described in section 2 with the simplest type of friction, namely that of (2.2a). Fig. 4 shows local densities along the z-axis, evaluated at different phases as described in the experimental paper: (a) is the experiment [16] and (b) the simulation [44]. The numerical curves were obtained by averaging the local density over the 15 ms following each phase φ and over 30 shaking cycles after discarding 30 "transient" cycles. As can be seen from this figure, the model correctly reproduces the experimental behavior of the beads, producing the same ϕ-independent, smoothly varying density profile as function of the height z. To check whether the present model is at all able to display a transition from a solid- to a fluid-like state both frequency f and amplitude A of the oscillations were varied. The trajectory of a selected "tracer" particle was monitored [44]. In the solid-like case the tracer particle remains confined to a very small region while in the fluid-like case the trajectory seems to explore the entire box. It is important to note that both situations can occur for the same value of Af^2 which means that even close to the onset of fluidization Af^2 is not a good scaling variable. It has in fact been shown [45] that in the fully fluidized regime the data can be scaled with the energy ($(Af)^2$ in one and $(Af)^{3/2}$ in two dimensions).

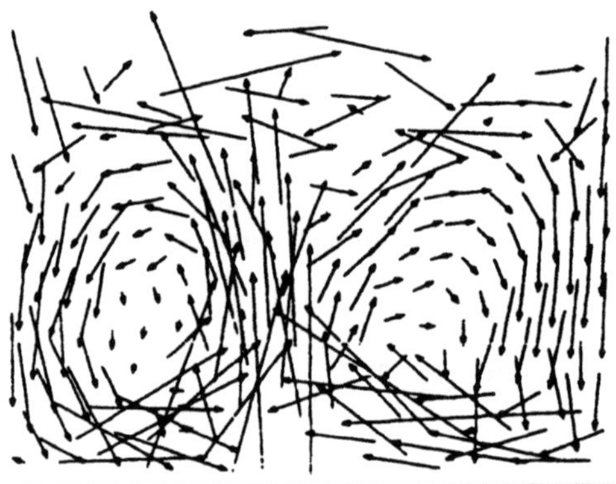

Fig. 5. Displacement of the particles after 15 cycles for $f = 70$Hz using 200 particles in a box with periodic boundary conditions of size $L/d_0 = 20$ with $A_0 = 1.5d_0$, $B = 0.5$, $w = 0$, $k_n = 5000/d_0$, $\gamma_n = 20g$, $\gamma_s = 200g/f$, $d_0 = 1mm$ (from Ref. [47]).

Let us next consider the case of a spatial modulation in the amplitude of the vibration, i.e. $B \neq 0$ in (3.1b), using periodic boundary conditions [47] and again a friction given by (2.2a). In Fig. 5 we see the displacements of the particles after 15 cycles for $B = 0.5$. Clearly the particles flow upwards in the center where the amplitude of the vibration is larger and form two convection cells. If the dissipation coefficient γ_n is increased by a factor of ten the convection is completely suppressed while it is quite insensitive to γ_s, even if $\gamma_s = 0$. The elastic modulus also has only a very weak influence as long as it remains larger than $10^3 g/sec^2$. The initial condition plays no noticeable effect showing that convection is no transient effect. The polydispersity w of the particles only slightly distorts the shape of the convection cells.

The strength of the convection was measured quantitatively by recording the average vertical components of the velocities of the particles in the center and at the edges. These quantities have also been measured experimentally by Rátkai [18]. The strongest convection for the aforementioned parameters is obtained around 60 Hz and it increases dramatically with the amplitude A_0 as was also seen in the experiment [18]. This resonance seems to be the driving force of the convective motion.

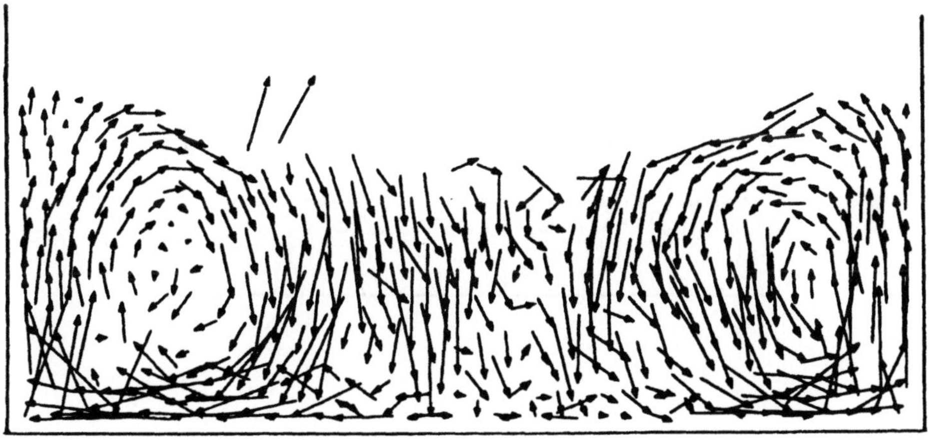

Fig. 6. Displacements after 10 cycles in a system with fixed vertical walls for $B = 0$, $w = 0.5$, $k_n = 5000g/sec^2$, $f = 20\text{Hz}$, $N = 400$, $L = 40d_0$, $A_0 = 3.0d_0$, $d_0 = 1mm$, $\gamma_n = 80g$, $\gamma_s = 0$ averaged over 10 cycles.

A completely different type of convection can be caused by the existence of fixed vertical walls without any modulation of the amplitude [47], i.e. for $B = 0$. One sees in Fig. 6 for $\gamma_s = 0$ convection cells where the motion of the particles at the wall is upward. On the other hand, when $\gamma_s \neq 0$ there is a very strong downward drag at each wall giving rise to a convection in the opposite sense. The two convection cells remain attached to the walls showing that the walls are at the origin of these cells. One also recognizes a slight heap formation close to the wall which might be a first sign of the sand heaps discovered by Faraday [9 – 11].

Let us analyze the origin of the convection due to fixed vertical walls. In the case of no shear friction the vertical walls do not transfer any vibrating motion of the container but represent only a steric hindrance to the flow. In this case, the following scenario applies: When, after levitating from the plate, the packing falls back on the bottom of the container only the horizontal component of the velocities of the particles arriving first will survive collisions with the downwards vertical motion of the rest of the packing that follows behind. So flow parallel to the bottom plate will spontaneously appear and is reinforced at each cycle. This parallel flow will only survive in regions where it is coherent and the size of these regions will grow due to the reinforcement. When one of these regions collides with a vertical wall the flow must go upwards since it cannot go anywhere else. This explains not only the orientation of the convection but also why the convection cells are attached to the walls as seen in Fig. 6. The driving force for these cells are therefore the horizontal flows along the bottom plate.

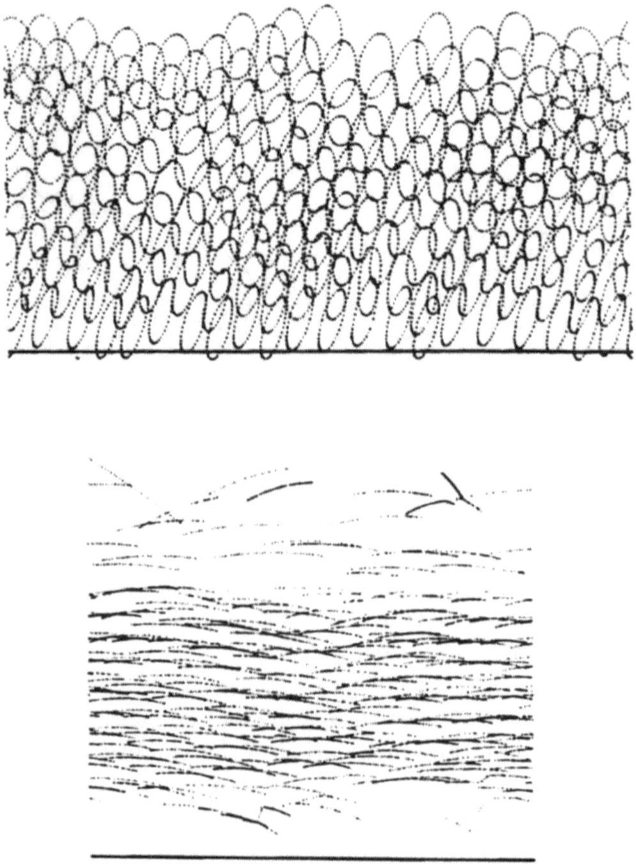

Fig. 7. Trajectories of the particles in steady state during one single cycle. The position of each particle is plotted after every 50 time-steps. The plots were obtained for $A_z = d_0$, $\alpha = \pi/4$, $\gamma_n = \gamma_s = 50g$ and (a) $f = 10$ Hz and (b) $f = 80$ Hz, top and bottom respectively.

When shear friction with the wall is present a different mechanism sets in: While the particles are pushed up and start to levitate, the packing is still quite compressed and a strong pressure is exerted on the walls giving rise to strong

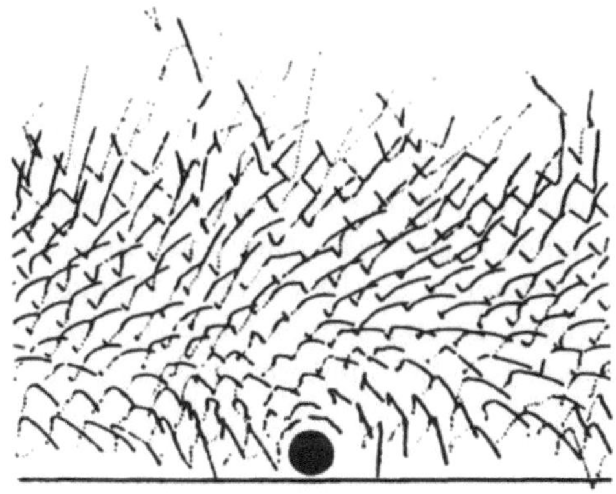

Fig. 8. Trajectories of the particles flowing in the presence of an obstacle for $A_z = d_0$, $\alpha = \pi/4$, $\gamma_n = \gamma_s = 50g$. The frequency was 80 Hz. The obstacle of diameter $d_1/d_0 = 1.5$ is given by the full circle.

shear friction while the relative motion of the particles with respect to the walls is upward. When afterwards the particles fall back and have downward relative motion with respect to the wall the packing is much looser and the shear friction much less efficient. Therefore the upward motion of the particles with respect to the wall is slowed down stronger, resulting in a net down drag along the wall. If γ_s is strong enough this effect can overcome the effect described in the above paragraph and the convection can reverse its orientation.

Let us next discuss the behavior of vibrating conveyor belts [48], i.e. granular material under harmonic vibrations having a given angle with respect to the direction of gravity as described in (3.2). Vibrating conveyor belts as a means of transportation are very typical for granular media, since neither solids nor fluids can be moved on them and are used for instance in the pharmaceutical industry to transport pills [73]. Let us consider the trajectories of the particles during one cycle of shaking. When the frequency is low enough all the beads move synchronously along elliptic trajectories curves as shown in Fig. 7a. The tilting angle of these ellipses increases with the angle of vibrations. For smaller shear friction coefficients γ_s the tilting angle tends to $\pi/2$, provided the vibration frequency is low enough. When the beads start to flow, the character of their trajectories changes: at not too high frequencies they move along sinusoidal. With increasing frequency, the trajectories become flatter and at the highest frequencies we observe a nearly horizontal flow (see Fig. 7b). A decrease of the vibration angle makes the horizontal motion more pronounced. A similar effect occurs when the friction coefficients are increased. For vanishing shear friction γ_s the beads move essentially vertically.

It is also interesting to check how a circular obstacle inserted into the system influences the flow. To this end, a fixed circular body can be inserted at $x_1 = L/2$, $z_1 = A_z$. The diameter of this obstacle has been varied from $d_1 = 0.1d_0$ to $d_1 = 2.5d_0$. The parameters characterizing the interactions of the obstacle with the particles are the same as in the case of particle-particle interactions. Note that due to the periodic boundary conditions, the obstacle is repeated along the belt. Even the presence of a rather small obstacle rapidly slows down the flow. In Fig. 8 we see the trajectories of the particles for an obstacle of $d_1 = 1.5d_0$. Figs. 7b and 8 have the same parameters so that without the obstacle Fig. 8 would look like Fig. 7b. Clearly the presence of the obstacle changes the trajectories of all the particles considerably. So, we cannot treat the obstacle as only locally influencing the flow, because the stiff repulsion between particles generates long-range correlations.

5. Simulating Size Segregation

One of the most puzzling phenomena encountered in granular matter is size segregation: When a mixture of grains of the same material (equal density) but different size is shaken in a container the larger particles rise to the top. This effect has been extensively studied experimentally [5, 74, 75] and has much importance in numerous industrial processes [76]. Recently this so called "Brazil nut effect" has attracted much interest among physicists [2].

Size segregation inevitably seems to contradict equilibrium statistical mechanics since the density of the overall packing increases with polydispersivity and so gravity makes situations with larger particles on the bottom energetically more favorable. Rosato et al. [6] proposed a Monte Carlo algorithm and put forward a kinetic argument to explain segregation using the fact that smaller particles are more mobile. In the same year Haff and Werner [8] did Molecular Dynamics simulations of rather small systems and claimed that segregation was essentially a consequence of solid friction and the rotation of the particles. Jullien et al. [7] used a piling technique which is non–stochastic as compared to the one of Ref. [6] and found a critical ratio $\mathcal{R}$ for the radia of spherical particles below which no segregation occurs. Based on these ideas Duran et al. [77] formulated a geometrical theory for segregation in which the small particles glide down along the surface of the larger particle, the critical ratio $\mathcal{R}$ of radia being between continuous and discontinuous gliding. They also claimed experimental evidence for two types of dynamics and visualized the discontinuous ascent of the larger particle through stroboscopic photos. Jullien et al. [78] reproduced the discontinuous dynamics by including horizontal random fluctuations into their model.

Parallel to these local theories there has been the "convection connection": It is known experimentally [18] and numerically [46, 47] that shaken assemblies of spheres form convection rolls which are attached to the walls of the container. For weak shaking the convection rolls only appear on the surface. Knight et al. [79] showed experimentally that segregation was due to this convection and the fact that larger particles have a harder time entering again into the bulk once they are on the surface. They also verified an exponential decay of the convection strength as function of depth for weak shaking [80]. Duran et al. [81] verified segregation due to convection in two dimensions for strong shaking and claimed that above mentioned local mechanisms are at work at weak shaking. Their stroboscopic pictures showed convection cells above the particles.

Large scale Molecular Dynamics simulations show [37] that also for weak shaking convection is responsible for segregation but in a more intricate way: Under certain conditions the larger particle is able to pull down the convection rolls due to the more efficient momentum transfer and then rise within its flow. Because of the exponential decay of convection with depth the ability to rise critically depends on the vertical position of the larger particles.

Throughout the simulation the parameters were $k_n = 3 \cdot 10^6\ g/sec^2$ (elastic modulus), $\gamma_N = 100\ Hz$, $\gamma_S = 1\ Hz$ (phenomenological normal and shear friction coefficients) and $\mu = 0.5$ (Coulomb friction parameter). We considered $N = 950$ particles were considered with diameters uniformly distributed in the interval $d_i \in [1.7, 2.3]\ cm$ and with masses $m_i = 2\pi R_i^2 \rho$, $\rho = 1\ g/cm^2$. The particles are put into a two–dimensional box having walls made of particles with the same material characteristics as the grains and which vibrates vertically according to $z_0(t) = A \cdot \sin(2\pi f t)$ with $A = 2\ cm$. Gravity acts in negative z–direction $F_i^{gr} = -m_i\ g$, $g = 981\, cm/sec^2$. The time step for the numerical integration of the Gear predictor–corrector scheme of 5th order was $\Delta t = 5 \cdot 10^{-5} sec$.

Segregation and convection behavior have been investigated [37] as function of the vibration frequency f in two different systems, either all particles are small or we add one single big particle of radius $R_1 = 4\ cm$ located in the center of the box close to the bottom. In order to investigate closer what happens at the onset of segregation we keep all the other parameters fixed. Fig. 9 shows the convection cells without the larger particle (left) and with the larger particle (right) for the frequency $f = 2.8\ Hz$, i.e. at the onset of segregation. The convection cells with and without the big particle, differ significantly while they are quite similar for larger or smaller frequencies. This indicates that at the onset the presence of the big sphere triggers the onset of convection which finally leads to segregation. Indeed we find that convection is always present when segregation happens. It is important to note that if one is close enough to the onset of segregation just putting the larger particles one row lower can entirely suppress the effect of segregation. This dependence of segregation on the height is quite strong and has not been discussed in the literature. By changing the frequency f very slowly and measuring the convection flow through a plane at a certain height we observed that the transition from the fluctuation regime to the convection regime is very sharp within the numerical precision ($\Delta f = 0.05\ Hz$). Moreover, when increasing the frequency the transition occurs almost exactly at the same frequency as when decreasing the frequency, i.e. there is no hysteresis.

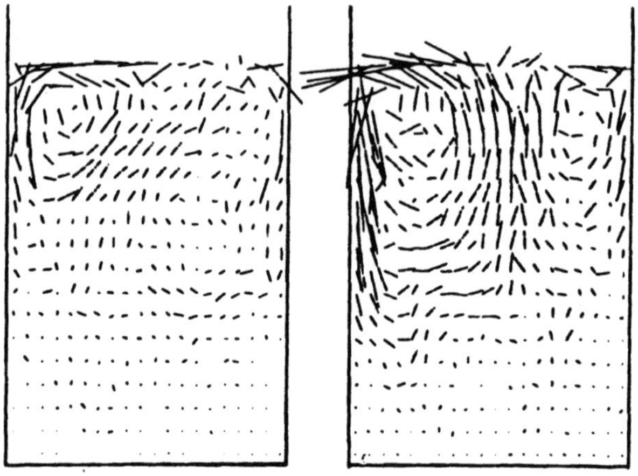

Fig. 9. Convection rolls in systems without (left) and with the larger particle (right) for $f = 2.8\ Hz$. The big particle triggers convection rolls. The velocities were obtained by averaging over 50 shaking periods of the box (from Ref. [37]).

The triggering of convection cells by the big particle is investigated more quantitatively by calculating the convective flux Φ defined as the sum of material (mass) flow in the center of the box j_{top} and the flow close to the walls j_{bot} by considering that these flows have opposite signs. The flows j_{top} and j_{bot} are defined by adding the number of particles which move in one direction minus the ones moving in the opposite direction and is measured in particles per cycle time. In fact we measure for each particle if the positions at subsequent nodes of the vibration are on different sides of a height line, where the height of the box was divided into 80 height lines between the bottom of the box and the surface of the packing. Fig. 10 shows the convective flux Φ through planes at different heights D for both systems and for the three different frequencies. For $f = 2.6\ Hz$ we find almost no directed flow but only fluctuations. For $f = 3\ Hz$ both systems, with and without large particle, behave similarly, as could also be observed in Fig. 9. For $f = 2.8\ Hz$, however, the convection cells clearly extend deeper due to the existence of the larger particle. Apparently the larger particle is able to pull the convection cells down.

This effect can be explained by the fact that in the region around the large particle the accelerations are higher since the momentum is transferred with less dissipative loss through the larger particle than through a corresponding pile of smaller particles of the same volume. We measured the sum of the absolute values of the forces of all the particles in a region around the position of the large particle and averaged it over time. The region was ring shaped with the inner border of radius 4.5 cm and the outer border of radius 8 cm. At the onset

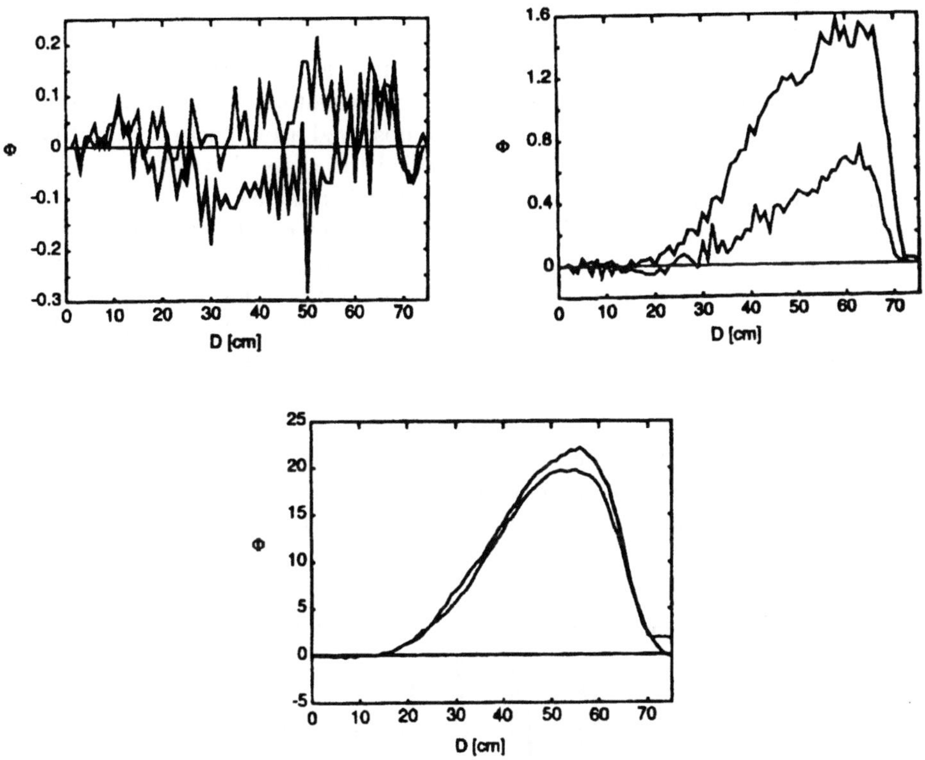

Fig. 10. Strength of the convection rolls measured through the flux Φ as a function of the height D, $f = 2.6\ Hz$ (upper left figure), $f = 2.8\ Hz$ (upper right figure) and $f = 3\ Hz$ (lower figure). The height D is measured such that the bottom is at the origin of the axis. The flux Φ is measured in units of particles per period. In the central case the convection rolls are stronger and reach deeper inside the material in the presence of the big particle. This triggered convection roll catches the big particle and forces it to rise to the top. (from Ref. [37])

of segregation ($f = 2.8\ Hz$) the average force of the small particles around the large particle is about 15% larger than that of the small particles in the same region in a system containing no large particle. This effect is strongest in the lower part of the ring shaped region. For $f = 2.6\ Hz$ the difference is only 5%. Therefore the accelerations in the region around a large particle are larger than if no particle would be present. We believe that this increase in high frequency oscillations is responsible for pulling the convection rolls down. It is, however, interesting to note that the granular temperatures (kinetic energy) in this region is roughly the same in the two systems.

One can see from Fig. 9 that the convection cells decay very sharply in strength but that even in the deep regions some essentially horizontal motion occurs. This is reminiscent of the stroboscopic pictures of [81] implying that even in the low acceleration regime some particles move inward horizontally. Within our framework this motion could be interpreted as the exponentially weak tail of the convection rolls.

Next we investigated the dependence of the onset of convection on the ratio of radia $\mathcal{R}$. Note that in our case $\mathcal{R} = R_1/1cm$ because the mean radius of the small particles is $1cm$. For $f = 3.2\ Hz$ we observed a big particle of radius $\mathcal{R} = 4.0$ moving up immediately. We also studied the cases $\mathcal{R} = 3.5$, $\mathcal{R} = 3.0$, $\mathcal{R} = 2.5$ and $\mathcal{R} = 2.0$. In these cases the large particle remains a certain waiting time on the bottom before it suddenly moves up quite rapidly. Fig. 11 shows a typical evolution of the vertical position of the big particle with $\mathcal{R} = 2$. The waiting times do not noticeably depend on $\mathcal{R}$ being of the order of 30 *sec*. Once the large particle comes to the top it performs an oscillating motion going up and down (whale effect) that has also been observed experimentally [82]. This motion seems due to the convection rolls: In the case where $\mathcal{R}$ is smaller the oscillating motion is more regular because the larger particle has less difficulty reentering into the bulk from the surface and follow the convective motion. For larger $\mathcal{R}$ the larger particle has more difficulty reentering thus leading to a more erratic horizontal motion. Particles with smaller $\mathcal{R}$ also seem to dip deeper into the bulk showing that the convection cells can move them more efficiently.

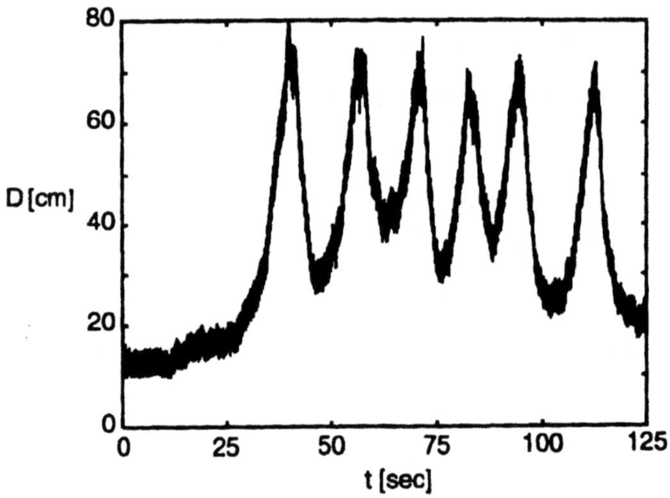

Fig. 11. Evolution of the vertical position of the big particle as function of time, $\mathcal{R} = 2.0$ for $f = 3.2\ Hz$ (from Ref. [37]).

Concluding, it seems that segregation of granular media in a vibrating box in two dimensions is intimately connected to convection. The larger particles, surrounded by a region of higher acceleration, loosen the material thus deepening the penetration of the convection rolls from the top into the medium. After some waiting time the larger particles are caught by the lower part of the rolls and pulled up. This triggering effect is only relevant close to the characteristic onset frequency of segregation which is sharply defined and strongly dependent on the initial depth of the large particle. Once the large particle is on the top it periodically goes up and down (whale effect) driven by the convection cells.

In addition to convection arching and geometry are also important as pointed out by many authors. How the large particles move in the exponentially weak convection field before being lifted upwards and the $\mathcal{R}$-dependence of the mobility are probably best described by Duran et al.'s local arching mechanisms [77, 81]. The fact that the whale effect of larger particles is less pronounced is certainly due to their lower mobility because of steric hindrance effects as formulated by Rosato et al. [6].

Many of the details of segregation are still not completely clear and in particular in three dimensions additional geometrical effects might play a role. This as well as other questions are difficult to study conclusively with a numerical technique due to the excessive requirements in computer time. It would for instance be interesting to see what happens when the box is so wide that the walls of the box are much farther away from the large particle than the height of the packing. In this experimentally relevant case the walls would not be able to stabilize the convection rolls. Simulations with periodic boundary conditions have, however, provided rather similar results as with fixed boundaries [83]. It would also be interesting to study larger ratios $\mathcal{R}$ in order to verify predictions made about a characteristic value of $\mathcal{R} = 12$ [7, 77] but for that case one would need to consider substantially larger systems. Three dimensional calculations have been made [83] but there the length scales are even smaller. The limitations in observation time due to the computational requirements also puts limits on the determination of the segregation velocity and we cannot exclude that particles rise on time scales much larger than the ones accessible numerically.

6. Simulating the FlowThrough Hoppers and Pipes

The outflow of granular materials from hoppers and silos is an important technological problem although it seems to be so commonplace and low-tech. After many years of uninterrupted service a silo might one day suddenly succumb under a "siloquake" or similar shock phenomena of surprising violence causing considerable harm. In fact silos and similar recipients are by orders of magnitude the industrial structures most susceptible to collapse. The reason for these catastrophic events are that the forces exerted by the flowing granular material against the wall of the container can fluctuate in magnitude by many orders.

Using similar techniques as in the last section but including the Coulomb (dynamic) friction of (2.2c) and rotations of particles as in Ref. [8] simulations were made for the flow out of a hopper [38, 56] and flow through a pipe [52].

The simulations for the flow out of a hopper [38] find the existence of a minimal outlet diameter below which clogging occurs due to arching which is larger for equal sized particles than for randomly distributed radia. In fact the rheology of the particles is very different when they are monodisperse: For a random distribution of radia the acceleration occurs on a ramified structure that has strong temporal fluctuations. Particles of equal size form regular, crystal-like domains and the motion occurs between the blocks. This bloc motion has been described in detail in the experiments of Drake [84].

Fig. 12. Vertical pipe plotted at regular time steps next to each other. Time goes from left to right. Gravity acts from top to bottom. For more details see Ref. [52].

In Fig. 12 we see a space-time diagram of the density inside a pipe with 600 particles and periodic boundary conditions. The particles initially have randomly distributed initial positions and velocities (left pipe). After some time spontaneously various patterns appear in the density: On one hand one has very dark regions, nearly constant in time. Then one sees black diagonal stripes of constant velocity down the pipe. Finally there are also some lighter horizontal lines. In the following we want to try to explain these rather complex structures.

Similar effects have also been observed in hoppers of opening angles of $\theta = 30°$ and the density at a position six particles diameters above the outlet has been measured as a function of time [56]. In Fig. 13 we see the Fourier transformation of this density in a log-log plot. Clearly the data fall on a straight line over nearly two decades. The slope is about $\alpha = -1.35 \pm 0.1$ obtained by a least square fit. That means that we have a power law spectrum of the form $1/f^{\alpha}$.

When particles of equal size are taken we observed equally well developed density patterns and find roughly the same power law decay of the spectrum. The effect is reduced when the diameter D of the outlet becomes too large. If it is too small the flow of particles can entirely stop due to arching. The critical diameter D_0 when this arching sets in has been studied before with similar techniques [38] where it was found that D_0 is larger when the particles have the same size. When we consider smooth walls, i.e. all wall particles having the same radii, we do not find density waves and the power spectrum looks significantly

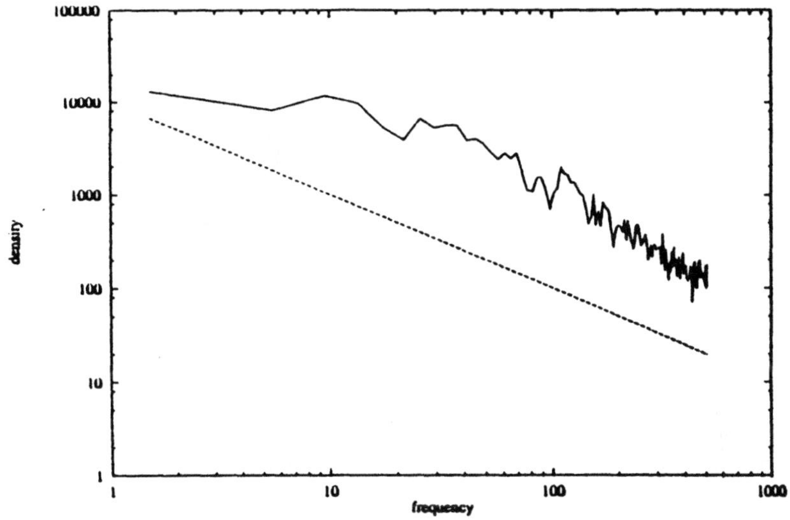

Fig. 13. Log-log plot of the spectrum of the density fluctuations in a hopper [56]. The elastic modulus is $k_n = 10^6 \mathrm{g/s^2}$ and the time step $\Delta t = 2 \cdot 10^{-4} sec.$

different. It shows an upwards curved slope with increasing frequency which one also finds when configurations block during the outflow. A similar effect was also found in simulations of flow on an inclined plane [41].

Baxter et al. [53] also measured the stresses acting on the walls of a three dimensional hopper with an opening angle of 45° during the outflow of sand. They observed a power law decay in the spectrum of the time dependence of the normal stress (pressure) where the exponent strongly depended on the run.

The stresses acting against the walls have been measured [59] by taking from our algorithm the normal p and tangential τ components of the force acting against each of the particles of the walls and monitor the time dependence of the stresses at a given position on the wall. The fluctuations of the first invariant of the stress tensor, defined as $\sqrt{p^2 + \tau^2}$, are of the same order as the values themselves. No apparent correlations are visible. In order to get a more quantitative information we show in Fig. 14 the power spectrum of these time sequences, i.e. we took the square of the amplitude of its Fourier transformation. For sufficiently open hoppers, like for $\Theta = 55°$, the spectrum decays with a power law over at least one order of magnitude with an exponent of about 1.3. This is in very good agreement with the experimental results of Baxter et al. [53] which worked at the same opening angle; (they define the angle as $90° - \Theta$). Their power law, however, extends over a much larger range than in the simulation because the computer requirements limit the observation times to several minutes while the real experiments can be carried out over many hours.

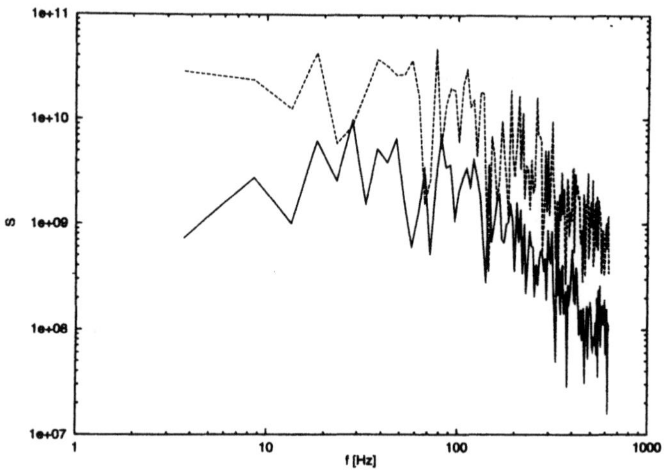

Fig. 14. Log-log plot of the power spectrum of the stresses for $\Theta = 60°$. The upper curve corresponds to the shear stress and the lower curve to the pressure. The slopes are -1.4 (top) and -1.7 (bottom). The opening had a diameter of ten particles.

It is very interesting to note that if the hopper walls become more inclined the spectrum changes abruptly and becomes white noise. Savage [85] found a similar situation in numerical calculations of the wall stresses in shear cells as function of density: Only at rather high densities the power spectrum showed a power law while for lower densities he observed white noise. In our case although the density cannot be varied a change in the opening angle determines if there are stagnation zones or not and evidently the density in the flowing regions is lower than in stagnations zones.

The observed shapes of stagnation zones in real three dimensional silos are curved, either cusplike downwards for primary flow or S-shaped for secondary flow [86]. Indeed the shape of the stagnation zone as found numerically become S-shaped after some transient time. Only close to the outlet the shape remains linear. The acceleration zones inside the flowing part have elliptic shape also in agreement with the observations of the inner core of the flow [86]. It seems that the fact that these observations were made on three dimensional hoppers while the numerical simulations are two dimensional (with exception of the Hertzian contact law) does not influence the qualitative picture very much.

It is interesting to see how the shape of the stagnation zone depends on the opening angle of the hopper Θ. For that sake also the case $\Theta = 0$, i.e. a box with a flat bottom was simulated. In this case the S-shape of the stagnation zone reaches down to the edge of the outlet in even closer agreement with the typical observations.

Investigating with large scale MD simulations the force distribution and the shape of the stagnation zones of outflowing hoppers one therefore has an interesting dependence on the opening angle: In the case of funnel flow ($\Theta = 75°$) there are no stagnation zones and the power spectrum of the stresses against the walls has white noise. Opening the angle one finds stagnation zones and a power law spectrum. This leads one to suspect that the power law in the stress spectrum does not originate from the power law in the density fluctuations [12, 56] because the density fluctuations also follow a power spectrum in the case of funnel flow. It seems more likely that the stagnation zones act like "noise transformers" in which essentially uncorrelated random kicks coming from the outflowing core are transmitted through the complex contact network to the wall and arrive power-law correlated. Similar observations have been made with the propagation of shock and sound waves in dense packed boxes [13].

7. Traffic Models

We have shown that similar to the avalanches that one observes on the surface of a sand pile also inside the bulk of granular material one has avalanche behavior which like the ones on the surface shows self-organized criticality on small scales [54]. The mechanisms that generate the patterns are similar but not identical to the original sandpile models. While the static friction similarly generates waiting times with a threshold it is not the motion of the sand itself that constitutes the avalanches but it is the group velocity of the holes between them: An individual particle can easily go from one dense region to the other by flying fast through a region of low density. There is therefore a backflow of information similar to the jamming on highways [87].

Everybody knows of the seemingly erratic motions of cars jammed on highways. One wonders whether they are due to a random behavior of the individual drivers or if there is an intrinsic chaotic mechanism. In favor of the first hypothesis is the existence of regular kinematic waves in dissipative systems with excluded volume [88]. For this reason many traffic models include rather important statistical noise in time [87]. In favor of the second hypothesis are measurements performed on Japanese highways showing a $1/f$ spectrum in the Fourier transformed density fluctuations [89] which might stem from some self-organized criticality [54]. It is therefore interesting to see if a traffic model without noise is able to give the observed erratic behavior and its $1/f$ spectrum.

In Ref. [57] we consider a continuous one-dimensional model. The system has length L with periodic boundary conditions; and the velocity v_i and position x_i of a vehicle i are continuous variables. The update rule is defined as follows:

- If the velocity is high with respect to the gap, then the car slows down:

$$\text{if} \quad v > \Delta x - \alpha \quad \Rightarrow \quad v \to \max(0\,,\, \Delta x - 1\,)\;; \qquad (7.1a)$$

  (the "max" is only necessary to prevent negative velocities);
- else if the velocity is low with respect to the gap and slower than five, then the car accelerates:

$$\text{if} \quad v < \Delta x - \beta \;\text{ and }\; v < v_{max} \quad \Rightarrow \quad v \to v + \min(\,1\,,\, \gamma * \Delta x\,)\;. \quad (7.1b)$$

Arbitrarily one can choose the maximum velocity to be $v_{max} = 5$ so that this rule allows maximum speeds up to nearly six.

- After the velocity has been updated for all vehicles according to the last two rules, we move all vehicles simultaneously by a distance equal to their velocities.

In Ref [57] initially $N = [\rho \cdot L]$ vehicles were placed on equally spaced sites 1 to N, all with velocity zero, where the total number of cars ρ was chosen small enough to prevent the first car hitting the last one through the periodic boundary conditions. Starting from this totally ordered initial state, the system is allowed to evolve according to the above rules, with the restriction that the first vehicle has a speed kept fixed at $v_{lead} = 4.99999$. This is a simplification of the well known situation where fast vehicles follow a slower one which they cannot pass.

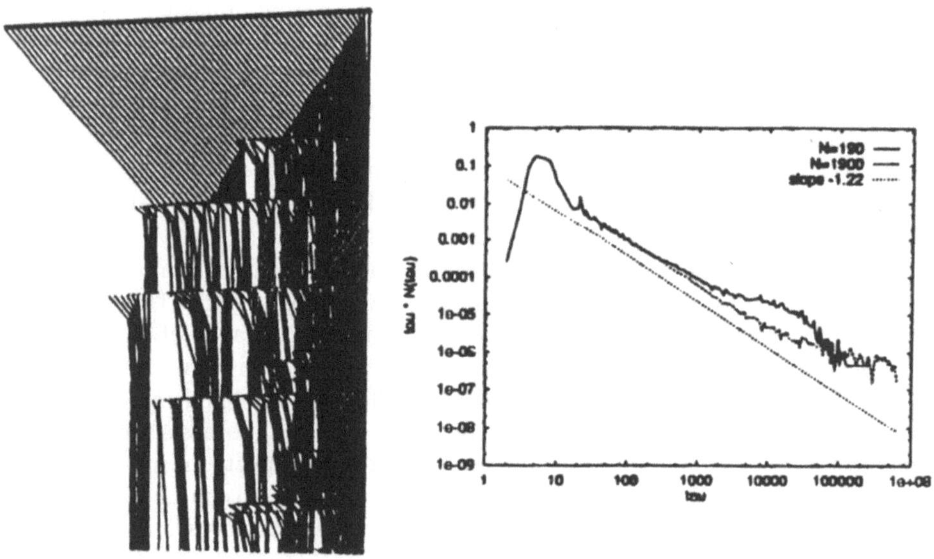

Fig. 15. Left: First 4200 time steps (from top to bottom) and flow of $N = 61$ cars from left to right on a lane of length $L = 1024$. Right: Distribution $n(\tau)$ of times τ between consecutive breaking events for $N = 190$ and $N = 1900$ cars; $\alpha = 0.5$, $\beta = 3.0$, and $\gamma = 0.1$ (from Ref. [57]).

In Fig.15(left) we see the time evolution of a system which has been transformed to the coordinates of the first vehicle, i.e. the positions of all cars are given relative to the first car. We see that equally spaced cars rapidly evolve into a fluctuating state (right hand side). In this new state density increases give rise to very short bursts (traffic jams) of very different sizes which redistribute the cars backwards such that in some cases they even start again in equally spaced patterns.

In order to ensure that this behavior is an intrinsic consequence of the dynamics and not just the enhancement of numerical round-off errors, single precision and double precision calculations have been compared to each other. The behavior of the model (i.e. the formation of the collective shocks) is robust with respect to parameter changes. More precisely, no qualitatively different behavior for changes of the parameters α, β, γ, and v_{lead} has been observed within the following range $0.1 \leq \alpha \leq 0.6$, $2.0 \leq \beta \leq 5.0$, $0.08 \leq \gamma \leq 0.12$, $4.5 \leq v_{lead} \leq 4.99999$.

One can also measure the distribution of times τ between consecutive "braking" events for the last vehicle (τ counts the time from the end of one braking to the beginning of the next). Braking is defined as a slowing down according to the rules for the velocity update.

For $N = 1900$ vehicles, after about $3 \cdot 10^5$ time steps to let the transients die out the distribution of τ was measured during about $1.1 \cdot 10^6$ further time steps. As seen in Fig. 15b, this distribution displays a remarkable straight line on a log-log plot, fulfilling a power law relation

$$n(\tau) \propto \tau^{-\alpha} \tag{7.2}$$

with $\alpha = -2.2 \pm 0.1$. This is an indication for the existence of self-organized criticality [54].

Many aspects of this model are remiscent of the so-called train model for earthquake dynamics [90]. Instead of pulling at one end, the slower car may be seen as pushing against the other cars which want to move faster. This leads to a slowly increasing average density, and at some time this density locally exceeds a critical threshold. The reaction is a more or less drastic slowing down of the corresponding vehicle, which may or may not force the next vehicle to slow down as well. By this mechanism, avalanches of all sizes are generated, which may propagate through the entire traffic jam.

8. Lattice Gas and Boltzmann Lattice Models

It seems natural to describe the flow of granular media using the concepts of fluid mechanics. One must, however, consider that as opposed to classical fluids there is local dissipation of energy. Taking into account the dissipation rate in the energy balance equations it has been possible [91, 92] to predict the existence of an instability: Slightly denser regions have more dissipation and therefore lower pressure which itself generates a flow that will enhance the density. So, the dissipation will be responsible for the formation of clusters of high density and these have been observed [91]. Also has it been possible to derive from this kinetic gas theory [92 – 94] that the viscosity increases very sharply with density.

One alternative to the direct solution of the equations of motion of fluids are the so called Lattice Gas (LG) [95] Lattice Boltzmann Models (LBM) [96]. These models are defined on a lattice with velocity vectors that can only point into few discrete directions and all have the same length. For the LBM this simplification is somewhat compensated by the fact that on each site one has more real degrees of freedom (six on a triangular lattice) than in the classical numerical techniques allowing for the definition of a local shear or a local rotation.

Let us first describe the Lattice Boltzmann Model as used in Ref. [58]. We consider a triangular lattice, and on each site $\boldsymbol{x}$ we have six real variables $N_i(\boldsymbol{x},t)$, $i = 1,..6$, representing (counted counter clockwise) the densities of the particles going in the direction i of the lattice. (For convenience we will in the following omit the site index $\boldsymbol{x}$ and denote by N_i' the value of the particle density after collision.) One updating of the system ($t \to t+1$) is given by two steps: (1) The collision step at which the six N_i are updated at each site through

$$N_i' = N_i + \lambda(N_i - N_i^{eq}) \tag{8.1}$$

and (2) the propagation step at which each N_i is shifted to the site of the nearest neighbor in direction i. Eq.(8.1) produces a relaxation towards the equilibrium densities N_i^{eq} which is numerically stable provided $-2 < \lambda < 0$. The value of λ sets the kinematic viscosity ν of the fluid. The equilibrium densities in (8.1) are given by

$$N_i^{eq} = \frac{\rho}{6}(1 + 2\boldsymbol{u}\cdot\boldsymbol{c}_i + 4(\boldsymbol{u}\cdot\boldsymbol{c}_i)^2 - 2\boldsymbol{u}^2) \tag{8.2}$$

where ρ is the mass density at site $\boldsymbol{x}$

$$\rho = \sum_i N_i \quad , \tag{8.3}$$

$\boldsymbol{c}_i$ the unit vector along direction i and $\boldsymbol{u}$ the velocity at site $\boldsymbol{x}$ defined through the momentum density per site

$$\rho\boldsymbol{u} = \sum_i \boldsymbol{c}_i N_i \quad . \tag{8.4}$$

The equilibrium distribution N_i^{eq} given in (8.2), is chosen to give mass and momentum conservation in the collision step. The flow will be forced into the direction of gravity $\boldsymbol{g}$, which is pointing parallel to the walls of the pipe. For that purpose an additional step is added after the collision step which is defined by $N_i'' = N_i' + \frac{1}{3}\boldsymbol{c}_i \cdot (\rho\boldsymbol{g})$. Periodic boundary conditions are imposed in the direction of gravity in which the system has a length of L. In the perpendicular direction one has walls separated by W lattice spacings. The lattice orientation is such that one of the lattice directions is parallel to the walls at which no-slip conditions are used. At the beginning of the simulation the average density $\bar{\rho}$ is fixed. It is an important parameter of the model which because mass is conserved.

The relaxation parameter λ depends on the material properties including the kinematic and the bulk viscosities. Since an exact relation between λ and the material constants is not known we will lean on some approximative arguments [95] that predict a vanishing bulk viscosity. In that case one can relate λ directly to the kinematic viscosity ν through $\lambda = -\frac{1}{2}(0.25 + 2\nu)^{-1}$. We will consider that ν is a function of the local density ρ.

A salient feature of granular media is the spontaneous formation of density waves, similar to traffic jams of the previous section. One possibility to explain the effect that generates these waves is to assume that the viscosity depends on density. Within the kinetic gas theory of granular media [92 – 94] the relation $\nu \propto (\rho - \rho_c)^{1/3}$ has been derived. Since the above relation imposes a maximum density ρ_c it is rather difficult to implement it directly within the context of the LBM where the particles do not have an excluded volume. We therefore chose a piecewise linear relation of the form $\nu = \nu_{min}$ if $\rho \leq \rho_t$ and $\nu = \nu_0 + \gamma(\rho - \overline{\rho})$ for $\rho > \rho_t$ (see Fig. 16a). $\overline{\rho}$ is the average density and the threshold density ρ_t is chosen to make ν a positive continuous function of the density. We initialize the system by having the same values of the N_i on each site and then let the system evolve to its steady state. In the case of stable flows steady state is reached after 2000 or 3000 time steps. In the case of the unstable flows that develop density waves, the simulations might take up to 20000 time steps to reach steady state.

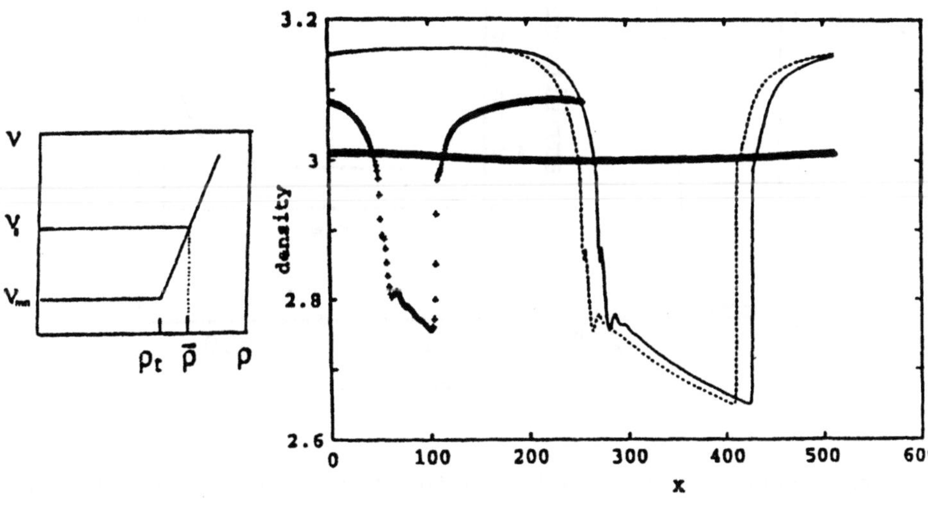

Fig. 16. Left: Density dependence of the viscosity chosen in the simulations. The slope is $\gamma = 6.25$ and the minimum viscosity is $\nu_{min} = 0.01$. Right: Density in the center of the channel as a function of the position X along the channel for $\rho_t = 2.962$, $\bar{\rho} = 3.0$, $g = 3.33 \times 10^{-5}$ and $W = 64$. The crosses are for $L = 256$ and 60,000 iteration steps after the initial perturbation. The other curves correspond to $L = 512$ for 5000 (thick line), 60,000 (full line) and 60,025 (dashed line) iterations after the perturbation was applied (from Ref. [58]).

In order to generate density waves it is necessary to introduce a small perturbation producing a 0.3% relative density difference on one line across the pipe, keeping the mass unchanged [58]. In Fig. 16b we see that this initially very weak perturbation dramatically builds up and develops into a density wave of over 10% density contrast. For a pipe of same width but half the length, i.e. a different aspect ratio the wave has a less pronounced profile. This dependence on the aspect ratio is not to be confounded with finite size effects.

Triggering the density wave by two spatially separated perturbations, rather than just a single one, one can convince oneself that the complex shape of the waves does not reflect the detailed way in which they were initiated. We also observe that there seems to be no characteristic wavelength: Fig. 16b shows that the waves have roughly the same shape and that their size scales with the dimension of the channel.

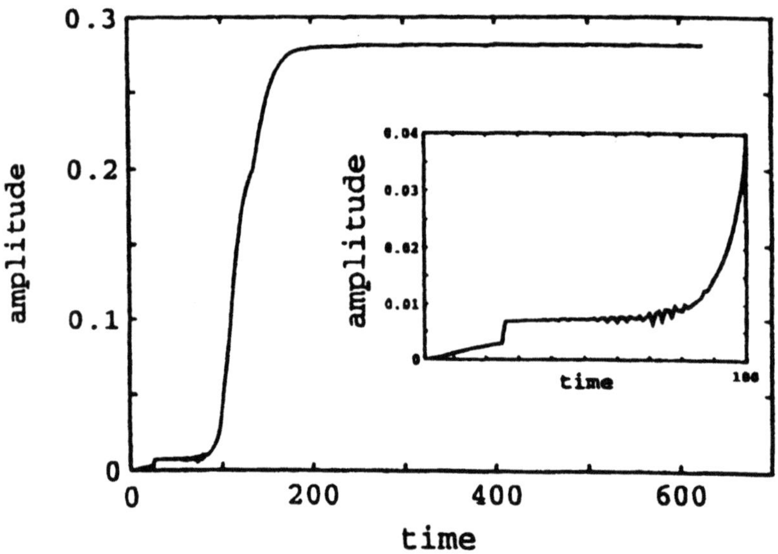

Fig. 17. Amplitude, i.e. difference between largest and smallest density, along the center of the pipe as a function of time measured in units of 100 iteration steps for $L = 256$ and otherwise the same parameters as in Fig. 16(right). The insert is a blow-up of the behavior at early times. (from Ref. [58])

Fig. 17 shows this amplitude as a function of time during 60,000 time steps. The insert shows the initial unstable phase leading to the rather drastic increase of the amplitude at the time 10,000. The first small increase in the density is due to the acceleration of the flow and can be understood from the velocity dependence in the pressure. The small jump at after 2500 time steps results from the perturbation. Before the instability is triggered at time 10,000, small oscillations in the amplitude are observed. It was checked that the amplitude indeed has its steady state value at time 60,000 by running the simulations ten times longer. The complicated relaxation towards the fully developed density wave indicates that strong non-linear effects come into play rendering a linear stability analysis meaningless. It would be interesting to understand this behavior further.

A Lattice Gas model [95] can be formulated such as to include local dissipation of energy by introducing rest particles and special collision rules as done in Ref. [25]. Let us consider N particles located on the sites of a two dimensional triangular lattice which is L sites long and W sites wide. On each site there are seven Boolean variables i which are one or zero depending on whether there is a particle having velocity, $\boldsymbol{v}_i(i = 0, 1, 2, \cdots 6)$ or not. Here $\boldsymbol{v}_i(i = 1, 2, \cdots 6)$ are the nearest neighboring (NN) lattice vectors and $\boldsymbol{v}_0 = \boldsymbol{0}$ is the rest (zero velocity) state. So, each state can be either empty or occupied by a single particle and the number of particles per site has a maximal value of 7 and a minimal value of 0. The time evolution consists of a collision step and a propagation step. In the collision step particles change their velocities due to collisions and in the subsequent propagation step particles move in the directions of their velocities to the NN sites where they collide again except if there are rest particles. The system is updated in parallel. Only the specified collisions shown in Fig.18 can deviate the trajectories of particles. All collisions conserve mass and momentum.

For two- and three body collisions, one has the probabilistic rules shown in Fig. 18a. The probability that a configuration may take place is shown next to the configuration. If the parameter p is nonzero, energy is dissipated during the collision.

Two moving particles colliding with a rest particle from opposite directions can stop each other in accordance with momentum conservation. But on each site only one rest particle is allowed. Therefore, the two particles stay at rest on the NN sites where they originally came from. However, on these sites already other rest particles may exist. To make things simple, one can still use the on–site collision as defined and temporarily allow more than one rest particle on a site during the collision. Immediately after the collision, the extra rest particles randomly hop to NN sites until they find a suitable site with no rest particle already sitting there. This is one way to incorporate the existence of a maximum density. Another possibility which is in fact easier to implement on parallel computers consists of introducing also rest particles on bonds [26]. The collision rules with rest particles are shown in Fig. 18b.

The driving force of the flow is gravity. One simply incorporates its effect by giving with probability g to a rest particle a velocity along the direction of gravity and to any moving particle a change in velocity by a unit vector along gravity, if the resulting state is empty. (see Fig. 18c).

The sites at the walls of the system only have two directions into which the particles can move. So, a particle colliding with the wall from one direction can be bounced back with probability b and specularly reflected into the other direction with probability $1 - b$ as seen in Fig 18d. If $b = 0$, the walls are smooth (perfect no-slip condition). Otherwise, the walls have some roughness.

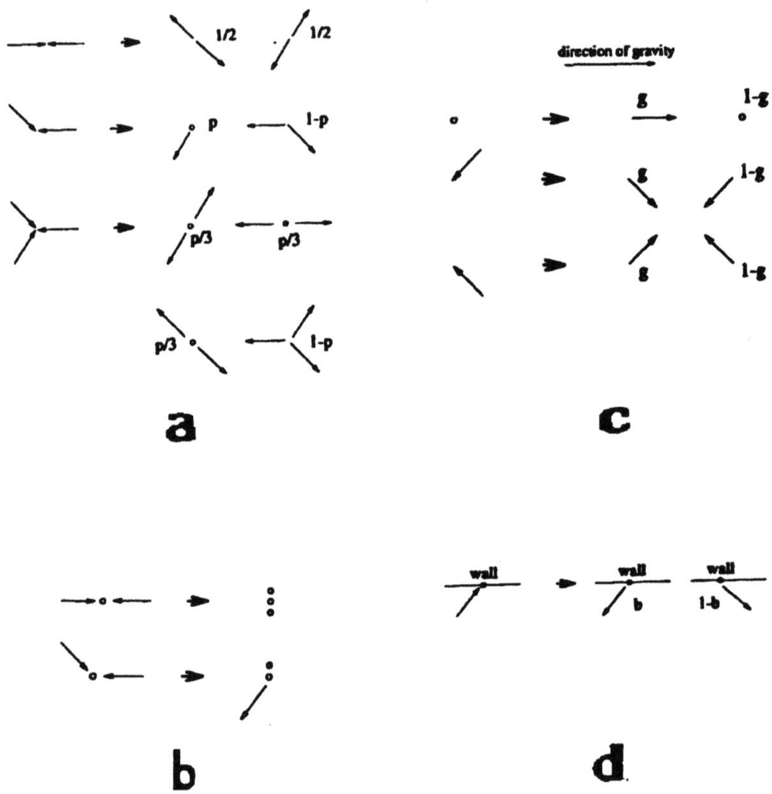

Fig. 18. (a) Probabilistic collision rules of the Lattice Gas of Ref. [25] (a) for two- and three body collisions, (b) for moving particles with a rest particle, (c) for gravity and (d) for a moving particle colliding with the wall. Thin arrows represent moving particles and small circles stand for rest particles. The number next to a configuration is the probability that the configuration takes place. Immediately after the collision, excessive rest particles on a site will randomly hop to the nearest neighboring sites until they find a suitable site with no rest particle already there.

Let us first consider a system with periodic boundary conditions in the direction of gravity. The initial configuration of the system is set to be random in the sense that every state (except the rest state) of each site is randomly occupied according to a preassigned average density ρ. Fig. 19 shows the time evolution of the density in the pipe during the early stage for $p = 0.1, g = 0.5, b = 0.5$. Here the system has length $L = 2200$ and width $W = 11$ with average density $\rho = 1.0$ The pipe was divided along the vertical direction into 220 bins with equal length of 10 and the number n_j of particles in the jth bin is counted. In Fig. 19 we plot n_i in the ith bin by a grayscale which is a linear in n_i. The n_i at a given time are plotted from bottom to top while the densities at different time steps are plotted from left to right as time increases. Gravity is from bottom to top (usually it is plotted in the opposite direction). We see that initially the density is rather uniform and gradually regions of high density are being formed out of the homogeneous system. A high density region may also die out and two high density regions may merge to form a single one.

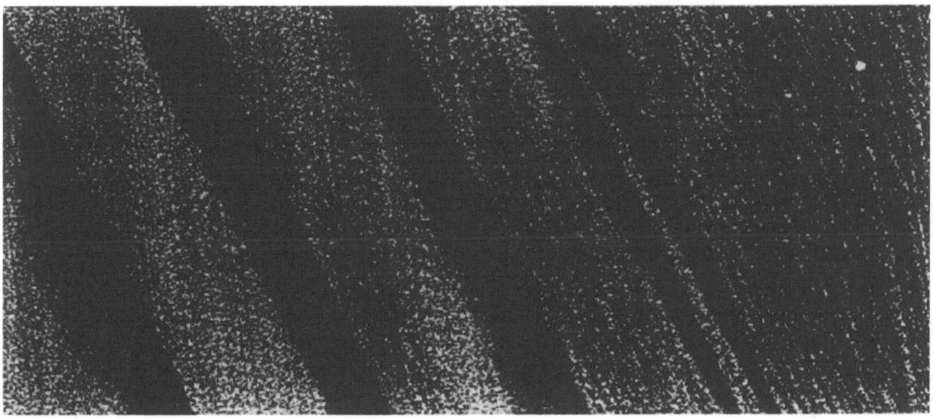

Fig. 19. Time evolution of the density $n_j\{j = 1, 2, \cdots 220\}$ divided in 220 bins along the pipe of $L = 2200$, $W = 11$ and $\rho = 1.0$. Densities at a given time are plotted from bottom to top (direction of gravity) while densities at different time steps are plotted from left to right (direction of time). Every 80 time steps are shown during 40,000 time steps. The grayscale of each bin is a linear function of n_i. Darker regions correspond to higher densities (from Ref. [25]).

To obtain the shape of the density waves one can record the density field in steady state at each time step and shift their spatial coordinates such that they overlap each other maximally. Since the density wave travels along the pipe, this shifted distance gives the average velocity. The maximal overlap rule can be applied hierarchically to obtain a clearer shape of the density wave. 64 simulation runs are then averaged to give the final density profiles which are presented in Fig.20. We notice that the density wave has a non–symmetric shape and its wave front is sharper than the backside of the wave. The width of the wave scales with the system length (almost linearly) and the amplitude of the wave only increases a little bit as the system length increases. Note the similarities to the profiles obtained with the LBG in Fig. 16.

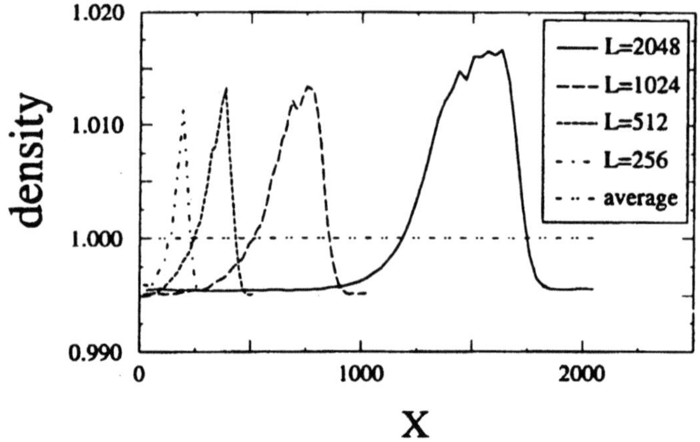

Fig. 20. Density as a function of position X along the pipe for various lengths L. The average has been made in the perpendicular direction. The model parameters are $\rho = 1.0$, $p = 0.1$, $g = 0.5$, $b = 0.5$ and $W = 11$ (from Ref. [25]).

To characterize the density fluctuations in a certain region with time, let us calculate their power spectra. The simulation is performed for very long time to obtain good statistics. One first subtracts the mean value from the data to remove the peak at $f = 0$ in the power spectra and calculates the spectra using a standard FFT routine. The power is the square of the amplitude of the Fourier Transformation of the time series. One representative power spectrum is shown in Fig. 21 for systems with $g = 0.5, b = 0.5, \rho = 1.0$ and $p = 0.8$. We observe a sharp peak due to the contribution of the density wave observed in Fig. 19. The frequency of this peak corresponds to a wave velocity of Lf/T_0 where L is the pipe length and T_0 is the time interval between two recordings (here $T_0 = 10$ time steps). The velocity measured in this way coincides with the direct measurement in real space. Apart from this peak one sees a background having a power law behavior where the spectrum falls off as $1/f^{\alpha}$. The exponent α is found to be around 1.33 in Fig. 21.

Both dissipation and the roughness of walls are necessary conditions for the presence of travelling density waves. If either $p = 0$ or $b = 0$ the spectrum has white noise. Neither the exponent of the power-law decay nor the velocity of the density wave do depend on dissipation p within the error bars. In fact, using a Kolmogorov–Obukhov approach revised for space–intermittent systems, Bershadskii [97] proposed that the exponent $\alpha = 1.33$ found for this model [25] might be a universal value of 4/3 for scalar fluctuations convected by stochastic velocity fields in dissipative systems. Our numerical results for the present LGA model show that the exponent does neither change with b nor with g but the average density ρ does affect the exponent. For very low density ($\rho \leq 0.8$) the exponent is zero, i.e. the density fluctuation in the pipe is just white noise since the interaction among the particles is so weak that no collective motion can

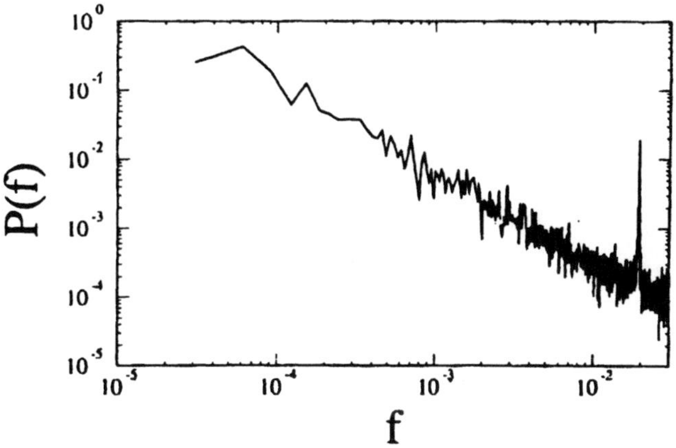

Fig. 21. Power spectrum $P(f)$ of the time series of the density fluctuation inside a region in a pipe of length L=220 and width W=11. The model parameters are $p = 0.8$, $b = 0.5$, $g = 0.5$, $\rho = 1.0$. The time series of the density fluctuation were recorded every 10 time steps (from Ref. [25]).

be formed. For average densities above 1.0, the exponent increases with ρ and becomes $\alpha \approx 1.86$ for $\rho = 2.0$. Beyond $\rho = 1.0$ the model cannot be taken very seriously anymore since when the number of rest particles exceeds the total number of sites there is an upper limit on the dissipation rate p that the system can bear.

From the experimental point of view, open boundary conditions are more realistic than the periodic boundary condition. Let us consider an open pipe which is initially empty. Particles are then injected from the upper boundary by a constant rate I and leave the system at the lower boundary without coming back. The injection rate is defined as follows. On each site of the upper boundary one considers the states with velocities pointing into the system. If such a state is not occupied, it can be filled with probability I. A time–evolution of the density in the pipe is shown in Fig. 22. Densities at a given time are plotted from bottom to top while densities at different time are plotted from left to right. Gravity acts from bottom to top. From this plot we observe that high density regions can also be formed in open systems. They travel along the pipe until they leave the system from the lower border or they may die out during their propagation. There are also more than one high density region at one time, in contrast to what we observed in Fig. 19 in periodic systems. It seems that all the density waves in Fig. 22 travel with a constant velocity.

The density fluctuations have been measured [25] in a bin very close to the bottom of the pipe. Its power spectrum is found to follow $1/f$ noise only around a critical injection rate $I_c = 0.52$ and white noise otherwise. Fig. 23 shows three power spectra for different injection rates, $I = I_c$, $I < I_c$ and $I > I_c$. The power spectrum for $I = I_c$ falls off with a slope close to -1 in the log–log plot. The

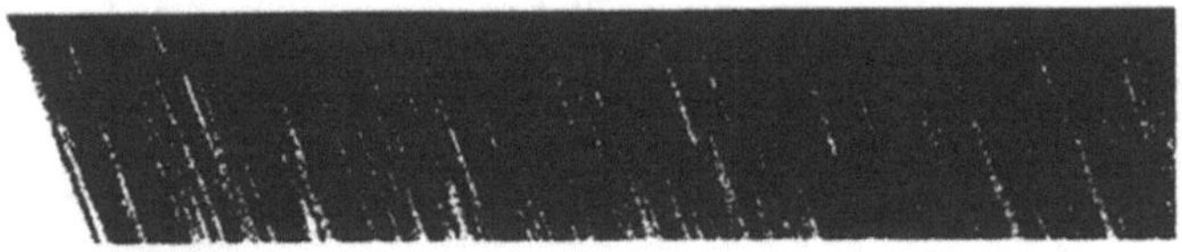

Fig. 22. Time evolution of the density $n_j\{j = 1, 2, \cdots 100\}$ in a pipe of $L = 1000$, $W = 5$ and $I = 0.5$. Other model parameters are $p = 0.5$, $b = 0.5$, $g = 0.2$. Densities at a given time are plotted from bottom to top (direction of gravity) while densities at different time steps are plotted from left to right (direction of time). Time goes from 0 to 40,000 time steps. The gray scale of each bin is a linear function of n_i. Darker regions correspond to higher densities (from Ref. [25]).

exponent α for the power spectra $1/f^\alpha$ is everywhere zero except around I_c. It seems that the power spectrum only has $1/f$ noise at the critical point. The critical injection rate is found to be independent on the model parameters and the system size. In fact one finds that for $I > I_c$ the system is clogging while for $I < I_c$ particles no clogging occurs. It seems that I_c is the maximal injection rate that the system can sustain without clogging.

9. A Thermodynamic Approach to Granular Media

Subjected to an external forces granular materials locally perform rather statistical motions due to their mutual collisions. For example on the loudspeaker the individual grains chaotically jump up and down forming a gas-like cloud of colliding particles. Also inside a shear-cell [15, 40, 98] or flowing down an inclined chute [41, 84] in addition to a laminar flow with a well defined (average) velocity profile one has Brownian-like motion of the particles perpendicular to the flow direction.

The above observations have inspired several authors to use thermodynamic concepts to describe granular media. On one hand a "granular temperature" T_{gr} has been defined [40, 99, 100] as $T_{gr} = \langle v^2\rangle - \langle v\rangle^2$, i.e. proportional to the kinetic energy surplus with respect to the global motion. Strictly speaking this definition is thermodynamically justified if an equipartition theorem exists which is not the case for granular particles since they dissipate energy at collisions. On the other hand, Edwards and collaborators [23] have put forward another idea: Based on

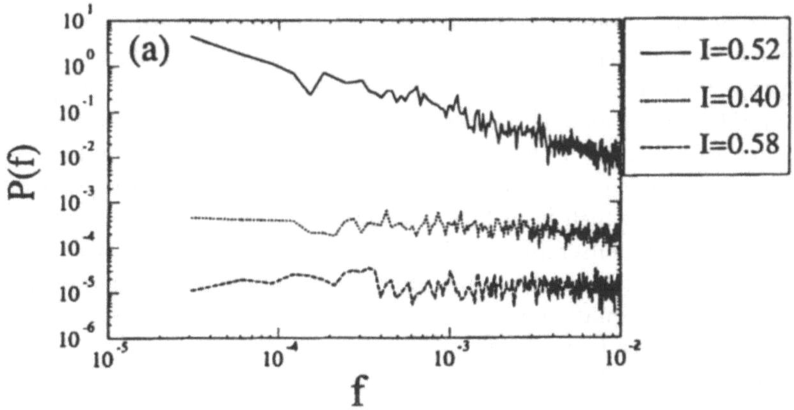

Fig. 23. Three typical power spectra for different injection rates I, $I = I_c = 0.52$ (top left), $I < I_c$ (top right), $I > I_c$ (bottom). The model parameters are $p = 0.5$, $b = 0.5$ and $g = 0.5$. The two lower curves have been shifted vertically for clarity (from Ref. [25]).

the important observation that granular materials do not conserve energy they proposed to consider the volume V to replace the internal energy in the usual thermodynamic formalism and define a temperature-like quantity $X = \partial V/\partial S$ which they called "compactivity". Although formally intact, this formalism is not easy to justify since in many real situations like on the vibrating table or on an inclined plane, the volume is not well limited at large heights. While Edwards's approach seems intuitively correct for dense packings and the definition of T_{gr} reasonable in the limit of strong internal motions or weak dissipation they fail in the corresponding opposite limit.

Let us present in the following a thermodynamic approach to granular materials [21] founded on similar principles as equilibrium thermodynamics which incorporates at least partially the intuitive pictures of previous work: We shall consider subsystems sufficiently small to have no velocity or density gradients and for which the energy flux into them is such that energy dissipation is homogeneous. Energy conservation implies that $\Delta I = \Delta E_{int} + \Delta D$ where ΔD is the energy dissipated in a given time and ΔI is the energy that was pumped into the system while ΔD was dissipated in order to maintain a steady state. The internal energy E_{int} is like in traditional thermodynamics the sum of kinetic and potential energy of all the degrees of freedom of the grains as elastic bodies (translation, rotation, elasticity, etc). One can now treat the excess dissipated energy $\Delta\mathcal{D} = \Delta D - \Delta I$ in a similar way as the heat in usual thermodynamics. Since the dissipated energy is proportional to the sum of normal forces f_n^i that push the particles together during collision i one can express changes in $\mathcal{D}$ as $\delta\mathcal{D} = \wp\delta C$ where $\wp$ is an internal pressure acting at collisions that we can define as $\wp = \rho\langle f_n^i/A_i\rangle$ where A_i is the area of contact of collision i and the average is performed over all collisions. We define the density ρ of collisions as the number of collisions per unit volume and unit time. The extensive conjugate quantity C has a geometrical interpretation and plays the role of a potential. It should in fact be proportional [21] to the overlap volume V_{ov} that one has for technical reasons in MD simulations which can be defined more precisely as the excluded volume that would arise if the centers of mass of the particles follow the real trajectories but one does not take into account the elasto-plastic deformation. The "equilibrium" - which is in fact a steady state driven by the energy flux - can be defined as the ensemble *minimizing* C and one can *postulate* in analogy to the second law of thermodynamics that C should decrease for any change of state at constant internal energy E_{int}: $\Delta C \leq 0$ driven by the elastic repulsion between colliding (overlapping) grains.

As in usual thermodynamics one can now work in different ensembles. Naturally one would work at fixed $\wp$ (granular ensemble) in which a granular potential G_r can be defined as $G_r = E_{int} + \wp C$ and where at constant $\wp$ the equilibrium is given by the minimum of G_r. The response function κ defined as $\kappa = \frac{\partial \mathcal{D}}{\partial \wp} = \wp \frac{\partial C}{\partial \wp}$ measures how much more energy can be dissipated if $\wp$ is increased. On top of the granular ensemble one can build up the traditional body of thermodynamics as if the grains were a gas of particles interacting elastically. One can fix or free the number N of particles, define a "granular" temperature T_g and entropy S or impose to the system either an external volume V or an external pressure p. Special for granular media is that one could also impose an external shear τ or the dilatancy V_d [14,15].

Considering a "state" given by the positions, orientations, linear and angular velocities of the grains as rigid bodies the entropy is well-defined as noted already in Ref. [23]. A reasonable definition for a "granular" temperature T_g would then be: $T_g = (\partial G_r / \partial S)_\wp$ which is, in fact, similar to the one defined previously [40,99,100] Experimentally $\wp$ and T_g are independent control parameters of the system: T_g is essentially driven by the amount ΔI of energy that is fed into the system per unit time. $\frac{\wp}{T_g}$ depends mainly on the density of collisions and can therefore increase by fragmenting the grains into smaller pieces. (Note that when a given grain is split into eight pieces, the cross section of each individual piece decreases by a factor four, so that $\wp$ will increase by two.)

10. Sound Propagation in Granular Assemblies

Recent experimental work shows that sound propagation in sand exhibits very complex behavior. Liu and Nagel [13] reported fluctuations in the acceleration-amplitudes measured at several points in the system. The spectrum of the time evolution of these amplitudes revealed a power law. They also measured the frequency response for their systems. The frequency response seems to be characteristic for the system, a slight rearrangement of the beads produces a completely different behavior. Leibig [101] calculated the frequency response for a system consisting of spherical particles interacting via linear springs on a square lattice. These numerical results are in agreement with the experimental ones [13].

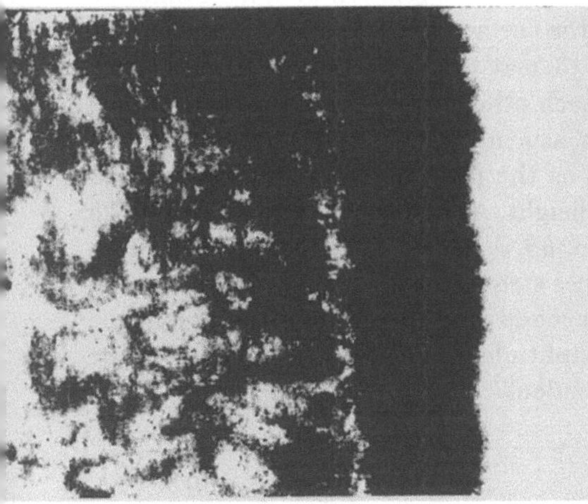

monodisperse

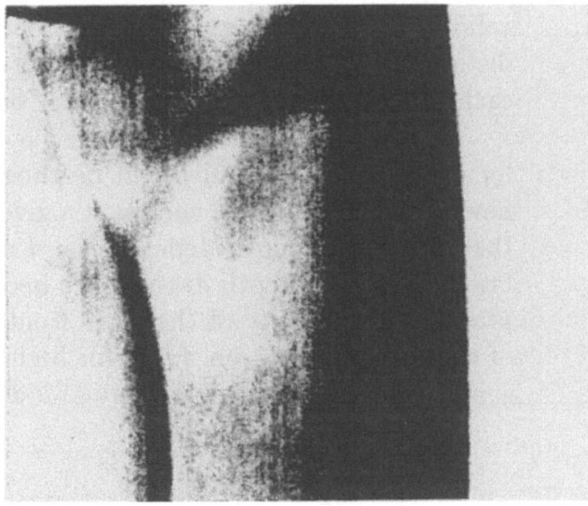

polydisperse

Fig. 24. Wave produced by a single compressional pulse of amplitude 0.01 d on the left wall. The grey scales code the kinetic energy of the particles. The left figure shows the monodisperse case and the right figure the polydisperse case. Note the secondary wave front clearly seen in the monodisperse case travelling (due to its lower amplitude) much slower behind the primary wave front (from Ref. [64]).

There have been predictions about the depth dependence of the sound velocity. Goddard made calculations [61] showing that the sound velocity v_c depends on pressure p like $v_c \propto p^{\frac{1}{6}}$ if the elastic interaction is Hertz's contact law. For high pressures this relation has been verified experimentally (cited in Ref. [61]). For low pressures, however, $v_c \propto p^{\frac{1}{4}}$ has been found experimentally. Both cases imply that the wave velocity approaches zero near the surface ($p = 0$), so that a vertical wave front will bend up, becoming increasingly horizontal which is known as the "mirage effect" [2, 13].

Melin [64] studied a two dimensional packing of grains confined within a rectangular box under gravity. The centers of the particles are situated on a triangular lattice. A sufficiently small perturbation (e.g. e weak polydispersity or a small vibration) can maintain the topology of this packing over regions much larger than the size of the container. So, since the neighbors are fixed throughout the entire simulation the time consuming part of determining the possible interaction partners can be avoided. This triangular lattice can be easily mapped onto a square lattice.

Melin [64] considers a system consisting of 256 * 256 monodisperse spheres under gravity and puts on the left wall a step shaped excitation, i.e. all spheres at this wall are instantaneously moved a small distance to the right. On the left part of Fig. 24 one can indeed recognize a rather weak dependence of the propagation velocity on depth. It is clear see that the relation $v_c \propto h^{\frac{1}{6}}$ cannot be valid in this case, because the wave near the surface has travelled a considerable distance. The reason for this behavior is that at the surface the particles touch each other and will exert some force on each other for any small perturbation.

This behavior can be quantified by measuring the position of the wavefront for every time step which can be chosen as the position of the particles which having the largest total energy at a given height. As one can clearly see in Fig. 25 the anticipated depth dependence of $v_c \propto h^{\frac{1}{6}}$ does not hold true for the simulated cases. The depth dependence becomes stronger with gravity. Moreover the propagation velocity of the wave front depends very much upon the amplitude of the initial excitation. Only for high amplitudes the depth dependence of the sound velocity becomes amplitude independent.

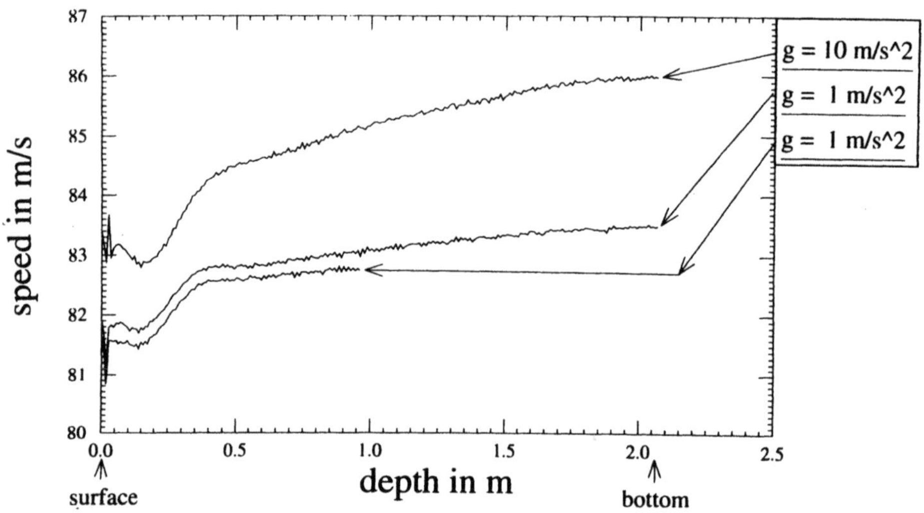

Fig. 25. Depth dependence of the sound velocity for two different values of gravity approximately 7.8 msec after the pulse. The third curve is for a system of only 256∗128 particles (i.e. half the height). It shows, that the boundary at the bottom only has a small effect. For better visualization this last curve has been shifted downwards by 0.25 m/sec. (from Ref. [64]).

It is interesting to note that the wave speed decreases with time. The explanation for this can be found in Fig. 24 where one can see that although the wavefront is very sharp the particles in the region behind the wavefront still have very much kinetic energy left. This means that the total energy in the wavefront decreases with time since some of energy is lost in the region it has traversed. Some of the kinetic energy is also lost due to its dissipation.

Introducing a polydispersity of $\Delta A_{max} = \pm 0.01A$ where $A = d^2$ smears the wave front out but otherwise no qualitative difference in the general behavior is visible (see right part of Fig. 24). The reason for this can be seen in Fig. 26.

Due to the polydispersity some of the horizontal bonds are already closed in the equilibrated state. Any excitation will travel faster along these closed bonds.

Fig. 26. Contact network for an equilibrated $256*256$ packing of spheres, the thickness of the lines is a measure for the stresses in the bonds. The particles have a polydispersity of 1% in the areas (from Ref. [64]).

Instead of looking at a single excitation, one can look at the behavior of a continuously excited system. Experimentally [13, 102] it is possible to measure the acceleration amplitudes and phases of a few test particles. The behavior turns out to be quite complex as seen in Fig. 27. The system has been driven sinusoidally at a frequency of 100Hz with an acceleration amplitude of $0.5g$. The particles forming the left wall are the sound source. Both figures have been taken 256 cycles (2.56 sec) after the beginning of the excitation. Fig. 27 shows the phases in x-direction for the monodisperse case.

One can see that the resulting patterns are far from regular. Note the very sharp phase shift on the right side of the system having an angle of 60°. These sharp phase shifts appear quite often. They are due to the confinement of the system in the rigid box [101].

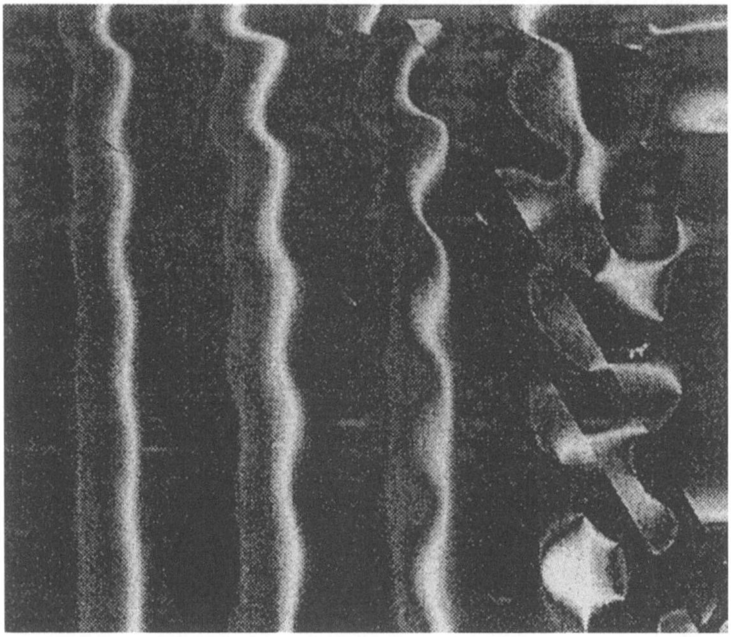

Fig. 27. The acceleration phases in x-direction for the monodisperse system 2.56 sec after starting the excitation (from Ref. [64]).

11. Plastic Deformation and Shear Bands in Dense Packings

The plastic yield criterion of granular material is quite different from that of metals because on one hand it linearly depends on pressure (Mohr-Coulomb) [103] and on the other hand because under shear a dilatation along an angle ψ is observed. The onset of plasticity is well described by a relation [103]:

$$\tau = c + \tan(\phi) \cdot \sigma_n \tag{11.1}$$

where τ and σ_n are the shear and the normal stresses and the material constants ϕ and c are the "friction angle" and the cohesion force, respectively. In general, the plasticity is "non-associate", i.e. the angle of dilatancy ψ under shear is different from ϕ.

Using an explicit Lagrangian technique similar to FLAC [104] an elasto-plastic rectangular medium was simulated in Ref. [65]. A compressive velocity in the horizontal direction was imposed. Besides the above mentioned material constants c, ϕ and ψ one can also change the elastic moduli or Lamé constants λ and μ. The simulations are performed on a grid which in its undeformed state is a square lattice of unit lattice constant with an additional diagonal giving thus a triangulation. The numerical technique has an intrinsic time step Δt. Measured in these units a compressional velocity of 10^{-5} is imposed per grid element.

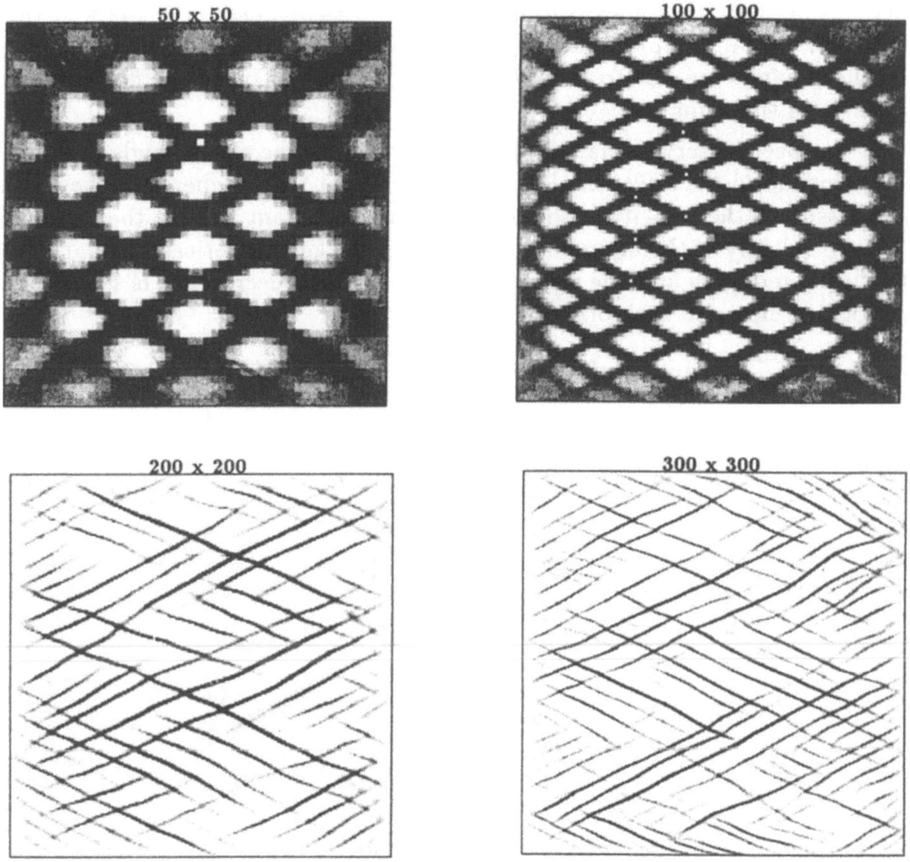

Fig. 28. Snapshot of the shear strain rate for systems of different size: (a) 50x50, (b) 100x100, (c) 200x200, (d) 300 x 300 for $\phi = 40$, $\psi = 0$, $c = 0$, $\lambda = \mu$, Poisson ratio $\nu = 0.25$ and $R = 10^{-9}$ (from Ref. [65]).

In Fig. 28 we see snapshots from the evolution of the systems with different mesh (lattice) sizes L but the same physical parameters. These results are shown after such a long time that the overall number of shear bands does not change with time (however the positions and activities of single shear zones is not constant).

The local changes of the second invariant of the strain rate: $e = \dot{\epsilon}_{II}$ determine the grey scale in Fig. 28. This means that the dark regions correspond to strong changes either in the direction or in the magnitude of the motion of the material. The white regions are elastic. We see that spontaneously shear bands are formed in which the plastic deformation occurs. These bands form an angle $45° - \phi/2 < \theta < (45° - \psi/2)$ with respect to the horizontal which is consistent with bifurcation theory [105, 106]. They have varying length and their position changes in time. Since our initial setup was completely homogeneous the random positions of the shear bands are due to minute effects in the round-off of the floating point numbers in the computer.

It has been observed that the pattern formed by the shear bands essentially only depends on the dimensionless parameter $R = pc/\mu V_{bc}$ where p is the confining pressure, c the velocity of sound, μ the Lamé constant and V_{bc} the externally imposed velocity of the boundary. This can be checked numerically by varying p, the elastic modulus and the size of the sample. With decreasing R shear bands are located closer to each other.

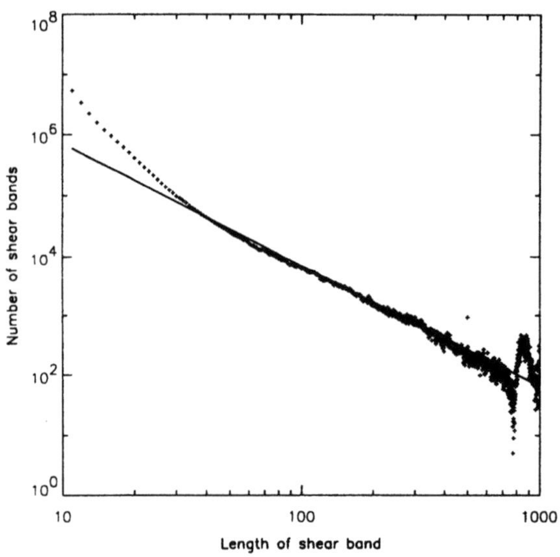

Fig. 29. Log-log plot of the length distribution for a system of size 300 x 300 for the same physical parameters as in Fig. 28. (from Ref. [65])

The scale invariance of the shear bands can be analyzed by changing the lattice size L or equivalently the resolution of the system. The geometrical fractal dimension of the shear band network is $d_0 = 1.82 \pm 0.1$ for $R = 10^{-9}$. The same result is obtained from the box-counting analysis of a single picture with the highest resolution [65].

The spontaneous appearance of a fractal set of shear bands starting out from a rather homogeneous situation suggests the existence of self-organized criticality [54]. Each shear band might be seen as a single internal avalanche on which the system can release stresses through larger displacements. Let us also look at the distribution of the length of shear bands. Numerically the length of the bands can be obtained using a burning algorithm [107]. In Fig. 29 we see a log-log plot of this distribution. For lengths smaller than the width of the bands the data are not useful. After that we see a power-law decay with an exponent of $\tau = 2.1 \pm 0.2$. This value is in reasonable agreement with length distributions measured on sand [62].

There is some resemblance between fracture of rocks and shear bands in granular media. It is, however, important to point out a crucial difference between the two: for cracks the most stressed regions are at the tips while the shear bands have their strongest strain rates in the center.

12. Discussion

With a rather simple description of a granular medium as an ensemble of inelastic spherical particles with shear friction we have shown that many interesting rheological properties can be reproduced. Various types of convection can occur on a vibrating plate which are strongly influenced by hindrances on the wall. Density waves appear during the flow through a pipe which can be explained as a consequence of dissipation: Due to the inelastic collisions between particles (also cars) an instability [91, 92] tends to form clusters of high density (dark horizontal stripes in Fig. 12). These clusters self-organize into a critical state giving a power law spectrum. As in classical traffic models the density dependence of the flux generates kinematic waves (dark tilted lines in Fig. 12). The density dependence of the viscosity as predicted by kinetic gas theory builds up waves of low density (light horizontal lines in Fig. 12). We see that simply introducing dissipation to a gas of particles produces several phenomena, that occur simultaneously in granular materials.

It is not always straightforward to determine the material constants corresponding to some parameters of the model, like γ_n and γ_s, and so a quantitative comparison with experiments in many cases still involves fitting some parameters. More realistic models including static and dynamic friction, rotations of particles, variations in the particle shapes, etc increase the number of parameters but are needed to establish a closer contact to reality. The most important restriction of the simulations presented here is that most have been performed in two dimensions. Three dimensional simulations must also be performed since many phenomena seem to be due to steric effects. The technique we presented can of course also be applied to three dimensions. It is just the available computer time that limits at present our possibilities to extend simulations to three dimensional systems of numerically reliable sizes.

The analogy of molecular systems for which Molecular Dynamics simulations were originally conceived gave us the incentive to describe within a thermodynamic formalism the fluctuations arising from the constant flux and dissipation of energy that drives a granular material's kinematic behavior. By separating the dissipative degree's of freedom (friction and plasticity) from the conservative ones (translation, rotation, elasticity) we define a "granular ensemble" coupled to a "dissipate bath" which is in fact the one in which experimental and numerical measurements are usually performed.

When the density of the granular medium exceeds the characteristic Reynolds dilatancy it behaves like a solid but retains very particular properties. One is the non-linear acoustics due to the complicated network on which the stresses are transmitted and another is the Hertzian contact law. As a consequence the sound velocity depends on pressure and amplitude and one finds secondary waves of lower velocity. Another effect is the non-associated plasticity due to the local effects of dilatancy during large scale deformations. Its consequence is the localization of the shear on planes which at least within a certain range form a fractal network.

I thank E. Flekkøy, J. Gallas, J. Lee, S. Melin, K. Nagel, G. Peng, A. Poliakov, T. Pöschel, G. Ristow and S. Sokołowski were my collaborators on the results presented in this course.

References

1. A. Hansen and D. Bideau (eds.) *Disorder and Granular Media* (North-Holland, Amsterdam, 1992)
2. H.M. Jaeger and S.R. Nagel, Science **255**, 1523 (1992) and S.R. Nagel, Rev. Mod. Phys. **64**, 321 (1992)
3. C.S. Campbell, Annu. Rev. Fluid Mech. **22**, 57 (1990)
4. A. Mehta, *Granular Matter* (Springer, Heidelberg, 1994)
5. J.C. Williams, Powder Techn. **15**, 245 (1976) and Fuel Soc. J. **14**, 29 (1963)
6. A. Rosato, K.J. Strandburg, F. Prinz and R.H. Swendsen Phys.Rev.Lett. **58**, 1038 (1987) and Powder Techn. **49**, 59 (1986); P. Devillard, J. Physique **51**, 369 (1990)
7. R. Jullien, P. Meakin and A. Pavlovitch, Phys. Rev. Lett. **69**, 640 (1992); R. Jullien and P. Meakin, Europhys. Lett. **22**, 523 (1993)
8. P.K. Haff and B.T. Werner, Powder Techn. **48**, 239 (1986)
9. M. Faraday, Phil. Trans. R. Soc. London **52**, 299 (1831)
10. P. Evesque and J. Rajchenbach, Phys. Rev. Lett. **62**, 44 (1989); C. R. Acad. Sci. Ser. 2, **307**, 1 (1988) and **307**, 223 (1988); C. Laroche, S. Douady and S. Fauve, J. de Physique **50**, 699 (1989); P. Evesque, J. Physique **51**, 697 (1990); J. Rajchenbach, Europhys. Lett. **16**, 149 (1991); E. Clément, J. Duran and J. Rajchenbach, Phys. Rev. Lett. **69**, 1189 (1992)
11. J. Walker, Sci. Am. **247**, 167 (1982); F. Dinkelacker, A. Hübler and E. Lüscher, Biol. Cybern. **56**, 51 (1987)
12. G.W. Baxter, R.P. Behringer, T.Fagert and G.A. Johnson, Phys. Rev. Lett. **62**, 2825 (1989)

13. C.-h. Lui and S.R. Jaeger, Phys. Rev. Lett. **68**, 2301 (1992)
14. O. Reynolds, Phil. Mag. Soc. **20**, 469 (1885)
15. Y.M. Bashir and J.D. Goddard, J. Rheol. **35**, 849 (1991)
16. E. Clément and J. Rajchenbach, Europhys. Lett. **16**, 133 (1991)
17. P. Evesque, E. Szmatula and J.-P. Denis, Europhys. Lett. **12**, 623 (1990); O. Zik and Stavans, Europhys. Lett. **16**, 255 (1991); O. Zik, J. Stavans and Y. Rabin, Europhys. Lett. **17**, 315 (1992)
18. G. Rátkai, Powder Techn. **15**, 187 (1976)
19. H.K. Pak and R.P. Behringer, Phys. Rev. Lett. **71**, 1832 (1993); S. Douady, thèse de doctorat, Ecole Normale Supérieure de Lyon
20. S.B. Savage, J. Fluid Mech. **92**, 53 (1979); G.M. Homsy, R. Jackson and J.R. Grace, J. Fluid Mech. **236**, 477 (1992); S.B. Savage and K. Hutter, J. Fluid Mech. **199**, 177 (1989)
21. H.J. Herrmann, J. Physique II **3**, 427 (1993)
22. J.T. Jenkins and S.B. Savage, J. Fluid Mech. **130**, 186 (1983)
23. S.F. Edwards and R.B.S. Oakeshott, Physica A **157**, 1080 (1989); S.F. Edwards, J. Stat. Phys. **62**, 889 (1991); A. Mehta and S.F. Edwards, Physica A **157**, 1091 (1989)
24. G.W. Baxter and R.P. Behringer, Phys. Rev. A **42**, 1017 (1990), Physica D **51**, 465 (1991); D. Désérable and J. Martinez, in *Powder & Grains* ed. C. Thornton (Balkema, Rotterdam, 1993), p. 345
25. G. Peng and H.J. Herrmann, Phys. Rev. E **49**, R1796 (1994) and preprint HLRZ 11/94
26. A. Károly and J. Kertész, Proc. 6th Joint EPS-APS Int. Conf. of Phys. Comp., eds. R. Gruber and M. Tomassini (EPS, Geneva, 1994)
27. H. Caram and D.C. Hong, Phys. Rev. Lett. **67**, 828 (1991) and Mod. Phys. Lett. B **6**, 761 (1992)
28. O. Walton, in *Two Phase Flow*, ed. M. Roco (Butterworth-Heinemann, Boston, 1992), chapter 25
29. G. Ristow, Ann. Rev. Comp. Phys., ed. D. Stauffer (World Sci. Publ. Co., Singapore, 1994)
30. P.A. Cundall, Report AD/A - 001 602, U.S. Nat. Tech. Information Service, (Springfield, Va, 1974)
31. P.A. Cundall and O.D.L. Strack, Géotechnique **29**, 47 (1979)
32. R.J. Bathurst and L. Rothenburg, J. Appl. Mech. **55**, 17 (1988)
33. Y. Zhang and P.A. Cundall, Proc. Symp. on the Mech. of Particulate Media, 10th National Congress on Applied Mechanics (Austin, Tex, (1986); J.M. Ting and B.T. Corkum, Proc. 6th Int. Conf. Num. Meth. in Geomech., Int. Commission for Num. Meth. in Geomech. **1**, 305 (1988); R. Dobry and T.T. Ng, Proc. 1st U.S. Conf. on Discrete Element Meth., Nat. Sci. Found. (Washington, DC, 1989); J.A. Issa and R.B. Nelson, idem; P.J. Hassan, thesis Univ. of Toronto (1990)
34. M.R. Kuhn and J.R. Mitchell, Proc. 1st U.S. Conf. on Discrete Element Meth., Nat. Sci. Found. (Washington, DC, 1989)
35. L.J. Lorig and B.H.G. Brady, ISRM British Geotech. Soc. (Cambridge, 1984), p.105

36. R. Barbosa and J. Ghaboussi, Proc. Workshop on Const. Laws for the Anal. of Fill Retention Struct., Canada Oil and Gas Adm. (Ottawa, 1987); M. Grabinsky, thesis Univ. of Toronto (1989); J.M. Ting, B.T. Corkum, C.R. Kauffman and C. Greco, J. of Geotech. Eng., ASCE **115**, 379 (1989)
37. T. Pöschel and H.J. Herrmann, preprint HLRZ 28/94
38. G. Ristow, J. Physique I **2**, 649 (1992) and Int. J. Mod. Phys. C **3**, 1281 (1992)
39. D.C. Hong and J.A. McLennan, Physica A **187**, 159 (1992)
40. C.S. Campbell and C.E. Brennen, J. Fluid Mech. **151**, 167 (1985); P.A. Thompson and G.S. Grest, Phys. Rev. Lett. **67**, 1751 (1991); D.M. Hanes and D.L. Inman, J. Fluid Mech. **150**, 357 (1985); O.R. Walton and R.L. Braun, J. Rheol. **30**, 949 (1986); M. Hakuno, K. Iwashita and Y. Uchida, Proc 1st U.S. Conf. on Discrete Element Meth., Nat. Sci. Found. (Washington, DC, 1989);
41. T. Pöschel, J. Physique II **3**, 27 (1993)
42. M.P. Allen and D.J. Tildesley, *Computer Simulation of Liquids*, Oxford University Press, Oxford, 1987
43. D. Tildesley, in *Computational Physics*, edited by R.D. Kenway and G.S. Pawley, NATO Advanced Study Institute, Edinburgh University Press, 1987
44. J.A.C. Gallas, H.J. Herrmann and S. Sokołowski, Physica A, **189**, 437 (1992)
45. S. Luding, E. Clément, A. Blumen, J. Rajchenbach and J. Duran, Phys. Rev. E **49**, 1634 (1994); S. Luding, H.J. Herrmann and A. Blumen, preprint HLRZ 77/93
46. Y-h. Tagushi, Phys. Rev. Lett. **69**, 1367 (1992) and preprints
47. J.A.C. Gallas, H.J. Herrmann and S. Sokołowski, Phys. Rev. Lett., **69**, 1371 (1992)
48. J.A.C. Gallas, H.J. Herrmann and S. Sokołowski, J. Physique II, **2**, 1389 (1992)
49. O. Cutress and R.F. Pulfer, Powder Techn. **1**, 213 (1967)
50. A. Drescher, T.W. Cousens and P.L. Bransby, Geotechn. **28**, 27 (1978)
51. R.L. Michalowski, Powder Techn. **39**, 29 (1984); J.D. Athey, J.O. Cutress and RF. Pulfer Chem. Eng. Sci. **21**, 835 (66); P.M. Blair-Fish and P.L. Bransby, J. Eng. for Industry **95**, 17 (73); J. Lee, S.C. Cowin and J.S. Templeton, Trans. Soc. Rheol. **18**, 247 (74); K.L. Schick and A.A. Verveen, Nature **251**, 599 (1974)
52. T. Pöschel, J. Physique I **4**, 499 (1994)
53. G.W. Baxter, R. Leone and R.P. Behringer, Europhys. Lett. **21**, 569 (1993); G.W. Baxter, R. Leone, G.A. Johnson and R.P. Behringer, Eur. J. Mech., B/Fluids **10**, #2 - Suppl., 181 (1991); G.W. Baxter, PhD thesis
54. P. Bak, Tang and Wiesenfeld, Phys. Rev. Lett. **59**, 381 (1987)
55. G.A. Held, D.H. Solina, D.T. Keane, W.J. Horn and G. Grinstein, Phys. Rev. Lett. **65**, 1120 (1990)
56. G. Ristow and H.J. Herrmann, Phys. Rev. E **50**, R5 (1994)
57. K. Nagel and H.J. Herrmann, Physica A **199**, 254 (1993)
58. E. Flekkøy and H.J. Herrmann, Physica A **199**, 1 (1993)
59. G. Ristow and H.J. Herrmann, preprint HLRZ 29/94
60. Experiments made in Cargèse in 1988
61. J.D. Goddard, Proc. R. Soc. London A **430**, 105 (1990)

62. A. Sornette, P. Davy and D. Sornette, Phys. Rev. Lett. **65**, 2266 (1990)
63. J.W. Rudnicki and Rice J.R., J.Mech.Phys. Solids, **23**, 371 (1975)
64. S. Melin, Phys. Rev. E **49**, 2353 (1994) and IJMPC **4**, 1103 (1993)
65. A. Poliakov and H.J. Herrmann, to appear in GRL and Fractals
66. K.L. Johnson, *Contact Mechanics* (Cambridge Univ. Press, 1989)
67. L.D. Landau and E.M. Lifshitz, *Elasticity Theory* (Pergamon Press, Oxford, 1975)
68. J. Lee and H.J. Herrmann, J. Phys. A **26**, 373 (1993)
69. T. Pöschel and H.J. Herrmann, Physica A**198**, 441 (1993)
70. T. Pöschel and V. Buchholtz, Phys. Rev. Lett. **71**, 3963 (1993); V. Buchholtz and T. Pöschel, Physica A **202**, 390 (1994)
71. W. Form, G.A. Kohring, S. Melin, H. Puhl and H.J. Tillemans, preprint HLRZ 75/93 for Comp. Meth. in Appl. Mech. and Eng.
72. F. Cantelaube, Y.L. Duparcmeur, D. Bideau and G.H. Ristow, preprint for J. de Physique
73. H. Schmidt and I. Peschl, Fördern u. Heben **15**, 606 (1965)
74. K. Ahmad and I.J. Smalley, Powder Technol. **8**, 69 (1973)
75. J. Bridgewater, Powder Technol. **15**, 215 (1976)
76. R.L. Brown, J. Inst. Fuel **13**, 15 (1939)
77. J. Duran, J. Rajchenbach and E. Clément, Phys. Rev. Lett. **70**, 2431 (1993)
78. R. Jullien, P. Meakin and A. Pavlovitch, Phys. Rev. Lett. **70**, 2195 (1993)
79. J.B. Knight, H.M. Jaeger and S.R. Nagel, Phys.Rev. Lett. **70**, 3728 (1993); E.E. Ehrichs, H.M. Jaeger, G.S. Karczmar, J.B. Knight, V. Kuperman and S.R. Nagel, preprint
80. J.B. Knight, C.G. Fandrich, C.N. Lau, H.M. Jaeger, S.R. Nagel, preprint
81. J. Duran, T. Mazozi, E. Clément and J. Rajchenbach, preprint
82. J. Duran, private communication
83. J.A.C. Gallas, H.J. Herrmann, T. Pöschel and S. Sokołowski, preprint HLRZ 92/93
84. T.G. Drake , J. Geophys. Res. **95**, 8681 (1990)
85. S.B. Savage, in *Powders and Grains*, ed. C. Thornton, Birmingham, 1993
86. J. Schwedes, *Fließverhalten von Schüttgütern in Bunkern* (Verlag Chemie, Weinheim, 1968)
87. K. Nagel and M. Schreckenberg, J. Physique I **2**, 2221 (1992)
88. M.J. Lighthill and G.B. Whitham, Proc. Roy. Soc. A **229**, 281 and 317 (1955); M. Leibig, Phys. Rev. E **49**, 184 (1994)
89. T. Musha and H. Higuchi, Jap. J. Appl. Phys. **15**, 1271 (1976)
90. M. de Sousa Vieira, Phys. Rew. A **46**, 6288 (1992)
91. I. Goldhirsch and G. Zanetti, Phys. Rev. Lett. **70**, 1619 (1993)
92. S. Savage, J. Fluid Mech. **241**, 109 (1992)
93. P.K. Haff, J. Fluid Mech. **134**, 401 (1983)
94. J.T. Jenkins, Arch. Rat'l. Mech. Anal., **87**, 355 (1985)
95. U. Frisch, B. Hasslacher and Y. Pomeau, Phys. Rev. Lett. **56**, 1505 (1986)

96. R. Benzi, S. Succi and M. Vergassola, Phys. Rep. **222**, 145 (1992); G.R. McNamara and G. Zanetti, Phys. Rev. Lett. **61**, 2332 (1988)
97. A. Bershadskii, Physica A **210**, 386 (1994)
98. R.A. Bagnold, Proc. Roy. Soc. London A **295**, 219 (1966) H.M. Jaeger, C.-h. Liu, S.R. Nagel and T.A. Witten, Europhys. Lett. **11**, 619 (1990)
99. S. Ogawa, *Proc. of US-Japan Symp. on Continuum Mechanics and Statistical Approaches to the Mechanics of Granular Media*, eds. S.C. Cowin and M. Satake (Gakujutsu Bunken Fukyu-kai, 1978), p.208
100. J.T. Jenkins and S.B. Savage, J. Fluid Mech. **130**, 186 (1983); H.M. Jaeger, C.-h. Lui and S.R. Nagel, Phys. Rev. Lett. **62**, 40 (1989)
101. M. Leibig, Phys. Rev. E **49**, 1647 (1994)
102. C. Liu, S.R. Nagel, Phys. Rev. B **48**, 15646 (1993)
103. J.D. Byerlee, J. Geophys. Res. **73**, 4741 (1968)
104. P.A. Cundall, Ingenieur Archiv **58**, 148 (1989)
105. P.A. Vermeer, Géotechnique **40**, (1990) 223; P.A. Vermeer and R. de Borst, Heron, **29**, 1 (1984)
106. I. Vardoulakis, Int. J. Numer. Anal. Meth. Geomech. **4**, 103 (1980)
107. H.J. Herrmann, D.C. Hong and H.E. Stanley, J. Phys. A **17**, L261 (1984)

Some Physical and Computational Aspects of Self-Organized Criticality

A. Díaz-Guilera

Departament de Física Fonamental, Universitat de Barcelona,
Diagonal 647, E-08028 Barcelona, Spain.
Electronic address: **albert at ulyses.ffn.ub.es**

1. Introduction

In the last few years, much attention has been paid to the phenomenon known as self-organized criticality (SOC). Bak, Tang, and Wiesenfeld (Bak et al., 1988) introduced a cellular automaton model, currently known as the sandpile model, as a paradigm for the explanation of two widely occurring phenomena in nature: $1/f$ noise and fractal structures. $1/f$ noise means that the spectral density of the fluctuations of a certain quantity behaves like $1/f^{\alpha}$ with $0.6 < \alpha < 1.4$ and it is due to a superposition of time scales. Fractals are understood as systems with spatial self-similarity, which means that the system has the same structure independent from the scale we look at (Mandelbrot, 1982). Additionally, one of the main applications of SOC models has been in the study of earthquake dynamics since the distribution of magnitudes of earthquakes is a power-law, known as the Gutenberg-Richter law (Gutenberg and Richter, 1954), over more than ten orders of magnitude. These phenomena have a lack of characteristic scales in common.

The sandpile model is a cellular automata where each site on a lattice is characterized by an integer variable, and this variable changes in time due to external perturbations and to interaction between different sites. With very simple rules the system reaches a state which is characterized by the lack of any characteristic scale. This scale invariance suggests that the system is critical in analogy with equilibrium critical phenomena, but in SOC one deals with dynamical nonequilibrium statistical properties and the system evolves naturally to the critical state without any tuning of external parameters, criticality is an intrinsic property of the dynamics of the system.

The importance of this new concept lies in the fact that the complex behavior of the extended dynamical system arises from very simple local rules. Such dynamical systems can provide an explanation of the wide occurrence of the above mentioned phenomena. There is also a close analogy with the concept of chaos, but the idea of SOC refers to the behavior of an extended system with many degrees of freedom.

The notes herein are organized as follows: in Sect. 2 we will examine the original sandpile model in one and two dimensions and discuss how to characterize the SOC state. In Sect. 3 we will briefly review some of the experimental evidences of SOC, models related to specific systems, generic cellular automata models, and some theoretical approaches, respectively. Our aim in these section is not to give an exhaustive review of this subject but to demonstrate to the reader the wide range of physical phenomena which can be explained or understood in terms of these new ideas. Finally, in Sect. 4 we will deal with the computational aspects of different models exhibiting SOC.

2. Sandpiles, Cellular Automata, and Self-Organized Criticality

2.1. The Bak, Tang, and Wiesenfeld sandpile model

First of all, we will analyze the original one-dimensional sandpile model introduced by Bak et al. (1988) in detail. The model is a cellular automata where each site i is characterized by an integer variable h_i which gives the height of the pile at a given point (see Fig. 1). We define the slope of a site as

$$z_i = h_i - h_{i+1}. \tag{2.1}$$

The addition of a sand grain on a randomly chosen site i results in the following changes in the slopes

$$\begin{aligned} z_i &\rightarrow z_i + 1 \\ z_{i-1} &\rightarrow z_{i-1} - 1. \end{aligned} \tag{2.2}$$

We proceed by dropping grains at random until one site reaches a slope which is larger than some critical value, $z_i > z_c$, ($z_c = 2$ usually in one-dimensional models) and the site topples by transferring one grain to its neighboring site on the right. In this case the slopes change accordingly to (see Fig. 1)

$$\begin{aligned} z_i &\rightarrow z_i - 2 \\ z_{i\pm1} &\rightarrow z_{i\pm1} + 1 \end{aligned} \tag{2.3}$$

unless at the rightmost site where the sand grains fall off the pile.

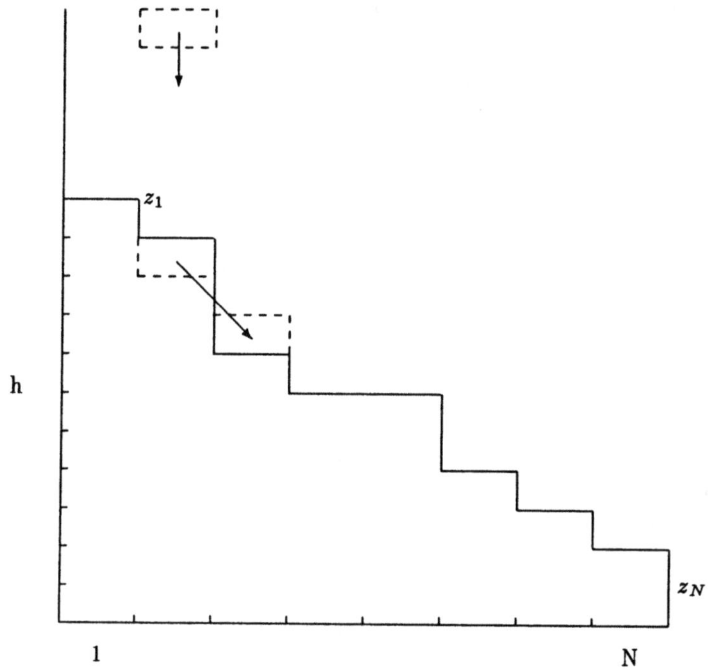

Fig. 1. Bak, Tang, and Wiesenfeld sandpile model with closed and open boundary conditions for the left and right boundaries, respectively.

The neighbors that are affected by the toppling can then topple in turn generating a chain reaction or *avalanche*. During the avalanche no more grains are added to the pile thus separating the two time scales involved in the dynamic evolution of the pile, one for the addition of the grains and another for the relaxation processes. When the avalanche is over we add more grains following (2.2) until a new avalanche starts and so on. After some transient, whose duration depends on the initial conditions, the system reaches a critical state in which $z_i = z_c \quad \forall i$. This state is stable, since after any perturbation the system relaxes returning to the stable state; the addition of a sand grain to the pile results in an avalanche that makes this added grain simply fall off the pile. Defining the size of the avalanche as the number of sites that topple, we can easily see that in this case it is equal to the distance from the site where sand is added to the open boundary; this state is critical in the sense that the avalanches have no characteristic size. The critical state is an attractor for the dynamics, which means that it does not depend on how the pile is built up (adding grains at random from scratch or relaxing from a supercritical state). However, this state has no spatial structure, and correlation functions are trivial.

Now we will take a look at the two-dimensional model where the relation to sandpiles is not so clear, but the critical state is not so trivial. In this case we describe the state of the system by the slope at each site and this slope is assumed to be a scalar variable. The rules are a simple generalization of the one-dimensional ones; thus the addition of a sand grain makes the slopes change to

$$\begin{aligned} z_{i,j} &\to z_{i,j} + 2 \\ z_{i-1,j} &\to z_{i-1,j} - 1 \\ z_{i,j-1} &\to z_{i,j-1} - 1. \end{aligned} \tag{2.4}$$

Usually this rule is modified such that the perturbation on the system is made directly in terms of the slope; thus we just add a unit of slope

$$z_{i,j} \to z_{i,j} + 1. \tag{2.5}$$

These rules make the analogy with sandpiles extremely unclear since it is difficult to imagine how to add a unit of slope to a two-dimensional pile.

We proceed with this rule until the slope of a site reaches the critical value $z_{i,j} > z_c = 4$, then

$$\begin{aligned} z_{i,j} &\to z_{i,j} - 4 \\ z_{i\pm1,j} &\to z_{i\pm1,j} + 1 \\ z_{i,j\pm1} &\to z_{i,j\pm1} + 1. \end{aligned} \tag{2.6}$$

For the two-dimensional sandpile model we can also think of a critical state with all sites having exactly the critical slope but now this state is *unstable* with respect to small perturbations, since an avalanche would propagate in two directions on the lattice with more than one grain falling off the pile, and it is not an attractor for the dynamics. In any case, after some transient time, the system reaches, through a self-organizing process, a *stable critical state* with a non-homogeneous distribution of slopes. This state responds to the addition of a sand grain with an avalanche of unpredictable size (number of sites that topple) and duration (number of iterations needed to reach the stable state). Statistically, we measure the distribution of avalanche sizes and lifetimes averaged over a large number of perturbations and we can observe the following power laws

$$D(s) \sim s^{\tau-1} \tag{2.7}$$

$$D(t) \sim t^{-b} \tag{2.8}$$

with no characteristic time or length scales.

In Fig. 2 we plot an example of the distribution of avalanche sizes in a log – log scale. There we can observe a clear power-law behavior for about two orders of magnitude; however, for large values of the size there is a poor statistics from where no clear conclusion can be extracted. In Sect. 4.3 we will show how to overcome these difficulties.

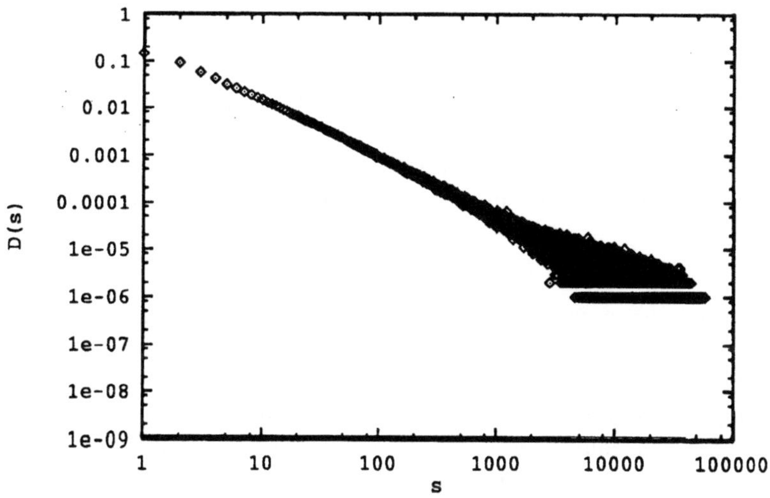

Fig. 2. Distribution of avalanche sizes.

Another statistical feature of the critical state is that after averaging the local values of the slopes it turns out to be perfectly homogeneous and isotropic, showing again the lack of any characteristic length scale.

2.2. Critical behavior

The critical state studied previously is characterized by a set of critical exponents in analogy with equilibrium critical phenomena.[2] The critical exponents relate different magnitudes describing the system, and in the sandpile model they can be introduced in terms of the variables describing the avalanches or in terms of the correlation function of the height fluctuations as is done, for instance, in nonequilibrium surface growth phenomena (Family and Vicsek, 1991).

In (2.7) and (2.8) we have already defined the exponents of the avalanche sizes and lifetimes distributions. The exponent that relates the duration and the characteristic length of the avalanche is called the dynamical exponent (z)

$$t \sim l^z \tag{2.9}$$

and the relation between the size and the length gives the fractal dimension (d_F)

$$s \sim l^{d_F} . \tag{2.10}$$

The dynamical exponent has a clear meaning in many physical systems. For instance, in a diffusion process, where the particles perform independent random

[2] See Tang and Bak, (1988b) for a complete list.

walks, this exponent characterizes the relaxation of a perturbation and its value is 2. On the other side, a signal propagating with constant velocity corresponds to $z = 1$. For the growth of the avalanches we expect an intermediate behavior between these two limits, except for the one-dimensional pile, where clearly this exponent is equal to one. On the other hand, if we expect the avalanches to be fractal objects its fractal dimension has to be smaller than the euclidean dimension of the lattice.

Other definitions concern the correlation function of the height fluctuations; we expect it to be of the following form, if there are no characteristic length or time scales,

$$\langle ||h(\mathbf{r},t) - h(\mathbf{0},0)||^2 \rangle \sim r^{2\xi} g(tr^{-z}) \tag{2.11}$$

where ξ is called the roughening exponent, z is again the dynamical exponent and g is a scaling function. We should note that the dynamical exponent has been defined in two different ways; but in both cases it relates time and length variables, and both definitions are expected to be equivalent.

As it happens in equilibrium critical phenomena, the values of the critical indices do not depend on the microscopic details of the dynamical rules, they effect only the prefactors in the power laws or the form of the scaling function. Therefore, different models with the same values of the exponents are said to belong to the same universality class.

2.3. What is new in SOC?

As we stated in the introduction, the sandpile model has tried to explain the wide occurrence of $1/f$ noise and fractal structures in nature. However, it has been shown that avalanches have a compact structure (Pietronero and Schneider, 1991) and the spectrum of the noise needs a careful investigation (Jensen et al., 1989). Nevertheless, the inclusion of a threshold condition in the dynamical rules is what leads the system to a critical state by means of a self-organizing process; and this happens in many physical systems as we will see in the next sections.

Noise and conservation are also key points in SOC phenomena. The original models were noise driven models; this means that the sand grains are added to the pile at randomly chosen sites. It was shown that conservative interaction rules [in (2.2) and (2.6) the slope is conserved] were crucial to obtain SOC in these models (Manna et al, 1990). More recently, however, the interest has been focused on systems with a deterministic and uniform driving,[3] such as stick-slip models of earthquakes, which exhibit SOC without a conservation law starting from a random initial distribution (Feder and Feder, 1991; Olami et al., 1992). All these models have in common the existence of some kind of noise, either in the way the system is driven or in the initial conditions; in principle, this seems to be one of the key ingredients of SOC but some authors have reported power-law distributions for completely deterministic automata (Wiesenfeld et al., 1990;

[3] In the one-dimensional sandpile picture, a deterministic and uniform driving would correspond to raise the left end of the platform.

Vieira, 1992; Nagel and Herrmann, 1993). However, for all these deterministic models there is a single site with a special behavior; this inhomogeneity, either in the initial conditions or in the external loading, is what gives rise in these cases to the complex behavior characteristic of SOC.

2.4. Relationship to real sandpiles

Real sandpiles present very complex behavior which is difficult to explain in terms of simple interaction rules, but due to the resemblance to the original models, and our intuitive idea of a sandpile, they have been studied in this context. However, there is still some controversy about the existence of SOC in real sandpiles because it also depends on how the pile is built, either by rotating a semicylindrical drum partially filled with sand or by dropping grains on a circular platform (Nagel, 1992). In order to explain some of the features that are observed in the experiments, more sophisticated computer models have been used (Mehta and Barker, 1994).[4]

3. Experiments, Computers and Theories

3.1. Physical systems exhibiting SOC

Although the relationship that SOC has with real sandpiles is still an open question, these ideas have been applied to other physical systems, mainly earthquakes, where the distribution of magnitudes decays as a power law for several orders of magnitude (Gutenberg and Richter, 1954). In the context of earth sciences the acoustic emission from volcanic rocks exhibits power-laws for the amplitude and time distributions (Diodati et al., 1991).

One of the first systems where SOC behavior was reported was in the evolution of magnetic-domain patterns (Babcock and Westervelt, 1990); these cellular patterns coarsen in response to an increasing magnetic bias field. By subsequent increment in the field, the pattern undergoes a sequence of topological changes until a new stationary configuration is reached. The size of the avalanche is the total number of cells eliminated, and its lifetime is given by the time between the first topological change and the cessation of the domain motion. In both cases power-law distributions are observed for about one decade. Also the Barkhausen effect in a ferromagnetic metallic glass shows some of the characteristics of SOC (Cote and Meisel, 1991).

The growth and coalescence of fluid droplets has also been a good candidate for the appearance of SOC (Janosi and Horvath, 1989; Plourde et al., 1994). A droplet grows in size on a sprayed surface and eventually reaches a critical mass, at which time it runs down the surface, triggering other stationary droplets in its path. The distributions of avalanche sizes and lifetimes show a power-law, whereas the power spectrum exhibits three approximate power-law regimes.

[4] For more details on this relationship see the Chapter in this book by H.J. Herrmann.

Other systems where SOC has also been investigated are, for instance, fragmentation (Oddershede et al., 1993), the acoustic emission generated during a martensitic transformation (Vives et al., 1994), and in the daily fluctuations of the air humidity (Vattay and Harnos, 1994).

3.2. Simple numerical models

In these paragraphs we want to deal with numerical models constructed to explain some of the features of specific physical problems like, for instance, the dynamics of earthquake faults (Carlson et al., 1994). Most of these works are based on the Burridge-Knopoff model (Burridge and Knopoff, 1967), which consists of a one-dimensional chain of blocks connected by springs that is pulled across a rough surface. The equation of motion for the moving blocks is given by

$$m\ddot{x}_j = k_c(x_{j+1} - 2x_j + x_{j-1}) - k_p(x_j - vt) - F(\dot{x}_j) \tag{3.1}$$

where x_j is the displacement of the j-th block from its equilibrium position, m is the mass of the blocks, k_c and k_p are the spring constants of the springs connecting the blocks to one another and to the pulling rod moving with velocity v, respectively, and F accounts for a nonlinear frictional force which introduces the threshold condition. Starting from random initial conditions and deterministically driving the springs, the size distribution of events is consistent with the Gutenberg-Richter law. A variant of this model that presents SOC even without random initial conditions consists of pulling the chain at one end (Vieira, 1992). Also different two-dimensional cellular automata models have been proposed in the context of the Burridge-Knopoff model (Feder and Feder, 1991; Olami et al., 1992; Nakanishi, 1990) and models where the springs break causing a redistribution of elastic forces (Chen et al., 1991).

In the previous subsection we mentioned that one of the physical systems where SOC has been observed is in the dynamics of magnetic domains. Numerical models related to this system have been also investigated (Che and Suhl, 1990; Bak and Flyvbjerg, 1992).

Some works have been devoted to studying SOC behavior in the vortex state of type-II superconductors (Pla and Nori, 1991; Pan and Doniach, 1994; Wang and Shi, 1994) since it presents a close analogy to sandpiles; in this case the local current density plays the role of the local slope of the pile.

Other numerical models with SOC behavior introduced in the literature are very diverse, as examples we can quote: the dynamics of river networks (Rinaldo et al., 1993), cloud formation (Nagel and Raschke, 1992), quaternary climatic fluctuations (Grieger, 1992), market prices with a threshold condition (Takayasu et al., 1992), forest-fire models (Drossel and Schwabl, 1992), ecosystems (Sol'e et al., 1992), models of evolution (Bak and Sneppen, 1993), the "Game of Life" (Alstrom and Leao, 1994), neural networks (Miranda and Herrmann, 1991), traffic jams (Nagel and Herrmann, 1993), or a model of biological oscillators (Corral et al., 1994).

3.3. Cellular automata and coupled map lattice models

Several cellular automata models exhibiting SOC have been reported in the literature. In the original sandpile model of Bak et al. (Bak et al., 1988), the system is perturbed externally by a random addition of sand grains. Once the slope between two contiguous cells has reached a threshold condition, sand is transferred to its neighbors in a fixed amount. Taking this model as a reference, different dynamical rules have been investigated leading to a wide variety of universality classes: directed flows (Kadanoff et al., 1989), threshold condition imposed on the height, on the gradient or even on the laplacian (Manna, 1991), continuous variables with a full transfer from a cell instead of a fixed discrete amount (Zhang, 1989; Pietronero et al., 1991; Fodor and Janosi, 1991; Díaz-Guilera, 1992),[5] deterministic perturbations in a nonconservative system (Feder and Feder, 1991; Christensen et al., 1992; Olami and Christensen, 1992; Socolar et al., 1993), anisotropy (Janosi, 1990), correlations in the deposition process (Ohtsuki and Iwata, 1992), are some examples.

These rules can be written in the form of a set of algebraic equations, one for each site. Thus for the Bak, Tang, and Wiesenfeld model we have

$$E(i,t+1) = E(i,t) - \Theta(E(i,t)-E_c)E_c + \sum_{j=nn} \Theta(E(j,t)-E_c)E_c/q + \eta(i,t) \quad (3.2)$$

$$\eta(i,t) = \eta(i) \prod_{j,\mathrm{allsites}} \Theta[E_c - E(j,t)] \quad (3.3)$$

where $\eta(i)$ is a random number at a randomly chosen site, q is the coordination number of the lattice, and $\Theta(x)$ is the Heavside step function. For a model where the variable is reset to zero they write

$$\begin{aligned} E(i,t+1) =& [1-\Theta(E(i,t)-E_c)]E(i,t) \\ &+ \sum_{j=nn} \Theta(E(j,t)-E_c)E(j,t)/q + \eta(i,t) \end{aligned} \quad (3.4)$$

and for a continuously driven model we have

$$E(i,t+1) = [1-\Theta(E(i,t)-E_c)]E(i,t) + \sum_{j=nn} \Theta(E(j,t)-E_c)E(j,t)/q + \eta(t) \quad (3.5)$$

where

$$\eta(t) = \eta \prod_j \Theta[E_c - E(j,t)] \quad (3.6)$$

is a global perturbation. Note that if $\alpha < 1$ the interaction rules are nonconservative.

A few of these models try to connect with some physical systems but the main goal is to study which conditions the SOC behavior is obtained under

[5] If the variable describing the state of the cell is continuous, it is more appropriate to call the model a coupled map lattice.

and to classify the different universality classes they belong to. In principle, one expects that the critical exponents do not depend on the microscopic details of the system but, for instance, in the deterministically driven models, it has been observed that the exponents do change as a function of the degree of conservation and, hence, the universality classes.

3.4. Theoretical studies

Of course, SOC has also been the object of several theoretical approaches. As it happens in general in critical phenomena, one of the first approaches is a mean-field treatment. In such a situation one neglects the effects of the fluctuations replacing the local variables by their averaged values. Thus we can find this analysis in the original BTW model (Tang and Bak, 1988a) and in other specific models such as the forest-fire (Christense et al., 1993) or the model for evolution (Flyvbjerg et al., 1993). A concept which is closely related to the mean-field is that of the upper critical dimension, which is the spatial dimension above which the fluctuations are irrelevant, and the mean-field results are exact. In most of the SOC models, this dimension seems to be 4 (Zhang, 1989; Obukhov, 1988; Díaz-Guilera, 1994).

Another approach is to investigate concrete models where the rules are more suitable for obtaining analytic results, although there is no clear relationship to physical systems. In this sense, perhaps the most successful model has been the so-called abelian sandpile model (Creutz, 1994). Exact results are also known for the original models by studying the scaling of the avalanches (Zhang, 1989) or by means of the renormalization group, widely used in critical phenomena, where the lack of characteristic scales is used (Pietronero et al., 1994).

There has also been a lot of interest in constructing differential equations which keep the scale invariance of the automata models. Anomalous diffusion equations with singularities in the diffusion coefficient have been considered in order to study the deterministic dynamics of the avalanches generated in the critical state (Carlson et al., 1990; Bantay and Janosi, 1992; Chau and Cheng, 1994; Kadanoff et al., 1992). Nonlinear stochastic differential equations which are built according to the symmetries of the discrete models have been analyzed in the framework of the dynamic renormalization group (Díaz-Guilera, 1994; Hwa and Kardar, 1989; Grinstein et al., 1990; Hwa and Kardar, 1992; Díaz-Guilera, 1993).

4. Computational Aspects of Self-Organized Criticality

In this section we deal with the concrete computational aspects of SOC phenomena. First of all we have to decide which kind of system we are going to simulate in our computer: dimension, lattice type, variables, initial and boundary conditions, and the dynamic evolution of the automata. We also have to know which kind of information we want to obtain and how to do it.

4.1. Preparing the system

Most of the simulations have been performed in a two-dimensional square lattice. The original sandpile model was described by a discrete scalar variable representing the height of the pile in a site. More recently, however, a continuous variable has been more widely used. This variable will be called energy but can have different physical meanings (Manna et al, 1990; Pietronero et al., 1991) (stress, neutron density, slope, ...) and it has a threshold value E_c above which a cell transfers energy to its neighboring cells (it fires). Also vector variables have been taken into account in different works (McNamara and Wiesenfeld, 1990). One usually starts with a subcritical system, i.e. $E_{ij} < E_c\ \forall i, j$, although the relaxation from a supercritical state has also been investigated.

Concerning the boundary conditions on the systems, they are usually assumed to be open, this is needed in a system with interaction rules that conserve the energy, otherwise an avalanche would never stop, either by considering reflecting boundaries or periodic boundary conditions. However, if there is dissipation in the interaction rules, periodic boundary conditions have been also considered since they are more appropriate to study the behavior of large systems. Another drawback of these conditions is that if an avalanche is very large, two or more propagating fronts of the same avalanche can collide and this can give rise to misleading conclusions. Open boundary conditions have a clear physical meaning in the context of sandpiles, since the sand grains reaching the boundaries fall off the system.

Initial conditions play a crucial role in the dynamics of the system if all the cells are driven continuously and deterministically. In this case random initial conditions are needed in order to get SOC. Nevertheless, if the cells are randomly driven the initial conditions are irrelevant.

4.2. Time evolution

Once we have prepared the initial state of the system we let it evolve according to some rules. The energy of each cell changes independently from one another until one of them reaches a critical value and the interactions between cells take place. The first mechanism is related to the external driving of the system; physical examples are, for instance, the pulling of the blocks in the model of earthquakes, the addition of grains to the sandpile or the change in the state of the oscillators. The second mechanism is the relaxation process by which the energy is redistributed through the system. This happens, for instance, when the force on a block exceeds the static friction force, then the block slips causing the redistribution of the forces in its neighbors, or when the local slope of a sandpile becomes too steep.

4.2.1. Perturbation rules

In the original models, the system was stochastically driven but later on the emphasis was put in continuously driven models since these seem to be more appropriate for describing, for instance, earthquake dynamics (Olami et al., 1992). In the first case one chooses a lattice cell at random (i, j) and increments its energy by an amount ΔE

$$E_{i,j} \rightarrow E_{i,j} + \Delta E \tag{4.1}$$

whereas in the second case all the cells are updated simultaneously

$$E_{i,j} \rightarrow E_{i,j} + \Delta E_{i,j} \quad \forall i, j. \tag{4.2}$$

If the driving is uniform, the energy increment is the same for each cell but in some systems, for instance in the models of integrate-and-fire oscillators (Corral et al., 1994) and other biological systems (Csilling et al., 1994), the energy increment depends on the state of the cell, giving rise to a richer behavior.

These rules are applied until one of the cells has an energy equal to the critical value $E_{i,j} = E_c$. But it would be very CPU time consuming to check all the lattice sites for each time step. For a stochastic dynamics one has just to check the lattice site where the energy has been incremented and for the deterministic rules one has to find the site with the largest energy $E_{\max}$ and then increment the energy at all sites

$$E_{i,j} \rightarrow E_{i,j} + E_c - E_{\max} \quad \forall i, j \tag{4.3}$$

and this site will start the avalanche (seed). This procedure implies two passes through the lattice: find the largest value of the energy and update the energies. The first one can be avoided by maintaining a list with the highest values (Grassberger, 1994; Middleton and Tang, 1994) and the second one by changing the threshold instead of changing the energies (Grassberger, 1994). [6] One has to be very careful with the possibility that two or more cells have the same $E_{\max}$. In such a situation, we would choose one of them at random as the seed and the remaining are kept slightly below the threshold. The reason lies in the fact that otherwise avalanches could overlap and then the interaction rules would need to be modified (Corral et al. , 1994; Middleton and Tang, 1994). [7]

[6] These algorithms become inefficient when there is a high degree of synchronization and when the driving is not uniform.

[7] Depending on the interaction rules we will analyze in the next paragraphs, this kind of dynamical noise is necessary in order to obtain SOC, otherwise the system presents a macroscopic degree of synchronization.

4.2.2. Interaction rules

When one of the cells, and only one, reaches the threshold ($E_{ij} > E_c$) due to its own dynamics, it transfers energy to its neighbors. Among different choices we take these rules to be continuous, nonconservative, short-ranged (nearest neighbors) and deterministic:

- The continuous version of the original sandpile Bak, Tang, and Wiesenfeld (BTW) rules:

$$\begin{aligned} E_{ij} &\to E_{ij} - E_c \\ E_{nn} &\to E_{nn} + \varepsilon E_c \end{aligned} \tag{4.4}$$

  where ε is the degree of conservation, corresponding $\varepsilon = 0.25$ to conserved interaction rules.
- Zhang model (Z):

$$\begin{aligned} E_{ij} &\to 0 \\ E_{nn} &\to E_{nn} + \varepsilon E_{ij} \end{aligned} \tag{4.5}$$

  where again conservation corresponds to $\varepsilon = 0.25$.
- Feder and Feder (FF) model that is also a short-range version of the rules used to model the synchronization of pacemaker cells (Peskin, 1975; Mirollo and Strogatz, 1990) :

$$\begin{aligned} E_{ij} &\to 0 \\ E_{nn} &\to E_{nn} + \varepsilon E_c \end{aligned} \tag{4.6}$$

  These rules are essentially nonconservative since the energy of a cell can be larger than E_c when it fires.

When the energy is transferred to the neighbors following these rules, they can, in turn, become critical and generate in this way an avalanche. Due to the underlying square lattice structure, two neighbors never fire at the same time; we can think of a chess board where black cells fire at even times and white ones at odd times.

At each avalanche step we have to search for the critical cells; instead of sweeping all the lattice again, we maintain a list with all the cells receiving energy in the previous step and those are checked. It is also important to note that the energy of the cells is updated simultaneously (parallel).

Concerning the interaction rules, one of the main ingredients is the role played by conservation; we already noted that rules (4.6) are nonconservative for any value of ε. These rules have another special characteristic: after all the cells on the lattice have fired at least once, there is no effect of the random initial conditions. Actually it has been shown that for these rules it is necessary to add some kind of dynamical noise in order to get SOC, otherwise the system presents synchronization (Christensen, 1994). For Zhang's rules (4.5) the transfer of energy depends on the energy of the cell, whereas it is a fixed quantity for the other rules.

4.2.3. Time scales

These dynamical rules involve two time scales, one for the intrinsic dynamics of the units or the external driving, and another for the interactions. In models of earthquakes, the first scale corresponds to the motion of the tectonic plates and the second one to the duration of the earthquake; since the former is orders of magnitude larger than the second one, the avalanches are assumed to be instantaneous. This assumption has also been made in recent studies on integrate-and-fire oscillators, but in order to discuss more realistic models it would be necessary to take into account some further factors such as the time the signal needs to propagate through the lattice or the refractory time associated to the response of a cell, just to mention a few. The effect of a refractory time has also been studied in the context of earthquake models (Lomnitz-Adler et al., 1992).

When both time scales are about the same order of magnitude, the intrinsic dynamics starts when the avalanche is still propagating and thus new avalanches can initiate. In this situation we are faced again with overlapping avalanches and there is no physical meaning in the measure of size and lifetime distributions of the avalanches.

4.3. Extracting information

Now we will focus on the kind of information we want to extract from our simulations. SOC is characterized by the absence of length and time scales in the propagation of the avalanches. These avalanches are characterized by their distributions of size and lifetimes; if there is a power law behavior it is characterized by an exponent. We can also look at the relationship between different magnitudes describing a single avalanche (size and length, for instance) in order to compute other critical exponents. This set of exponents is obtained by averaging over a large number of perturbations (time average), which should be compared with an average over different realizations of the stochastic source.

There are some cases where a change in the parameters characterizing the dynamical rules can give rise to a transition from a SOC behavior to a periodic behavior. In such a situation the time evolution of the variables describing the system gives information on how the stationary state is reached. Also in these models the degree of synchronization is a useful quantity in our understanding of the problem.

4.3.1. Distributions

The magnitudes whose distributions can be studied are the size and the lifetime or duration of the avalanches:

- *Avalanche size*:

Usually the size of the avalanche is defined as the number of cells that fire; from a geometrical point of view this is the size of the avalanche. However, some authors use the number of firings as the definition for the size, which is more closely related to the energy transfer during the avalanche, but the more appro-

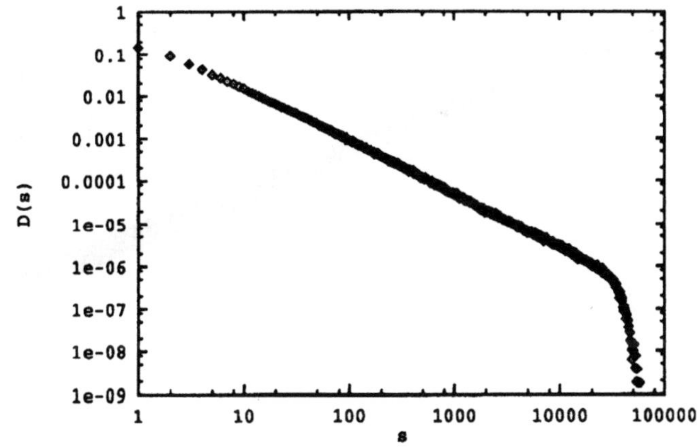

Fig. 3. Distribution of avalanche sizes averaged over bins of exponentially increasing size.

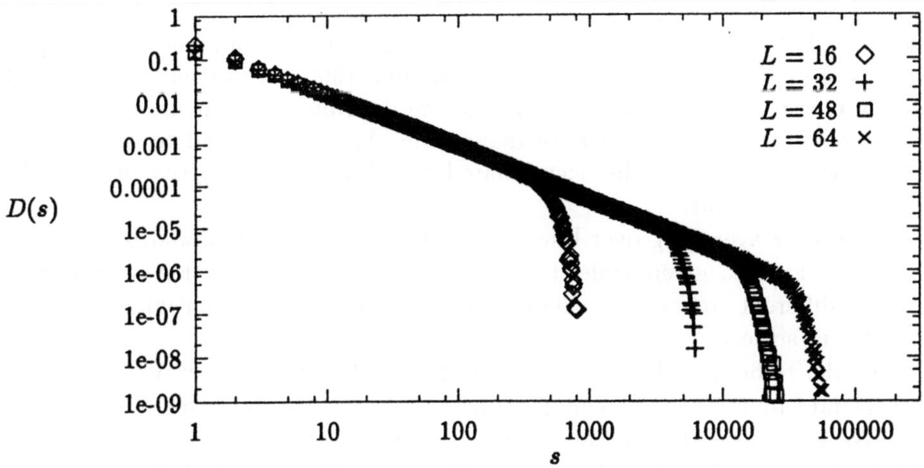

Fig. 4. Distribution of avalanche sizes averaged over bins for different values of L, where L is the linear dimension of the lattice.

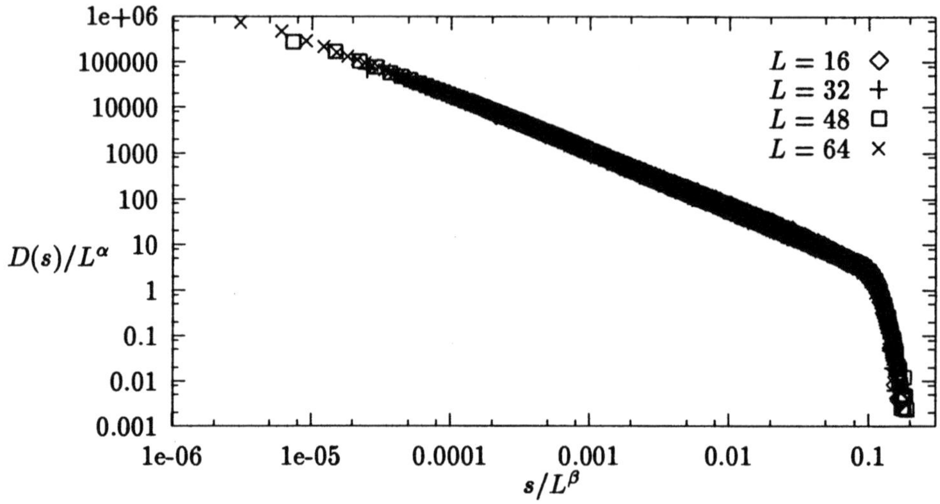

Fig. 5. Finite size scaling ansatz of the distribution of avalanche sizes.

priate name for this magnitude would be the *activation number* (Robinson, 1994). In Fig. 2 we have already plotted an example of such distribution in a log – log scale, having defined the size as the number of firings. Note that the statistics for the tail of the distribution is very poor. This can be improved by averaging on bins of exponentially increasing size (Manna, 1990) as this plotted in Fig. 3. In this plot we can identify the slope of the distribution with the exponent of the power-law for a few orders of magnitude. Another method to get a better representation is to plot the accumulated distribution which gives an exponent increased in one unit.

Even after averaging over bins we obtain a power-law followed by an exponential decay at a given scale. In Fig. 4 we plot the distribution of avalanche sizes for different lattice sizes where we can see that the exponential decay starts at different scales.

In order to be sure that this is effectively a finite-size effect, we try to collapse all this data by the proper scaling of the coordinate axes in terms of the system size. We have done this in Fig. 5 with $\alpha = -3.7 \pm 0.1$ and $\beta = 3.0 \pm 0.1$ from which we can conclude that the distribution function obeys the following scaling law

$$D(s) = L^\alpha F(s/L^\beta) \tag{4.7}$$

where F is a scaling function. At this point one could state that this finite-size scaling property is one of the main characteristics of SOC. If the distribution

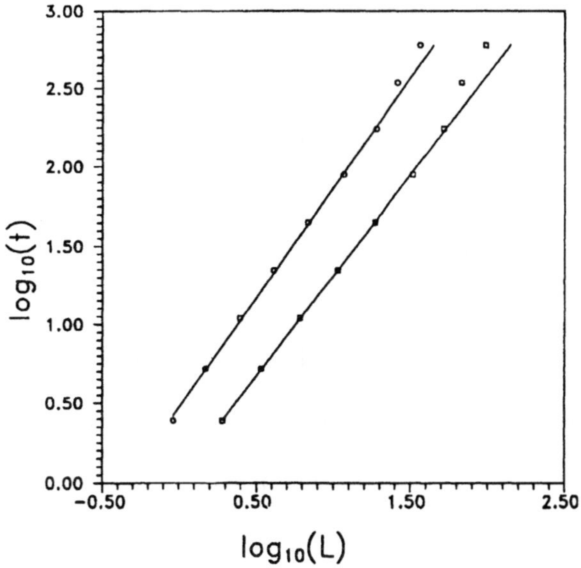

Fig. 6. The duration of an avalanche versus its mean characteristic length in logarithmic scales; for two different definitions of the length: o) length is defined as the radius of gyration with respect to the center of mass, □) the length is the maximum distance to a perimeter site from the seed. Straight lines are best linear fits to the sets of data with slopes $z = 1.40$ and $z = 1.28$, respectively.

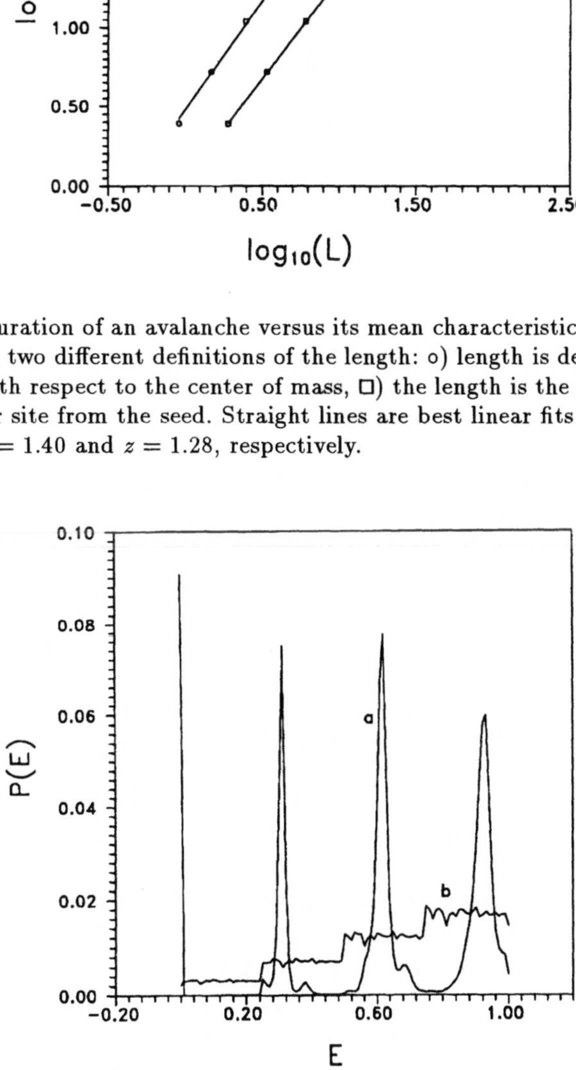

Fig. 7. Distribution of energies for different interaction rules: (a) Zhang, and (b) BTW.

is a power law (2.7), it is easy to show the following relationship between the exponents

$$\tau - 1 = \alpha/\beta \tag{4.8}$$

If we denote by s_c the avalanche size where the exponential decay starts, the exponent β is related to the growth of this quantity with the linear dimension of the lattice, L, i.e.

$$s_c \sim L^{\beta}. \tag{4.9}$$

This exponent has to be larger than zero to ensure the power-law behavior in the thermodynamic limit ($L \to \infty$); for $\beta = 0$ there is a characteristic length in the system $s_c^{1/d}$ and hence the scale invariance disappears, whereas for negative values of β the exponential dominates the large scale behavior.

- *Lifetime of the avalanche*:

There is no ambiguity in defining this magnitude since it is clearly the number of iterations needed for the system to get a stable configuration, i.e. all cells below the threshold. We obtain the same kind of distributions as for the avalanche size with different exponents.

4.3.2. Exponents

In Sect. 2.2 we defined the two exponents characterizing the behavior of the average growth of single avalanches: the dynamic exponent and the fractal dimension. Both exponents involve the characteristic length of the avalanche. There have been several definitions of this magnitude:

- The maximum distance from a perimeter cell to the cell where the avalanche originates (the seed) (Christensen et al., 1991).
- The so-called "taxicab" metric (Robinson, 1994), which corresponds to an average between the maximum and minimum distances to the seed.
- The radius of gyration with respect to its center of mass (Díaz-Guilera, 1992). For the first definition, the length is measured along the direction where the avalanches grow faster in time; this choice can make the dynamic exponent (the exponent relating the duration of the avalanche to its characteristic length) closer to unity. In Fig. 6 we plot the duration of the avalanches as a function of their characteristic length for two different definitions, each extracted from the same set of simulations. The last points separate from the straight line due to a finite-size effect of the open boundary conditions.

We also mentioned in Sect. 2.2 that different models that have the same values of the exponents belong to the same universality class. In order to show this, we have plotted the distribution of time and lattice averaged energy after the avalanche in Fig. 7 for two different dynamical interaction rules, Zhang's (4.5) and BTW (4.4) when the system is randomly driven. Note the very different distribution of energies in both cases. However, they share at least one of the critical exponents, the dynamical exponent, as can be seen in Fig. 8 where we have plotted the duration of an avalanche versus its mean characteristic length in logarithmic scales.

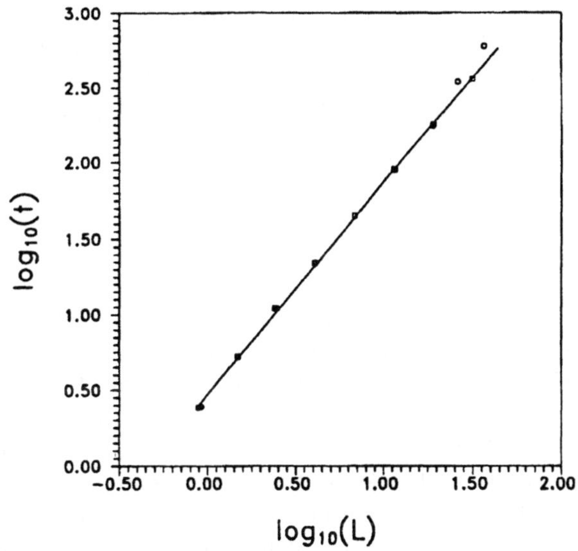

Fig. 8. Same as in Fig. 6 for different interaction rules: ∘, Zhang; □, BTW.

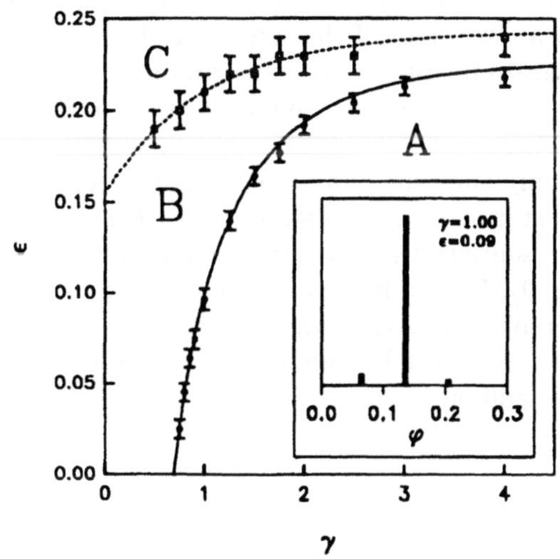

Fig. 9. Schematic phase diagram in terms of γ (the convexity of the driving) and ϵ (the level of conservation) for model (4.5). The symbols correspond to the phase transitions observed in the simulations. The solid line corresponds to the analytical result (Corral et al., 1994) and the dashed line is an exponential fit to the numerical data. The inset displays the distribution of phases after an avalanche in region A for a system of size $L = 64$ with $\gamma = 1.00$ and $\epsilon = 0.09$.

The other exponent is the fractal exponent; although, in principle, one of the initial goals of SOC was to explain the appearance of fractal structures it soon became clear that the avalanches have a compact structure rather than fractal (Pietronero and Schneider, 1991). This compactness does not mean that they are spherical, but rough in the sense of some models describing the growth of surfaces (Family and Vicsek, 1991).

4.3.3. Entrainment

Another way to characterize these coupled map lattice models is by means of the degree of synchronization which can be visualized in the distribution of energies. This can give an idea of the degree of entrainment between different units. In Fig. 7 we plotted it for two different interaction rules when the system is randomly driven. For Zhang's rules it has been reported that the width of the peaks in the distribution shrinks as the system size increases, and in the thermodynamic limit this would correspond to a quantization in the values of the energy. However, a close look at the distribution of energies in the lattice after any avalanche shows no correlations and, hence, no entrainment between neighboring cells.

In recent models that are driven continuously and nonlinearly (Corral et al., 1994), it has been reported that synchronization in a population of oscillators and SOC can be viewed as two sides of the same coin. In Fig. 9 we have plotted the schematic phase diagram for rules (4.5) in terms of the level of conservation, ε in (4.4)-(4.6), and the convexity of the driving, γ, starting from random initial conditions.

In region A there is only one type of avalanche that sweeps the whole system when it has reached the stationary state (see Fig. 10 where several snapshots of such an avalanche are presented); this gives rise to a high degree of synchronization (see the inset in Fig. 9). Region B is characterized by periodic behavior with a discrete distribution of avalanche sizes whereas region C presents the behavior characteristic of SOC, i.e. a power law distribution of avalanche sizes with its corresponding finite-size scaling.

4.3.4. Time evolution

Finally, the evolution of the different macroscopic variables describing the state of the system gives complementary information about its behavior. Thus, in Figs. 11-13 we plot the time evolution of the avalanche size for the three regions discussed in the previous paragraph, corresponding to nonlinearly driven oscillators with interaction rules (4.5). In the first case we can see how the system synchronizes in such a way that the avalanche that persists is system-wide. In the second case, we can observe the periodic behavior, and in the third case the absence of any predictability is manifest as one of the characteristics of SOC.

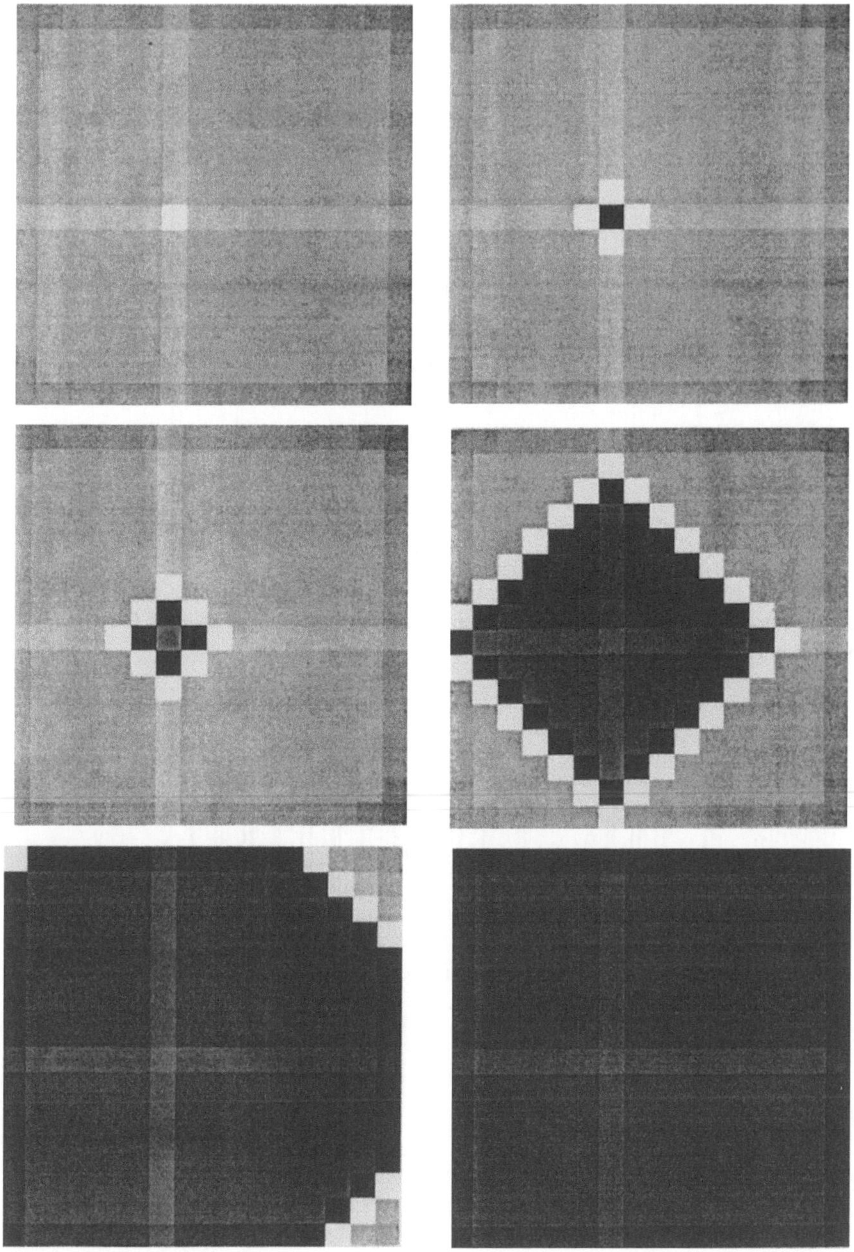

Fig. 10. Several snapshots of a system-wide avalanche in region A for $L = 16$. White squares are critical cells, black are zero energy sites, intensity of grey is a measure of the energy. Time goes from left to right and from top to bottom.

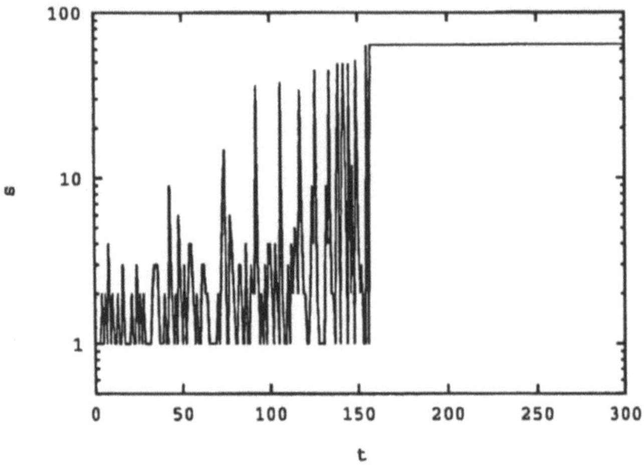

Fig. 11. Time evolution of the avalanche size in region A of Fig. 9 for $L = 8$. Each time step corresponds to an avalanche.

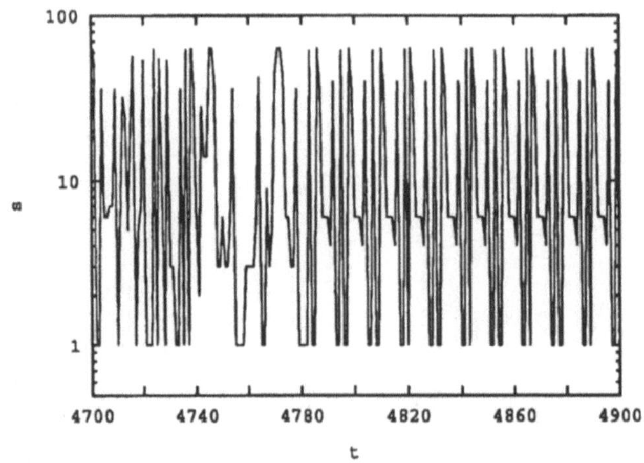

Fig. 12. Same as Fig. 11 in region B.

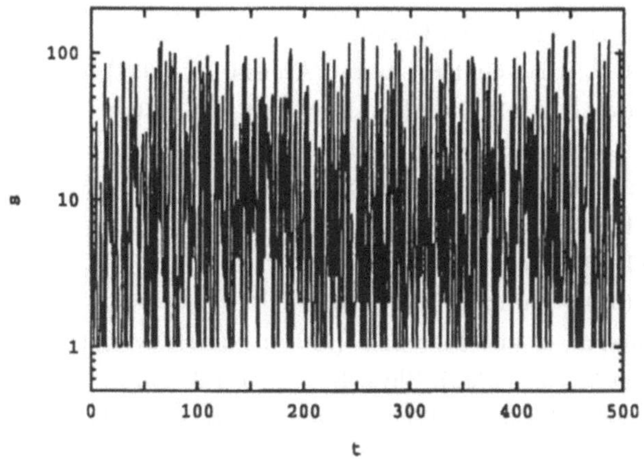

Fig. 13. Same as Fig. 11 in region C.

This work has benefited from discussions with several people, mainly A. Corral, C.J. Pérez, A. Arenas, and K. Christensen. Special thanks to A. Corral for his help with the figures and to C.J. Pérez for a critical reading of the manuscript. F.inantial support from CICyT, projects # PB89-0233 and # PB92-0863, is also acknowledged.

References

Alstrom, P. and Leao, J. (1994): Phys. Rev. E **49** R250.

Babcock, K.L. and Westervelt, R.M. (1990): Phys. Rev. Lett. **64** 2168.

Bak, P., Tang, C. and Wiesenfeld, K. (1988): Phys. Rev. A **38** 364.

Bak, P. and Flyvbjerg, H. (1992): Phys. Rev. A **45** 2192.

Bak, P. and Sneppen, K. (1993): Phys. Rev. Lett. **71** 4083.

Bantay, P. and Janosi, I.M. (1992): Phys. Rev. Lett. **68** 2058.

Burridge, R. and Knopoff, L. (1967): Bull. Seismol. Soc. Am. **57** 3411.

Carlson, J.M., Chayes, J.T., Grannan, E.R. and Swindle, G.H. (1990): Phys. Rev. Lett. **65** 2547.

Carlson, J.M., Langer, J.S. and Shaw, B.E. (1994): Rev. Mod. Phys. **66** 657.

Chau, H.F. and Cheng, K.S. (1994): Physica **208A** 215.

Che, X. and Suhl, H. (1990): Phys. Rev. Lett. **64** 1670.

Chen, K., Bak, P. and Obukhov, S. (1991): Phys. Rev. A **43** 625.

Christensen, K. (1994): Private communication.

Christensen, K., Fogedby, H.C. and Jensen, H.J. (1991): J. Stat. Phys. **63** 653.

Christensen, K., Olami, Z. and Bak, P. (1992): Phys. Rev. Lett. **68** 2417.

Christensen, K., Flyvbjerg, H. and Olami, Z. (1993): Phys. Rev. Lett. **71** 2737.

Corral, A., Pérez, C.J., Díaz-Guilera, A. and Arenas, A. (1994): unpublished.

Cote, P.J. and Meisel, L.V. (1991): Phys. Rev. Lett. **67** 1334.

Creutz, M. (1994): Computers in Physics **Mar/Apr** 198.

Csilling, A., Janosi, I.M., Pasztor, G. and Scheuring, I. (1994): Phys. Rev. E **50** 1083.

Díaz-Guilera, A. (1992): Phys. Rev. A **45** 8551.

Díaz-Guilera, A. (1993): Fractals **1** 963.

Díaz-Guilera, A. (1994): Europhys. Lett. **26** 177.

Diodati, P., Marchesoni, F. and Piazza, S. (1991): Phys. Rev. Lett. **67** 2239.

Drossel, B. and Schwabl, F. (1992): Phys. Rev. Lett. **69** 1629.

Family, F. and Vicsek, T. eds. (1991): Dynamics of fractal surfaces, World Scientific, Singapore.

Feder, H.J.S. and Feder, J. (1991): Phys. Rev. Lett. **66** 2669.

Flyvbjerg, H., Sneppen, K. and Bak, P. (1993): Phys. Rev. Lett. **71** 4087.

Fodor, Z. and Janosi, I.M. (1991): Phys. Rev. A **44** 1386.

Grassberger, P. (1994): Phys. Rev. E **49** 2436.

Grieger, B. (1992): Physica **191A** 51.

Grinstein, G., Lee, D.H. and Sachdev, S. (1990): Phys. Rev. Lett. **64** 1927.

Gutenberg, B. and Richter, C.F. (1954): Seismicity of the Earth and Associated Phenomena, Princeton Univ. Press., Princeton.

Hwa, T. and Kardar, M. (1989): Phys. Rev. Lett. **62** 1813.

Hwa, T. and Kardar, M. (1992): Phys. Rev. A **45** 7002.

Janosi, I.M. and Horvath, V.K. (1989): Phys. Rev. A **40** 5232.

Janosi, I.M. (1990): Phys. Rev. A **42** 769.

Jensen, H.J., Christensen, K. and Fogedby, H.C. (1989): Phys. Rev. B **40** 7425.

Kadanoff, L.P., Nagel, S.R., Wu, L. and Zhou, S. (1989): Phys. Rev. A **39** 6524.

Kadanoff, L.P., Chhabra, A.B., Kolan, A.J., Feigenbaum, M.J. and Procaccia, I. (1992): Phys. Rev. A **45** 6095.

Lomnitz-Adler, J., Knopoff, L. and Martínez-Mekler, G. (1992): Phys. Rev. A **45** 2211.

Mandelbrot, B.B. (1982): The fractal geometry of nature, Freeman, San Francisco.

Manna, S.S., Kiss, L.B. and Kertesz, J. (1990): J. Stat. Phys. **22** 923.

Manna, S.S. (1990): J. Stat. Phys. **59** 509.

Manna, S.S. (1991): Physica **179A** 249.

McNamara, B. and Wiesenfeld, K. (1990): Phys. Rev. A **41** 1867.

Mehta, A. and Barker, G.C. (1994): Rep. Progr. Phys. **57** 383.

Middleton, A.A. and Tang, C. (1994): unpublished.

Miranda, E.N. and Herrmann, H.J. (1991): Physica **175A** 339.

Mirollo, R.E. and Strogatz, S.H. (1990): SIAM J. Appl. Math. **50** 1645.

Nagel, K. and Raschke, E. (1992): Physica **182A** 519.

Nagel, S.R. (1992): Rev. Mod. Phys. **64** 321.

Nagel, K. and Herrmann, H.J. (1993): Physica **199A** 254.

Nakanishi, H. (1990): Phys. Rev. A **41** 7086.

Obukhov, S.P. (1988): In *Random Fluctuations and Pattern Growth*, H.E. Stanley and N. Ostrowsky eds., Kluwer, Dordrecht.

Oddershede, L., Dimon, P. and Bohr, J. (1993): Phys. Rev. Lett. **71** 3107.

Ohtsuki, T. and Iwata, H. (1992): Phys. Rev. A **45** 2228.

Olami, Z., Feder, H.J.S. and Christensen, K. (1992): Phys. Rev. Lett. **68** 1244.

Olami, Z. and Christensen, K. (1992): Phys. Rev. A **46** R172.

Pan, W. and Doniach, S. (1994): Phys. Rev. B **49** 1192.

Peskin, C.S. (1975): Mathematical Aspects of Heart Physiology. Courant Institute of Mathematical Sciences, New York University, New York.

Pietronero, L. and Schneider, W.R. (1991): Phys. Rev. Lett. **66** 2336.

Pietronero, L., Tartaglia, P. and Zhang, Y.C. (1991): Physica **173A** 22.

Pietronero, L., Vespignani, A. and Zapperi, S. (1994): Phys. Rev. Lett. **72** 1690.

Pla, O. and Nori, F. (1991): Phys. Rev. Lett. **67** 919.

Plourde, B., Nori, F. and Bretz, M. (1994): Phys. Rev. Lett. **71** 2749.

Rinaldo, A., Rodriguez-Iturbe, I., Rigon, R., Ijjasz-Vazquez, E. and Bras, R.L. (1993): Phys. Rev. Lett. **70** 822.

Robinson, P.A. (1994): Phys. Rev. E **49** 3919.

Socolar, J.E.S., Grinstein, G. and Jayaprakash, C. (1993): Phys. Rev. E **47** 2366.

Solé, R.V., López, D., Ginovart, M. and Valls, J. (1992): Phys. Lett. **172A** 56.

Takayasu, H., Miura, H., Hirabayashi, T. and Hamada, K. (1992): Physica **184** 127.

Tang, C. and Bak, P. (1988a): Phys. Rev. Lett. **60** 2347.

Tang, C. and Bak, P. (1988b): J. Stat. Phys. **51** 797.

Vattay, G. and Harnos, A. (1994): Phys. Rev. Lett. **73** 768.

Vieira, M.S. (1992): Phys. Rev. A **46** 6288.

Vives, E., Ortin, J., Mañosa, L., Rafols, I., Perez-Magrane, R. and Planes, A. (1994): Phys. Rev. Lett. **72** 1694.

Wang, Z. and Shi, D. (1994): Solid State Comm. **90** 405.

Wiesenfeld, K., Theiler, J. and McNamara, B. (1990): Phys. Rev. Lett. **65** 949.

Zhang, Y.C. (1989): Phys. Rev. Lett. **63** 470.

Critical Percolation with Moving Disks

I. M. Polo[1] and J. F. Fernández [1,2]

[1]Instituto Carlos I de Física Teórica y Computacional and
Departamento de Física Aplicada, Universidad de Granada,
18071-Granada, Spain.
[2]Centro de Física, Instituto Venezolano de Investigaciones Científicas,
Apartado 21827, Caracas-1020A, Venezuela

Abstract: Our aim is to study diffusion of a point particle within a system of moving disks in the critical percolation region. We simulate this system by Molecular Dynamics. All disks, of mass M each, and the point particle, of mass m, are in thermal equilibrium. The critical percolation point corresponds to some disk density and the $m/M \to 0$ limit. By varying m/M, the cross-over between standard critical percolation and a regime dominated by dynamics can be studied. We propose for that purpose an *open* system of particles (disks) that do not interact with each other and report some preliminary results.

1. Introduction

Brownian motion is usually approached by either of two separate independent paths: (a) diffusion within fluids [8], and, more recently, through diffusion in percolating systems [9] . Whereas in the former case the phenomenon is dominated by *dynamics*, diffusion in a percolating system is controlled by *geometry*. We propose a model for simulations that can bridge the gap between these two regimes.

Using Molecular Dynamics, we simulate the diffusion of a point particle of mass m within a system of non interacting disks of mass M each that do not interact with each other. All disks and the diffusing particle are in thermal equilibrium. In the limit $m/M \to 0$, all disks are stationary (since the diffusive particle moves infinitely faster than the disks then). The particle becomes trapped for disk densities ρ above the percolation threshold value ρ_c, and there is then no diffusion. It is our purpose to study how percolation effects carry over into the regime dominated by dynamics for $m/M \neq 0$. In particular, we wish to study critical percolation both as $m/M \to 0$ and as $\rho \to \rho_c$.

We propose for that purpose an *open* system of particles (disks) that do not interact with each other. Some preliminary results are reported. We next explain why it is relevant that the system be open. Critical behavior of the

[8] For a review and further references, see (Gomer, 1990)

[9] For a recent application and further references, see (Hunt, 1994)

diffusion constant D depends on long time effects. They come in through the correlation function $\langle v_i(t_1)v_i(t_2)\rangle$, since

$$D = \int_0^\infty dt \langle v_i(0)v_i(t)\rangle, \tag{1.1}$$

as follows from the definition of D,

$$D = \frac{1}{2}\lim_{t\to\infty}\frac{d\langle x_i^2(t)\rangle}{dt}, \tag{1.2}$$

and $x_i(t) = \int_0^t v_i(t')dt'$. The long (non-exponential) time tails, first discovered by Alder and Wainwright (Alder and Wainwright, 1970), occur only because of the memory of previous collisions between the diffusive particle and system disks [10]. Both reflecting and periodic boundary conditions force disks that would otherwise go off to infinity to come back for further collisions with the diffusive particle. That enhances the memory effect beyond its true value. (Reflecting sound waves would have a similar effect.). The model we propose errs by *under*estimating memory effects. It is therefore complementary to *closed* models.

2. The Model

The open model we propose is as follows. Consider a circle of radius R centered on a point particle as it diffuses through a gas of disks. (We choose the circle to be such that there are approximately 1000 disks within it.) The model keeps memory of previous collisions between the diffusive particle and a given disk only if the disk has not gone beyond the circle boundary in between collisions. That is achieved as follows: we keep track of particle trajectories only as long as they remain *within* the circle defined, and we let particles that are uncorrelated with the diffusive particle stream into the circle from an outside bath.

In order to estimate the errors introduced by the model, we find the probability that the distance between a disk and the diffusive particle increases beyond R in between two succesive collisions between them. That can only happen if the point particle first diffuses away from the disk at least a distance R and returns to collide with it at time t later (since disks move freely, except for collisions with the diffusive particle). The probability p for such an event is clearly largest for a non moving disk. A crude upper bound that is sufficient for our purposes can be obtained as follows: let $p_+(2R,t)$ be the probability for the diffusive particle to *be* beyond $2R$ *at* time t; a bit of reflection shows that $p < 2p_+(2R,t)$. Assuming that diffusion takes place according to $(2\pi Dt)^{-1}exp(-r^2/4Dt)$ leads to $p_+(2R,t) = exp(-R^2/Dt)$, and hence,

$$p < e^{-R^2/Dt}. \tag{2.1}$$

[10] See for instance, the review (Pomeau and Résibois, 1975)

One can expect therefore to obtain accurate results for $\langle v_i(t_1)v_i(t_2)\rangle$ and for $\langle [x(t)-x(0)]^2\rangle$ only as long as $Dt << R^2$; we have always fulfilled the inequality $Dt < 10R^2$. Moreover, we have checked that the results obtained are, within statistical errors, independent of R for R larger than the value we have used ($\rho\pi R^2 = 900$, where ρ is the disk density).

3. Results

We have run about 10^7 collisions of the diffusive particle for several values of m/M in the range $10^{-5} \leq m/M \leq 1$, and for several values of the disk density ρ near ρ_c. An outline of some of our preliminary results follows. For sufficiently small values of the diffusive mass (as small as $m/M \sim 10^{-5}$) we find for $\rho < \rho_c$ that

$$D \sim \omega^{\mu}, \tag{3.1}$$

where $\omega \equiv \rho_c - \rho$, $\rho_c r^2 = 0.358$, r is the disk radius, and $\mu \approx 1.3$. That is, as expected, in agreement with results obtained for critical percolation with static disks (Lowe and Masters, 1993).

Preliminary results suggest that, for $\rho = \rho_c$,

$$D \sim (m/M)^{\alpha}, \tag{3.2}$$

and that $\alpha \approx 0.2$.

We have had interesting discussions with Prof. J. Marro and Prof. J. Stankiewicz. Support from DGCIT, of Spain, through grant # PB91-0709, is gratefully acknowledged.

References

Alder, B.J. and Wainwright, T.E. (1975): Phys. Rev. A **1**, 18.

Gomer, R. (1990): Rept. Prog. Phys. **53**, 917.

Hunt, A. (1994): J. Phys. C **6**, 1239.

Lowe, C.P. and Masters, A.J. (1993): Physica A **195**, 149.

Pomeau, Y. and Résibois, P. (1975): Phys. Rept. **19**, 63.

Fourier Techniques in Numerical Methods for Evolutionary Problems

J.M. Sanz-Serna

Departamento de Matemática Aplicada y Computación,
Facultad de Ciencias, Universidad de Valladolid,
Valladolid, Spain

1. Introduction

1.1 Scope

These lecture notes contain a summary of the application of Fourier analysis to the numerical solution of time-dependent partial differential equations. The presentation emphasizes two topics: how to use Fourier techniques to analyze and understand finite-difference and finite-element methods and how to derive and code pseudospectral Fourier numerical methods.

The first topic is essential for anyone who wishes to use numerical methods in partial differential equations. While the Von Neumann stability analysis is found in virtually all introductions to the subject, many textbooks do not discuss the ideas of stability and consistency from the Fourier space point of view. Similarly, most elementary texts do not provide an adequate coverage of the notions of numerical dispersion and numerical dissipation. I have tried to fill these gaps.

Pseudospectral methods, our second main topic, are very useful when simulating partial differential equations arising in physics. These methods, being younger than finite-difference/finite-element methods, are not as widely known as they deserve. It is sad that many published papers report finite-difference simulations in situations where a pseudospectral method would have been far more efficient and not more difficult to code.

In order to cater for as wide an audience as possible, virtually no previous knowledge on Fourier analysis or numerical methods has been assumed. The text has been supplemented with exercises and anyone who is really interested in the subject should try and solve most of them. Many exercises contain useful material not covered in the main text. Sometimes the exercises ask for a computer program to be written and MATLAB is an ideal environment for those programs. To help the reader, we have used throughout a set of mathematical conventions that follow those used in MATLAB.

There are eight sections. Sects. 6–8 are devoted to Fourier analysis of finite-difference schemes and Sect. 9 to the pseudospectral method. The first four sections are introductory and serve as a foundation for the last four. Sect. 2 contains a short summary of Fourier series and Sect. 3 deals with partial differential equations. Discrete Fourier analysis, including the Fast Fourier Transform algorithm, is the subject of Sect. 4. I have devoted a full section (Sect. 5) to carefully discussing the relations between Fourier series and discrete Fourier analysis, a point where many expositions are perhaps too terse.

In view of the number of available pages, some important topics have not been covered. Broadly speaking, I only consider problems in one space dimension under periodic solutions. Multidimensional periodic problems can of course be dealt with by Fourier techniques; the adjustments to be made to the material here are minor. On the other hand, problems in the whole space are treated by Fourier integrals rather than by Fourier series; the change from Fourier series to Fourier integrals is sometimes a delicate business (for example, the effects of sampling become more subtle). Homogeneous Dirichlet boundary conditions (vanishing function) or homogeneous Neumann boundary conditions (vanishing normal derivative) on an interval $0 \leq x \leq L$ (or on its multidimensional equivalent $0 \leq x_n \leq L$, where n numbers the space variables) are also amenable to Fourier techniques; they require sine or cosine series, which also have a discrete version with a fast implementation.

1.2 Some mathematical preliminaries

It is convenient to list here some basic linear algebra results (Golub and Van Loan 1989; Horn and Johnson 1985) that will be used later. The reader should skip this section now and return to it when referred from other sections.

The *norm* or length of a real or complex vector $\mathbf{X} = (X_1, \ldots, X_\nu)$ with ν entries is given by $|\mathbf{X}| = (\sum_n |X_n|^2)^{1/2}$. The *norm* of a $\nu \times \nu$ real or complex matrix A is defined by $||A|| = \max\{|A\mathbf{X}|/|\mathbf{X}| : \mathbf{X} \neq \mathbf{0}\}$. It follows that $|A\mathbf{X}| \leq ||A||\,|\mathbf{X}|$, for each $\nu \times \nu$ matrix A and ν-vector $\mathbf{X}$, while $||AB|| \leq ||A||||B||$ if A and B are $\nu \times \nu$ matrices.

If A is a real or complex square matrix, $\rho(A)$, the *spectral radius* of A, is the maximum modulus of the eigenvalues of A. Since the eigenvalues of a power A^m are the powers of the eigenvalues of A, it holds that $\rho(A^m) = \rho(A)^m$. For each square matrix $\rho(A) \leq ||A||$. On the other hand, $||A||$ may be computed by the formula $||A|| = \rho(A^*A)^{1/2}$, where A^*, the *adjoint* of A, is the conjugate of the transposed of A. The *spectral abscissa,* $\alpha(A)$, of A is the maximum real part of the eigenvalues of A.

A *unitary matrix* is a matrix Q for which $Q^* = Q^{-1}$. Unitary matrices preserve vector norms, $|Q\mathbf{X}| = |\mathbf{X}|$, and, by the definition of $||Q||$, this implies that $||Q|| = 1$. Furthermore, unitary matrices preserve matrix norms, in the sense that for unitary Q and arbitrary square A, $||QA|| = ||AQ|| = ||A||$.

A *normal* matrix is a matrix that commutes with its adjoint A^*, i.e, $AA^* = A^*A$. Unitary matrices, real symmetric matrices and real skew-symmetric matrices are obviously normal. A matrix is normal if and only if there exists a unitary matrix Q so that Q^*AQ is the diagonal matrix Λ of the eigenvalues λ_n of A. Since multiplication by unitary matrices does not change matrix norms, for a normal matrix $||A|| = ||\Lambda||$ and therefore $||A|| = \max_n |\lambda_n| = \rho(A)$. From here, if A is normal, $||A^m|| = ||A||^m$. (For nonnormal matrices, $\rho(A) \leq ||A||$ and $||A^m|| \leq ||A||^m$.)

The *exponential* of a square matrix A is, by definition, $\exp(A) = I + A + (1/2)A^2 + \cdots$. It holds that $\exp(A)^{-1} = \exp(-A)$. Also $(d/dt)\exp(tA) = A\exp(tA)$, if t is a real variable. Then $\exp(tA)\mathbf{X}^0$ is the solution of the differential system $(d/dt)\mathbf{X}(t) = A\mathbf{X}(t)$ with initial condition $\mathbf{X}(0) = \mathbf{X}^0$. The eigenvalues of $\exp(A)$ are the numbers $\exp(\lambda)$, with λ an eigenvalue of A. If A is normal, then $\exp(A)$ is also normal and therefore $||\exp(A)|| = \rho(\exp(A))$. It then follows trivially (the modulus of the complex exponential is the exponential of the real part) that, if A is normal, then $||\exp(A)|| = \exp(\alpha(A))$. For nonnormal matrices $\exp(\alpha(A))$ may be smaller than $||\exp(A)||$.

If $P(z)$ is a complex polynomial and A is a matrix, the matrix $P(A)$ is defined in an obvious way. For instance, if $P(z) = 3+2z+z^2$, then $P(A) = 3I+2A+A^2$, with I the identity matrix. A *rational function* $R(z)$ of the complex variable z is the quotient of two complex polynomials $R(z) = P(z)/Q(z)$. If A is a square matrix and $R(z)$ is a rational function, then $R(A)$ is, by definition, the matrix $P(A)Q(A)^{-1}$ (or $Q(A)^{-1}P(A)$, because $P(A)$ and $Q(A)^{-1}$ commute). Note that $R(A)$ can only be defined when $Q(A)$ is an invertible matrix. For instance, $R(z) = (1+z/2)/(1-z/2)$ is a rational function and $R(A) = [I+(1/2)A][I-(1/2)A]^{-1}$; this is defined if 2 is not amongst the eigenvalues of A. The eigenvalues of $R(A)$ are $R(\lambda)$, with λ an eigenvalue of A. If A is normal, then $R(A)$ is normal and therefore $||R(A)|| = \rho(R(A))$ is given by the maximum modulus of $R(\lambda)$ as λ runs through all the eigenvalues of A.

2. Review of Fourier Series

2.1 The $\mathcal{L}^2$ theory

The literature on Fourier series is of course huge; we only consider the $\mathcal{L}^2$ theory (Strang 1986).

For a given, fixed $L > 0$, we deal with complex-valued functions $f(x)$ of a real variable x, $-\infty < x < \infty$, that are L-periodic, $f(x) \equiv f(x+L)$. The space $\mathcal{L}^2[0, L]$ consists of all L-periodic functions f for which the quantity

$$||f|| = \left(\int_0^L |f(x)|^2 \, dx \right)^{1/2} \tag{2.1}$$

is finite. For instance, if x represents time and f is the value of a periodic electric current, then the square of the right-hand side of (2.1) is (proportional to) the energy dissipated in a resistor during a period; $\mathcal{L}^2[0, L]$ contains all current functions with finite energy per period.

It is convenient to imagine that each f in $\mathcal{L}^2[0, L]$ is a 'vector' in a space with infinitely many dimensions and that $||f||$ is the *norm* or *length* of such a vector. For f, g in $\mathcal{L}^2[0, L]$,

$$\langle f, g \rangle = \int_0^L f(x) g(x)^* \, dx$$

($*$ denotes complex conjugate) is the *inner product* of f and g. If $\langle f, g\rangle = 0$ we say that f and g are *orthogonal;* we imagine that the vectors f and g are perpendicular.

The system of (L-periodic) functions $\ldots, \phi_{-2}, \phi_{-1}, \phi_0, \phi_1, \phi_2, \ldots$, where

$$\phi_n(x) = \exp\left(\frac{2\pi n i}{L}x\right), \quad n = 0, \pm 1, \pm 2, \ldots, \tag{2.2}$$

are pairwise orthogonal, $\langle \phi_n, \phi_m\rangle = 0$, $n \neq m$. Just as each geometric vector $\boldsymbol{v}$ in ordinary three-dimensional space can be written in the form

$$\boldsymbol{v} = v_1\boldsymbol{i} + v_2\boldsymbol{j} + v_3\boldsymbol{k} \tag{2.3}$$

in terms of a system $\{\boldsymbol{i}, \boldsymbol{j}, \boldsymbol{k}\}$ of pairwise perpendicular unit vectors, each f in $\mathcal{L}^2[0, L]$ can be referred to the system (2.2):

$$f = \sum_{n=-\infty}^{\infty} \hat{f}_n \phi_n, \tag{2.4}$$

with

$$\hat{f}_n = \frac{1}{L}\langle f, \phi_n\rangle = \frac{1}{L}\int_0^L f(x)\phi_n(x)^*\, dx, \quad n = 0, \pm 1, \pm 2, \ldots. \tag{2.5}$$

The series in (2.4) is the *Fourier series* of f; we use the notation

$$P_N(f) = \sum_{n=-N}^{N} \hat{f}_n \phi_n, \quad N = 0, 1, 2, \ldots, \tag{2.6}$$

for the corresponding *truncations*. Each term $\hat{f}_n\phi_n$, $n = 0, \pm 1, \pm 2, \ldots$ in (2.4) is said to be *a Fourier mode* of f.

The following properties are crucial.

1. For each f in $\mathcal{L}^2[0, L]$, (2.4) converges to f in the sense that, as $N \to \infty$, $\|f - P_N(f)\| \to 0$. This does not imply that, when the right-hand side of (2.4) is evaluated at a point x, the resulting series of complex numbers converges to the value $f(x)$ (pointwise convergence). In the electric current example, the current f is approximated by $P_N(f)$ in such a way that the energy in the residual $f - P_N(f)$ can be made arbitrarily small by taking N suitably large; this does not imply that at a given time x the value of $f(x)$ is the limit, as $N \to \infty$, of the values $(P_N(f))(x)$.
2. For each f in $\mathcal{L}^2[0, L]$, the sequence of Fourier coefficients $\hat{f}_n$ is *square summable*, i.e., $\sum_{-\infty<n<\infty} |\hat{f}_n|^2 < \infty$. Conversely, each square summable sequence of complex numbers corresponds to the Fourier coefficients of a function in $\mathcal{L}^2[0, L]$. Thus the Fourier series defines a correspondence between functions f and sequences of coefficients $\{\hat{f}_n\}$, just as (2.3) provides a correspondence between ordinary vectors $\boldsymbol{v}$ and sets of coefficients (v_1, v_2, v_3): the idea is to use the coefficients $\hat{f}_n$ instead of the function f. Furthermore *Parseval's identity* holds:

$$||f||^2 = L \sum_{n=-\infty}^{\infty} |\hat{f}_n|^2. \tag{2.7}$$

This is analogous to $|\boldsymbol{v}|^2 = v_1^2 + v_2^2 + v_3^2$ in (2.3).

3. If f is in $\mathcal{L}^2[0, L]$, then $P_N(f)$ is, among all the linear combinations of the form

$$S_N = \sum_{n=-N}^{N} g_n \phi_n, \tag{2.8}$$

(the g_n's are arbitrary complex numbers) the one that makes $||f - S_N||$ as small as possible. Functions of the form (2.8) are called *trigonometric polynomials of degree* N.

Exercise 1 Prove that the functions in (2.2) are pairwise orthogonal. Prove that each ϕ_n has length $\sqrt{L}$, so that the functions $\Phi_n = \phi_n/\sqrt{L}$ are pairwise orthogonal and have unit length. Substitute $\phi_n = \sqrt{L}\Phi_n$ in (2.4) to get $f = \sum \hat{F}_n \Phi_n$, with $\hat{F}_n = \langle f, \Phi_n \rangle$. This is directly analogous to the familiar formulae $v_1 = \langle \boldsymbol{v}, \boldsymbol{i} \rangle$, $v_2 = \langle \boldsymbol{v}, \boldsymbol{j} \rangle$, $v_3 = \langle \boldsymbol{v}, \boldsymbol{k} \rangle$ for the coefficients in (2.3). In terms of the $\hat{F}_n$'s, (2.7) becomes $||f||^2 = \sum |F_n|^2$, a formula to be compared with $|\boldsymbol{v}|^2 = v_1^2 + v_2^2 + v_3^2$. Thus the Φ_n's lead to formulae that are simpler to remember; some authors prefer to use it instead of the ϕ_n's in (2.2).

Exercise 2 Show that, with the definitions in (2.5) and (2.6), the coefficients $\hat{f}_n$ are such that $\langle f, \phi_n \rangle = \langle P_N(f), \phi_n \rangle$, $n = 0, \pm 1, \ldots, \pm N$. Denote by X_N the set of trigonometric polynomials (2.8) and prove that $P_N(f)$ is the unique element in X_N for which $f - P_N(f)$ is orthogonal to all functions in X_N. Thus $P_N(f)$ is the *orthogonal projection* of f onto X_N. This explains Property 3 above.

Exercise 3 Consider the L-periodic square wave function $f(x) = 1$, if $0 < x < L/2$, $f(x) = -1$, if $L/2 < x \leq L$. Show that $\hat{f}_n = -2i/(\pi n)$ for n odd and $\hat{f}_n = 0$ for n even.

Exercise 4 Consider the L-periodic saw-tooth function $f(x) = x$, if $0 < x \leq L/2$, $f(x) = L - x$, if $L/2 < x \leq L$. Show that $\hat{f}_0 = L/4$, $\hat{f}_n = -L/(\pi n)^2$, for n odd, and $\hat{f}_n = 0$, for $n \neq 0$ even.

2.2 The trigonometric version

It is sometimes useful to write ϕ_n in trigonometric form

$$\phi_n(x) = \cos \frac{2\pi n}{L} x + i \sin \frac{2\pi n}{L} x;$$

substitution in (2.4) leads to the following trigonometric form of the Fourier series

$$f = c_0(f) + \sum_{n=1}^{\infty} \left(c_n(f) \cos \frac{2\pi n}{L} x + s_n(f) \sin \frac{2\pi n}{L} x \right), \tag{2.9}$$

with

$$\begin{aligned} c_0(f) &= \hat{f}_0, \\ c_n(f) &= \hat{f}_n + \hat{f}_{-n}, \quad n = 1, 2, \ldots, \\ s_n(f) &= i(\hat{f}_n - \hat{f}_{-n}), \quad n = 1, 2, \ldots . \end{aligned}$$

Using (2.5), one arrives at the following formulae that allow the computation of $c_n(f)$, $s_n(f)$ without using the $\hat{f}_n$'s:

$$c_0(f) = \frac{1}{L} \int_0^L f(x)\, dx, \tag{2.10}$$

$$c_n(f) = \frac{2}{L} \int_0^L f(x) \cos \frac{2\pi n}{L} x\, dx, \quad n = 1, 2, \ldots,$$

$$s_n(f) = \frac{2}{L} \int_0^L f(x) \sin \frac{2\pi n}{L} x\, dx, \quad n = 1, 2, \ldots .$$

The form (2.9) has the advantage that, if f is real-valued, then $c_n(f)$ and $s_n(f)$ are real; the coefficients $\hat{f}_n$ are complex even if f is real.

In (2.9), f is decomposed into a constant $c_0(f)$ (which coincides with the average of f over one period, see (2.10)) and a sum of functions of the form ($c_n = c_n(f)$, $s_n = s_n(f)$)

$$c_n \cos \frac{2\pi n}{L} x + s_n \sin \frac{2\pi n}{L} x, \quad n = 1, 2, \ldots .$$

Assume for a moment that f (and hence c_n and s_n) are real. Then

$$c_n \cos \frac{2\pi n}{L} x + s_n \sin \frac{2\pi n}{L} x = A_n \cos \left(\frac{2\pi n}{L} x - \psi_n \right), \tag{2.11}$$

where $A_n = \sqrt{c_n^2 + s_n^2}$, $\psi_n = \arctan(s_n/c_n)$, so that $c_n = A_n \cos \psi_n$ and $s_n = A_n \sin \psi_n$. Thus, the function in (2.11), *the n-th harmonic of* f, $n = 1, 2, \ldots$, corresponds to a sinusoidal profile, whose amplitude A_n and initial phase $-\psi_n$ are determined by the Fourier coefficients c_n and s_n. The (smallest or basic) period of (2.11) as a function of x is L/n so that one period $0 \le x \le L$ of f is covered by n cycles of the sinusoid (2.11). Note that in the trigonometric format (2.9), n is nonnegative; both the coefficients $\hat{f}_n$ and $\hat{f}_{-n}$ contribute to the harmonic with period L/n, $n = 1, 2, \ldots$. If f is not real-valued this interpretation can be applied to its real and imaginary parts.

The form (2.4) is easier to handle mathematically. The trigonometric form possesses more meaning.

Exercise 5 Prove that if f is odd, $f(x) \equiv -f(-x)$, then $c_n(f) = 0$, $n = 0, 1, 2, \ldots$. Prove that if f is even, $f(x) \equiv f(-x)$, then $s_n(f) = 0$, $n = 1, 2, \ldots$. Prove the converse of these results.

Exercise 6 If all you know of f are the coefficients $\hat{f}_n$, how would you tell whether f is real-valued? How would you tell whether f is even, odd?

2.3 Fourier series and derivatives

We denote by ∂_x the operator of differentiation with respect to x. The system (2.2) is more advantageous than other orthogonal systems because each ϕ_n is an *eigenfunction* of the operator ∂_x:

$$\partial_x \phi_n = \lambda_n \phi_n, \quad \lambda_n = 2\pi n i/L. \tag{2.12}$$

On ϕ_n differentiation reduces to multiplication by the *eigenvalue* λ_n. Hence, from (2.4),

$$\partial_x f = \sum_{n=-\infty}^{\infty} (\lambda_n \hat{f}_n)\phi_n.$$

From Property 2 in Sect. 2.1, we see that $\partial_x f$ exists and belongs to $\mathcal{L}^2[0, L]$ if and only if $\sum |\lambda_n|^2 |\hat{f}_n|^2 < \infty$, i.e., $\sum n^2 |\hat{f}_n|^2 < \infty$. For the square wave in Exercise 3, $n^2|\hat{f}_n|^2 = 4/\pi^2$ and the series diverges: f has jump discontinuities at $x = 0, \pm L/2, \pm L, \ldots$ and correspondingly $\partial_x f$ has delta functions at those points; $\partial_x f$ is not in $\mathcal{L}^2[0, L]$. For the saw-tooth function in Exercise 4, $n^2|\hat{f}_n|^2 = 4/(\pi^2 n^2)$, which leads to a convergent series and to $\partial_x f$ in $\mathcal{L}^2[0, L]$. In fact, $\partial_x f$ is the square wave function.

For higher derivatives, $k = 1, 2, \ldots$,

$$\partial_x^k f = \sum_{n=-\infty}^{\infty} (\lambda_n^k \hat{f}_n)\phi_n,$$

and therefore $\partial_x^k f$ exists and belongs to $\mathcal{L}^2[0, L]$ if and only if the Fourier coefficients $\hat{f}_n$ of f decay as $|n| \to \infty$ fast enough for the series $\sum n^{2k}|\hat{f}_n|^2$ to be finite. Smooth functions have Fourier coefficients that decrease fast or, in other words, *the smoother the function the poorer in harmonics with large n.* This is intuitive. The function ϕ_n varies from $\phi_n = 1$ to $\phi_n = -1$ in an x-interval, $0 \le x \le L/(2|n|)$, whose length is small for $|n|$ large; therefore large coefficients $|\hat{f}_n|$ for large $|n|$ in (2.4) lead to sharp variations in f.

There is another interesting implication. By Parseval's identity (2.7) applied to the function $f - P_N(f)$,

$$\|f - P_N(f)\|^2 = L \sum_{|n|>N} |\hat{f}_n|^2;$$

smooth f corresponds to quickly decreasing $\hat{f}_n$ and hence to small lengths of the residuals $f - P_N(f)$: *the smoother the function, the faster the Fourier series converges.* See Figs. 1 and 2.

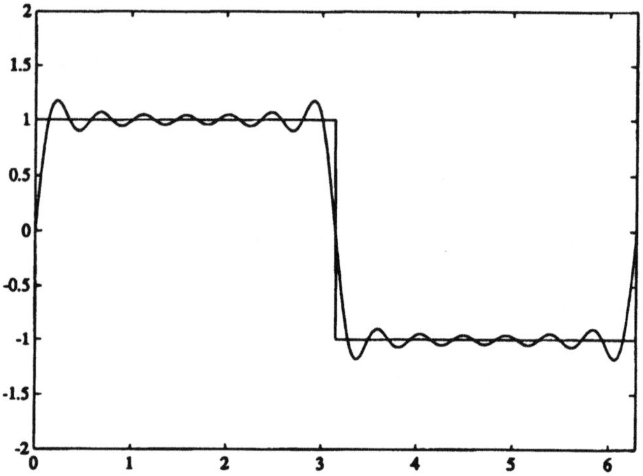

Fig. 1. The square wave function (Exercise 3) and the truncation P_{13} of its Fourier series

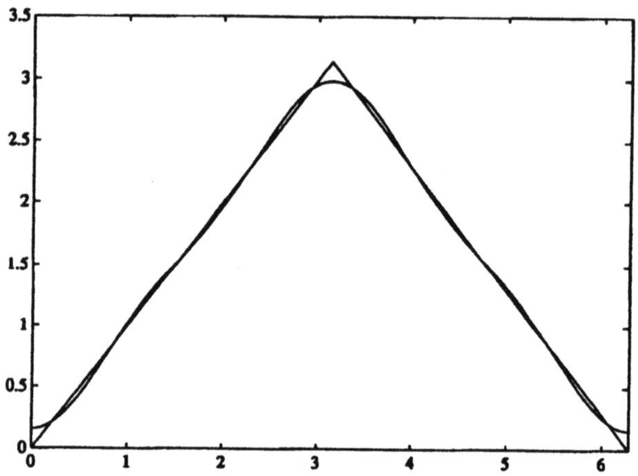

Fig. 2. The saw-tooth function (Exercise 4) and the truncation P_3 of its Fourier series. Compare with Fig. 1: here fewer harmonics give a better approximation

Exercise 7 Prove that if f has derivatives of all orders in $\mathcal{L}^2[0, L]$, then its Fourier coefficients decrease faster than any power of $|n|$, i.e., for each $k = 1, 2, \ldots$, $|n|^k|\hat{f}_n| \to 0$ as $|n| \to \infty$.

Exercise 8 Prove that the Fourier coefficients of (Canuto et al. 1988) $3/(5 - 4\cos x)$ are $2^{-|n|}$ (you may compute the needed integrals by the residue theorem of complex variables). Thus the Fourier coefficients decrease exponentially as $|n| \to \infty$.

Exercise 9 For the function in Exercise 3, write a program that draws the graphs of f and $P_N(f)$. (Use the trigonometric format to avoid complex quantities.) Run this program for several values of N. Do the same for the functions in Exercise 4 and Exercise 8. What are your conclusions?

3. Fourier Analysis of Initial Value Problems

3.1 Formal solution

Let $P(z) = a_0 + a_1 z + \ldots + a_d z^d$ be a polynomial with complex coefficients. We consider the following periodic initial value problem. We wish to find a complex-valued function $u = u(x,t)$, $-\infty < x < \infty$, $t \geq 0$, L-periodic in x, that satisfies the differential equation

$$\partial_t u(x,t) = P(\partial_x) u(x,t), \quad -\infty < x < \infty, \quad t > 0, \tag{3.1}$$

along with the initial condition

$$u(x,0) = u^0(x), \quad -\infty < x < \infty, \tag{3.2}$$

where u^0 is a given function in $\mathcal{L}^2[0,L]$. For $P(z) = az^2$, a a positive constant, we have $P(\partial_x) = a\partial_x^2$ and (3.1) is the *heat equation* $\partial_t u = a\partial_{xx} u$ with diffusivity constant a. The choice $P(z) = -cz$, c real, leads to the *advection equation* $\partial_t u = -c\partial_x u$; for $P(z) = iz^2$, we have the *Schroedinger equation* in nondimensional units $\partial_t u = i\partial_{xx} u$ (or $i\partial_t u = -\partial_{xx} u$), etc.

The problem (3.1)–(3.2) is easily solved by Fourier series (Strang 1986). The ϕ_n's are eigenfunctions of ∂_x (see (2.12)) and hence eigenfunctions of the operator $P(\partial_x)$ in (3.1):

$$\begin{aligned} P(\partial_x)\phi_n &= (a_0 + a_1\partial_x + \ldots + a_d\partial_x^d)\phi_n \\ &= (a_0 + a_1\lambda_n + \ldots + a_d\lambda_n^d)\phi_n = P(\lambda_n)\phi_n. \end{aligned}$$

It is convenient to introduce the notation

$$\mu_n = P(\lambda_n)$$

for the eigenvalues of $P(\partial_x)$. For each fixed value of t, the solution $u(x,t)$ of (3.1)–(3.2) is sought as a Fourier series

$$u(x,t) = \sum_{n=-\infty}^{\infty} \hat{u}_n(t)\phi_n(x), \tag{3.3}$$

i.e., as a *superposition of eigenfunctions.* Substitution of (3.3) in (3.1) yields

$$\sum_{n=-\infty}^{\infty} \frac{d}{dt}\hat{u}_n(t)\phi_n(x) = \sum_{n=-\infty}^{\infty} \mu_n \hat{u}_n(t)\phi_n(x)$$

or

$$\frac{d}{dt}\hat{u}_n(t) = \mu_n \hat{u}_n(t), \quad n = 0, \pm 1, \pm 2, \dots . \tag{3.4}$$

Thus the partial differential equation (3.1) for u is equivalent to the system of infinitely many ordinary differential equations (3.4) for the Fourier coefficients. This system is *diagonal* or uncoupled: the n-th equation only involves the n-th coefficient $\hat{u}_n(t)$. Diagonalization is of course what one looks for when using eigenfunctions. In turn, (3.2) provides the initial values for (3.4),

$$\hat{u}_n(0) = \hat{u}_n^0, \quad n = 0, \pm 1, \pm 2, \dots, \tag{3.5}$$

where $\hat{u}_n^0$ are the Fourier coefficients of u^0. From (3.4) and (3.5), $\hat{u}_n(t) = \exp(\mu_n t)\hat{u}_n^0$ and (3.3) reads

$$u(x,t) = \sum_{n=-\infty}^{\infty} \exp(\mu_n t)\hat{u}_n^0 \phi_n(x). \tag{3.6}$$

Each term in this series is called a *mode* of the solution.

Exercise 10 Particularize (3.6) to the heat, advection and Schroedinger equations mentioned above.

Exercise 11 Systems of ν equations with ν unknown functions can be cast in the format (3.1) by allowing u to be a vector with ν components and letting the coefficients a_j of $P(z)$ be $\nu \times \nu$ constant matrices (z remains a complex variable). Write $P(z)$ for the system

$$\partial_t v = c\partial_x w, \quad \partial_t w = c\partial_x v, \tag{3.7}$$

(c a positive constant). Prove that (3.6) remains valid for systems; $\mu_n = P(\lambda_n)$ and $\exp(\mu_n t)$ are now $\nu \times \nu$ matrices (see Sect. 1.2) and $\hat{u}_n^0$ a vector whose ν components are the Fourier coefficients of the components of u^0. Particularize (3.6) to the system (3.7).

Exercise 12 Prove that the wave equation $\partial_{tt}\zeta = c^2\partial_{xx}\zeta$ is equivalent to (3.7) through the change of variables $v = \partial_t\zeta$, $w = c\partial_x\zeta$. This illustrates the reformulation of equations involving ∂_t^k, $k > 1$, as systems of the first order in t (i.e., systems involving only first derivatives with respect to t). The resulting systems may then be solved as in Exercise 11.

3.2 Well-posed problems

The solution $u = u(x,t)$ of (3.1)–(3.2) is a function of two arguments. These arguments do not play a symmetric role; it is convenient to introduce the notation $u(\cdot,t)$ to refer to the function of the first argument obtained by giving the fixed numerical value t to the second argument. If for each fixed $t \geq 0$, $u(\cdot,t)$ belongs to $\mathcal{L}^2[0,L]$, then we may imagine that each $u(\cdot,t)$ is a vector in $\mathcal{L}^2[0,L]$; this vector changes with t starting from the initial vector u^0 at $t = 0$. It is this change that we wish to study.

A crucial quantity is given by ($\Re$ denotes real part)

$$\alpha = \alpha(P(\partial_x)) = \sup_{-\infty<n<\infty} \Re\mu_n,$$

the *spectral abscissa* of the operator $P(\partial_x)$ (cf. Section 1.2). Two situations may arise.

(i) $\alpha < \infty$. Then we use (2.7) in (3.6) to get, for $t \geq 0$,

$$\begin{aligned}
||u(\cdot,t)||^2 &= L \sum_{n=-\infty}^{\infty} |\exp(\mu_n t)|^2 \, |\hat{u}_n^0|^2 \\
&= L \sum_{n=-\infty}^{\infty} \exp(2\Re\mu_n t) \, |\hat{u}_n^0|^2 \\
&\leq Le^{2\alpha t} \sum_{n=-\infty}^{\infty} |\hat{u}_n^0|^2
\end{aligned}$$

or

$$||u(\cdot,t)|| \leq e^{\alpha t} ||u^0||. \tag{3.8}$$

Thus, if the initial datum u^0 is in $\mathcal{L}^2[0, L]$, the solution $u(\cdot,t)$ remains in $\mathcal{L}^2[0, L]$ at all later times $t > 0$. Furthermore, for each fixed $t > 0$, the length of the evolved vector $u(\cdot,t)$ can be bounded by a factor $e^{\alpha t}$ (independent of u^0) times the initial length $||u^0||$. Small u^0 lead to small solutions. The problem is said to be *well posed* (Richtmyer and Morton 1967; Kreiss and Oliger 1973; Sanz-Serna 1985; Sanz-Serna and Verwer 1989).

(ii) $\alpha = \infty$. Then, for the initial condition $u^0 = \phi_n$, with norm $\sqrt{L}$, the solution $u(\cdot,t) = \exp(\mu_n t)\phi_n$ has a length $\sqrt{L}\exp(\Re\mu_n t)$ that can be made arbitrarily large by varying n. It is therefore impossible to bound $||u(\cdot,t)||$ by a u^0-independent factor times $||u^0||$. The problem is said to be *ill posed.* Initial conditions close to 0 may result in arbitrarily large solutions. Such problems are not good candidates to become physical models.

Exercise 13 Prove that the equations in Exercise 10 give rise to well-posed initial value problems. How about the *backward heat equation* $\partial_t u = -a\partial_{xx}u$, *a* a positive constant?

Exercise 14 A system (cf. Exercise 11) leads to a well-posed initial value problem if there exist constants K and α such that

$$\sup_{-\infty<n<\infty} ||\exp(\mu_n t)|| \leq Ke^{\alpha t}, \quad t > 0.$$

(Here $||\cdot||$ denotes norm for $\nu \times \nu$ matrices, see Sect. 1.2.) For well-posed systems, derive an estimate similar to (3.8). Study the well-posedness of the initial value problem for (3.7).

3.3 Dissipation and dispersion

In this subsection we assume for simplicity that the initial datum u^0 in (3.2) is real-valued and write it in trigonometric form as in (2.11), i.e.,

$$u^0(x) = A_0 + \sum_{n=1}^{\infty} A_n \cos\left(\frac{2\pi n}{L}x - \psi_n\right).$$

We also assume that in (3.1) the variables x and t correspond to physical space and time.

Let us consider first the advection equation $\partial_t u = -c\partial_x u$ (c a real constant). The solution (3.6) written in trigonometric form is

$$u(x,t) = A_0 + \sum_{n=1}^{\infty} A_n \cos\left(\frac{2\pi n}{L}x - \frac{2\pi n}{L}ct - \psi_n\right). \tag{3.9}$$

Therefore u is a superposition of sinusoidal waves (Whitham 1974)

$$A_n \cos\left(\frac{2\pi n}{L}x - \frac{2\pi n}{L}ct - \psi_n\right); \tag{3.10}$$

each of these is constant on the lines in space-time with equations $x - ct = \xi$, ξ a constant. In other words, (3.10) propagates with velocity c without changing shape. Since this holds for each n, the same is true for the sum in (3.9) and in fact, $u(x,t) = u^0(x - ct)$, see Figs. 3–4.

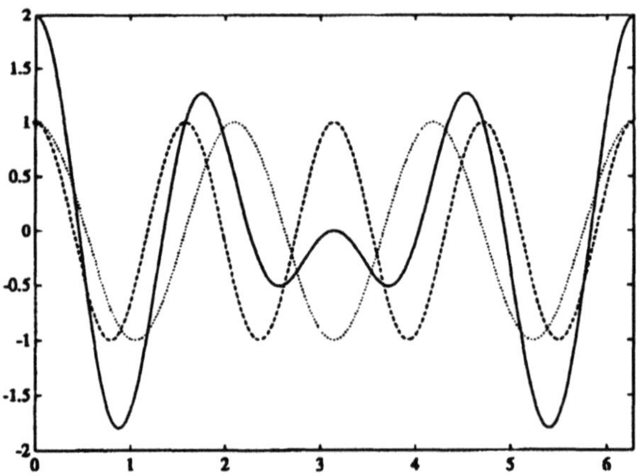

Fig. 3. Initial condition (solid line) $u^0(x) = \cos 3x + \cos 4x$ ($L = 2\pi$) obtained by superposing the wave numbers 3 (dotted line) and 4 (dashed line)

In (3.10), $\kappa_n = 2\pi n/L$ is the *wave number,* i.e., the number of complete cycles of the cosine function per 2π units of length, while $\omega_n = 2\pi nc/L$ is the *(angular) frequency,* i.e., the number of cosine cycles per 2π units of time. The quotient $\omega_n/\kappa_n = c$ provides the (phase) velocity of the wave. On the other

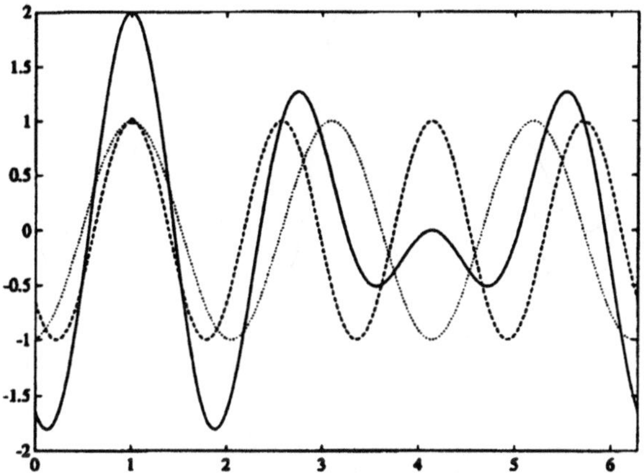

Fig. 4. Solution $u(\cdot, t)$ at time $t = 1$ of the advection equation $\partial_t u = -c\partial_x u$, $c = 1$, for the initial condition in Fig. 3. Both harmonics have travelled one unit to the right

hand, $\ell_n = L/n$ is the *wave length,* (distance between two consecutive maxima of the cosine function) and $T_n = L/(cn)$ is the period in time of the wave. Again $\ell_n/T_n = c$.

Let us now turn to the equation $\partial_t u = \partial_{xxx} u$ with solution

$$u(x,t) = A_0 + \sum_{n=1}^{\infty} A_n \cos\left(\frac{2\pi n}{L}x - \left[\frac{2\pi n}{L}\right]^3 t - \psi_n\right).$$

We are again superposing sinusoidal waves. But now the wave with wave number $\kappa_n = 2\pi n/L$ travels with a velocity $(2\pi n/L)^2$ that is not the same for all wave numbers. This behavior is known as *dispersion.* When dispersion is present the 'shape' of $u(\cdot, t)$ changes with t. This is illustrated in Fig. 5.

The equation $\omega_n = \kappa_n^3$ relating frequency and wave number is known as the *dispersion relation* of the equation (Whitham 1974).

Finally for the heat equation $\partial_t = a\partial_{xx} u$, a positive, the solution is

$$u(x,t) = A_0 + \sum_{n=1}^{\infty} \left[A_n \exp\left(-\frac{4\pi^2 a n^2}{L^2} t\right)\right] \cos\left(\frac{2\pi n}{L}x - \psi_n\right);$$

the sinusoidal components do not move with time; this is not a wave-like equation. It is the amplitude of the components that changes with t. The harmonics decay, which physically would correspond to a *dissipative* behavior. Higher wave numbers decay at a faster rate (see Fig. 6). A large diffusivity constant a results in a faster decay.

Exercise 15 Write in trigonometric form the solutions of the initial value problem for the equation $\partial_t u = \partial_{xx} u + \partial_{xxx} u$. Study the well-posedness of the initial

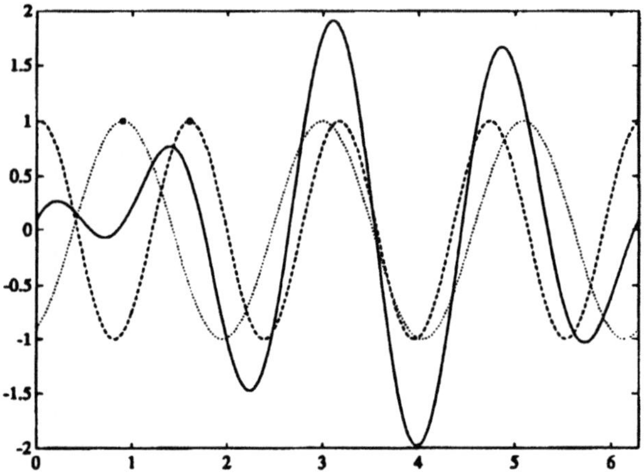

Fig. 5. Solution $u(\cdot,t)$ at time $t = 0.1$ of the equation $\partial_t u = \partial_{xxx} u$ for the initial condition in Fig. 3. The wave number 3 moves with velocity 9, so that the maximum that was initially at $x = 0$ is now at $x = 0.9$. The wave number 4 has velocity 16 and the maximum initially at $x = 0$ is now at $x = 1.6$. (These maxima are indicated by small circles.) Clearly the shape of the solution $u(\cdot,t)$ changes with t

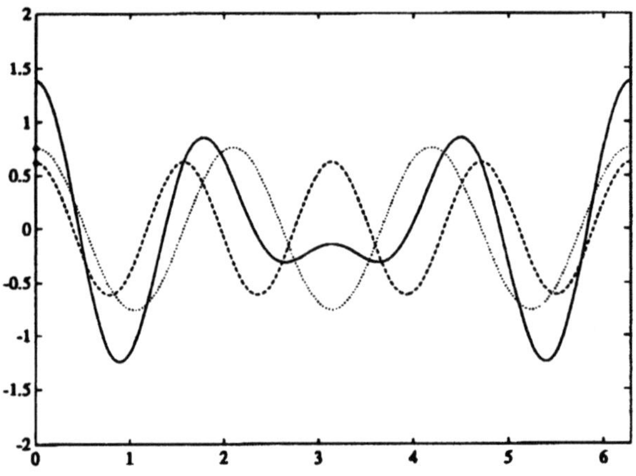

Fig. 6. Solution $u(\cdot,t)$, $t = 0.03$ of the heat equation $\partial_t = \partial_{xx} u$, for the initial condition in Fig. 3

value problem and discuss the solution behavior, that combines dissipation and dispersion.

Exercise 16 Write in trigonometric form the solutions of the initial value problem for the equation $\partial_t u = \partial_{xx} u + au$, a a real constant. Study the well-posedness of the initial value problem. Discuss the solution behavior for different values of a. Note that solution growth is compatible with well-posedness.

4. Discrete Fourier Analysis

4.1 The discrete Fourier transform

The discrete Fourier transform (Strang 1986) is one of the main tools of modern applied mathematics. For the time being, it is convenient not to think at all of the discrete transform as a discrete version of the Fourier series; the relation between discrete transforms and Fourier series is discussed in Sect. 5 below.

We work in the space $\mathbb{C}^M$ of column vectors $\mathbf{X}$ with M complex components $\mathbf{X} = [X_0, X_1, \ldots, X_{M-1}]^T$; note that subscripts run from 0 to $M-1$, rather than the standard 1 to M. The superscript T means transpose.

The (M-dimensional) *discrete Fourier transform* is the linear transformation in $\mathbb{C}^M$ that associates with each vector $\mathbf{X}$ the vector $F_M\mathbf{X}$, where F_M is the $M \times M$ complex matrix whose entry (ℓ, n), $\ell, n = 0, 1, \ldots, M-1$, is $w_M^{\ell n}$, ℓn-th power of the number

$$w_M = \exp\left(-\frac{2\pi i}{M}\right) = \cos\frac{2\pi}{M} - i\sin\frac{2\pi}{M}. \tag{4.1}$$

For instance, for $M = 2$, $w_2 = -1$ and

$$F_2 = \begin{bmatrix} w_2^0 & w_2^0 \\ w_2^0 & w_2^1 \end{bmatrix} = \begin{bmatrix} 1 & 1 \\ 1 & -1 \end{bmatrix};$$

for $M = 3$, $w_3 = -1/2 - i\sqrt{3}/2$ and

$$F_3 = \begin{bmatrix} w_3^0 & w_3^0 & w_3^0 \\ w_3^0 & w_3^1 & w_3^2 \\ w_3^0 & w_3^2 & w_3^4 \end{bmatrix} = \begin{bmatrix} 1 & 1 & 1 \\ 1 & -1/2 - i\sqrt{3}/2 & -1/2 + i\sqrt{3}/2 \\ 1 & -1/2 + i\sqrt{3}/2 & -1/2 - i\sqrt{3}/2 \end{bmatrix};$$

for $M = 4$, $w_4 = -i$ and

$$F_4 = \begin{bmatrix} w_4^0 & w_4^0 & w_4^0 & w_4^0 \\ w_4^0 & w_4^1 & w_4^2 & w_4^3 \\ w_4^0 & w_4^2 & w_4^4 & w_4^6 \\ w_4^0 & w_4^3 & w_4^6 & w_4^9 \end{bmatrix} = \begin{bmatrix} 1 & 1 & 1 & 1 \\ 1 & -i & -1 & i \\ 1 & -1 & 1 & -1 \\ 1 & i & -1 & -i \end{bmatrix}.$$

It is important to observe that $w_M^M = 1$, i.e., w_M is an M-th root of 1. In fact, w_M is the M-th root of unity whose argument is negative and as small as possible.

The idea behind the use of the discrete transform is that many vector operations are easier if performed on the transformed vectors. One then (i) transforms the data $\mathbf{X}$, (ii) operates with the transformed $F_M\mathbf{X}$ to find the transformed of the solution and (iii) transforms back to find the solution. The third step requires the knowledge of the inverse matrix F_M^{-1}, which is really simple:

$$F_M^{-1} = \frac{1}{M}\bar{F}_M. \tag{4.2}$$

Here $\bar{F}_M$ is the matrix obtained by conjugating all the entries of F_M. Since F_M and F_M^{-1} only differ in the factor M and in the change $i \to -i$, the algorithms used to compute discrete transforms (see Sect. 4.3) are easily adapted to compute inverse discrete transforms.

On the other hand, $F_M = F_M^T$ and hence (see Sect. 1.2) $F_M^* = \bar{F}_M = MF_M^{-1}$. If $\mathbf{Y} = F_M\mathbf{X}$, then

$$|\mathbf{Y}|^2 = \mathbf{Y}^*\mathbf{Y} = (F_M\mathbf{X})^*\mathbf{Y} = M\mathbf{X}^*F_M^{-1}F_M\mathbf{X} = M|\mathbf{X}|^2,$$

i.e.,

$$\sum_{n=0}^{M-1} |X_n|^2 = \frac{1}{M}\sum_{n=0}^{M-1} |Y_n|^2; \tag{4.3}$$

this is a discrete version of Parseval's identity (2.7). Except for the normalizing factor M, F_M is a unitary matrix and using $F_M\mathbf{X}$ instead of $\mathbf{X}$ does not alter the length of the vectors involved.

Exercise 17 Write explicitly F_6 and F_8. Use (4.2) to write explicitly $\bar{F}_M$ and F_M^{-1}, $M = 2, 3, 4, 6, 8$. Have you observed that $\bar{F}_M$ may be obtained by permuting the columns of F_M? Check that $F_4F_4^{-1}$ yields the unit matrix.

Exercise 18 Two vectors $\mathbf{X}_1$, $\mathbf{X}_2$ in $\mathbb{C}^M$ are said to be orthogonal if their inner product $\mathbf{X}_1^*\mathbf{X}_2$ vanishes. Show that the column vectors of F_M are pairwise orthogonal. Show that the same is true for the column vectors of F_M^{-1}.

Exercise 19 Prove that $F_M\bar{F}_M = MI$; this yields (4.2).

4.2 An application: systems of ordinary differential equations with circulant matrices

Many situations give rise to systems of the form

$$\frac{d}{dt}\mathbf{X}(t) = A\mathbf{X}(t), \tag{4.4}$$

with A a *circulant* constant complex matrix (Strang 1986)

$$A = \begin{bmatrix} a_0 & a_1 & a_2 & \cdots & a_{M-1} \\ a_{M-1} & a_0 & a_1 & \cdots & a_{M-2} \\ a_{M-2} & a_{M-1} & a_0 & \cdots & a_{M-3} \\ \vdots & \vdots & \vdots & \ddots & \vdots \\ a_1 & a_2 & a_3 & \cdots & a_0 \end{bmatrix}.$$

It is fortunate that, for $n = 0, 1, \ldots, M-1$, the n-th column of F_M^{-1} (see (4.2))

$$\mathbf{V}_n = (1/M)\begin{bmatrix} \bar{w}_M^0 \\ \bar{w}_M^n \\ \bar{w}_M^{2n} \\ \vdots \\ \bar{w}_M^{(M-1)n} \end{bmatrix} = (1/M)\begin{bmatrix} \exp\left(\frac{2\pi 0ni}{M}\right) \\ \exp\left(\frac{2\pi ni}{M}\right) \\ \exp\left(\frac{2\pi 2ni}{M}\right) \\ \vdots \\ \exp\left(\frac{2\pi(M-1)ni}{M}\right) \end{bmatrix} \tag{4.5}$$

is an eigenvector of A. More precisely

$$A\mathbf{V}_n = \sigma_n \mathbf{V}_n, \quad n = 0, 1, \ldots, M-1, \tag{4.6}$$

with

$$\sigma_n = a_0 \bar{w}_M^0 + a_1 \bar{w}_M^n + \ldots + a_{M-1} \bar{w}_M^{(M-1)n}, \quad \bar{w}_M = \exp\left(\frac{2\pi i}{M}\right).$$

We therefore look for the solution $\mathbf{X}(t)$ of (4.4) as a *superposition of eigenvectors*

$$\mathbf{X}(t) = \sum_{n=0}^{M-1} Y_n(t) \mathbf{V}_n, \tag{4.7}$$

(the $Y_n(t)$ are complex numbers) or in matrix notation (a product matrix-times-vector is the linear combination of the matrix columns whose coefficients are the components of the vector)

$$\mathbf{X}(t) = F_M^{-1} \mathbf{Y}(t). \tag{4.8}$$

The new vector $\mathbf{Y} = F_M \mathbf{X}$ is therefore the transform of $\mathbf{X}$. Substitution of (4.7) in (4.4) leads, in view of (4.6), to

$$\sum_{n=0}^{M-1} \frac{d}{dt} Y_n(t) \mathbf{V}_n = \sum_{n=0}^{M-1} \sigma_n Y_n(t) \mathbf{V}_n,$$

i.e.,

$$\frac{d}{dt} Y_n(t) = \sigma_n Y_n(t), \quad n = 0, 1, \ldots, M-1; \tag{4.9}$$

in terms of the Y_n's the system has uncoupled or diagonalized and is easily integrated:

$$Y_n(t) = \exp(\sigma_n t) Y_n(0), \quad n = 0, 1, \ldots, M-1. \tag{4.10}$$

We conclude from (4.7) and (4.9) that the solution is

$$\mathbf{X}(t) = \sum_{n=0}^{M-1} \exp(\sigma_n t) Y_n(0) \mathbf{V}_n. \tag{4.11}$$

In practice, to find $\mathbf{X}(t)$ at any given numerical value of t, one computes $\mathbf{Y}(0)$ by a discrete transform (see (4.8)), integrates in the Y_n variables as in (4.10) and returns to the X_n variables by performing the inverse discrete transform in (4.8) at time t.

The reader has certainly noticed the similarity between this material and the contents of Sect. 3.1; $u(x,t)$ corresponds to $\mathbf{X}(t)$, $P(\partial_x)$ to A, the $\hat{u}_n$'s to the Y_n's, (3.1) corresponds to (4.4), (3.3) to (4.7), (3.4) to (4.9) and (3.6) to (4.11). At each fixed value of t, $u(x,t)$ is parameterized by a continuum of values of x; here $\mathbf{X}(t)$ is parametrized by the discrete subscript $n = 0, 1, \ldots, M-1$. There is a Fourier coefficient $\hat{u}_n$ for each integer n, however there are only M variables Y_n. The basis functions ϕ_n in (3.3) are pairwise orthogonal and so are the vectors $\mathbf{V}_n$ in (4.7), see Exercise 18.

Other applications of the discrete Fourier transform involve the solution of algebraic systems of linear equations, see Exercise 21, and the computation of convolutions (Strang 1986).

Exercise 20 Prove (4.6).

Exercise 21 Solve the linear algebraic equations $A\mathbf{X} = \mathbf{B}$, where A is the matrix in (4.4) and $\mathbf{B}$ a known vector. (Hint: If $\mathbf{X} = F_M^{-1}\mathbf{Y}$ and $\mathbf{B} = F_M^{-1}\mathbf{C}$, then $Y_n = C_n/\sigma_n$, $n = 0, 1, \ldots, M-1$.) Prove that to solve such a system one needs one discrete transform, one inverse discrete transform and M divisions. This idea can be extended to more general matrices (Strang 1986).

4.3 The fast Fourier transform

4.3.1 Preliminary remarks

Once F_M has been formed as in Sect. 4.1, to find $F_M\mathbf{X}$ for a given vector $\mathbf{X}$ requires M^2 complex multiplications if one follows the standard recipe for matrix/vector products (each entry in F_M has to be multiplied by an element of $\mathbf{X}$). In 1965 Cooley and Tukey popularized an algorithm, *the Fast Fourier Transform, FFT,* that finds $F_M\mathbf{X}$ with less than $(1/2)M\log_2 M$ multiplications (the exact number depends on the details of the specific implementation used). This implies enormous savings. For $M = 2^{12} = 4096$, a dimension that is typical in many applications, $M^2 = 2^{24}$, and the FFT requires less than 6×2^{12} multiplications; this means that FFT is at least 600 times faster. Since $\log_2 M$ grows very slowly with M, the cost of the FFT grows for all practical purposes like $O(M)$; the straightforward matrix-times-vector algorithm has an $O(M^2)$ cost.

The idea behind the FFT is not difficult (Strang 1986). Suppose that M is even $M = 2N$ and we need to compute $\mathbf{Y} = F_M\mathbf{X}$. We begin by splitting $\mathbf{X}$ into two vectors of length N

$$\mathbf{X}' = [X_0, X_2, \ldots, X_{M-2}]^T, \quad \mathbf{X}'' = [X_1, X_3, \ldots, X_{M-1}]^T$$

and computing two transforms

$$\mathbf{Y}' = F_N\mathbf{X}', \quad \mathbf{Y}'' = F_N\mathbf{X}'', \tag{4.12}$$

whose dimension is only *one half* of that of the sought transform. We may recover $\mathbf{Y}$ from $\mathbf{Y}'$ and $\mathbf{Y}''$. In fact, it is straightforward to prove that the first N components of $\mathbf{Y}$ are given by

$$Y_n = Y_n' + w_M^n Y_n'', \quad n = 0, 1, \ldots, N-1, \tag{4.13a}$$

while the N last components are given by

$$Y_{N+n} = Y_n' - w_M^n Y_n'', \quad n = 0, 1, \ldots, N-1. \tag{4.13b}$$

Note that (4.13a)–(4.13b) only require N multiplications $w_M^n Y_n''$. This evaluation of M-dimensional transforms in terms of N-dimensional transforms is the essence of the FFT.

Even though in practice one implements directly (4.13), the idea above is more easily grasped by rewriting (4.13) in matrix form. For instance, with $M = 4$, $N = 2$, we have

$$\begin{bmatrix} Y_0 \\ Y_1 \\ Y_2 \\ Y_3 \end{bmatrix} = M_4 \begin{bmatrix} Y_0' \\ Y_1' \\ Y_0'' \\ Y_1'' \end{bmatrix},$$

where

$$M_4 = \begin{bmatrix} 1 & 0 & 1 & 0 \\ 0 & 1 & 0 & -i \\ 1 & 0 & -1 & 0 \\ 0 & 1 & 0 & i \end{bmatrix}.$$

Therefore, by (4.12), (O_2 denotes the 2×2 zero matrix)

$$\begin{aligned} \mathbf{Y} = M_4 \begin{bmatrix} F_2\mathbf{X}' \\ F_2\mathbf{X}'' \end{bmatrix} &= M_4 \begin{bmatrix} F_2 & O_2 \\ O_2 & F_2 \end{bmatrix} \begin{bmatrix} \mathbf{X}' \\ \mathbf{X}'' \end{bmatrix} \\ &= M_4 \begin{bmatrix} F_2 & O_2 \\ O_2 & F_2 \end{bmatrix} \begin{bmatrix} 1 & 0 & 0 & 0 \\ 0 & 0 & 1 & 0 \\ 0 & 1 & 0 & 0 \\ 0 & 0 & 0 & 1 \end{bmatrix} \mathbf{X}; \end{aligned}$$

the rightmost matrix, that we shall denote by P_4, permutes the entries of $\mathbf{X} = [X_0, X_1, X_2, X_3]^T$ to give $P_4\mathbf{X} = [X_0, X_2, X_1, X_3]^T$. To sum up

$$F_4 = M_4 \begin{bmatrix} F_2 & O_2 \\ O_2 & F_2 \end{bmatrix} P_4.$$

In the case $M = 8$, $N = 4$, the formulae (4.13) lead in the same way to

$$F_8 = M_8 \begin{bmatrix} F_4 & O_4 \\ O_4 & F_4 \end{bmatrix} P_8,$$

where

$$M_8 = \begin{bmatrix} 1 & 0 & 0 & 0 & 1 & 0 & 0 & 0 \\ 0 & 1 & 0 & 0 & 0 & w_8 & 0 & 0 \\ 0 & 0 & 1 & 0 & 0 & 0 & w_8^2 & 0 \\ 0 & 0 & 0 & 1 & 0 & 0 & 0 & w_8^3 \\ 1 & 0 & 0 & 0 & -1 & 0 & 0 & 0 \\ 0 & 1 & 0 & 0 & 0 & -w_8 & 0 & 0 \\ 0 & 0 & 1 & 0 & 0 & 0 & -w_8^2 & 0 \\ 0 & 0 & 0 & 1 & 0 & 0 & 0 & -w_8^3 \end{bmatrix}$$

and P_8 is the permutation matrix

$$P_8 = \begin{bmatrix} 1 & 0 & 0 & 0 & 0 & 0 & 0 & 0 \\ 0 & 0 & 1 & 0 & 0 & 0 & 0 & 0 \\ 0 & 0 & 0 & 0 & 1 & 0 & 0 & 0 \\ 0 & 0 & 0 & 0 & 0 & 0 & 1 & 0 \\ 0 & 1 & 0 & 0 & 0 & 0 & 0 & 0 \\ 0 & 0 & 0 & 1 & 0 & 0 & 0 & 0 \\ 0 & 0 & 0 & 0 & 0 & 1 & 0 & 0 \\ 0 & 0 & 0 & 0 & 0 & 0 & 0 & 1 \end{bmatrix}.$$

4.3.2 The algorithm

Assume now that M is a power of 2, $M = 2^\mu$, μ a positive integer. The idea just described can be successively applied to reduce the computation of transforms of dimension M to transforms of dimension $M/2$, $M/4$, $M/8$, etc. One goes all the way down to 2-dimensional transforms that are, of course, trivially computed. (A 2-transform requires just two additions.)

For instance, on combining the examples of Sect. 4.3.1, we get for $M = 8$

$$\begin{aligned} \mathbf{Y} = F_8\mathbf{X} &= M_8 \begin{bmatrix} F_4 & O_4 \\ O_4 & F_4 \end{bmatrix} P_8\mathbf{X} \\ &= M_8 \begin{bmatrix} M_4 \begin{bmatrix} F_2 & O_2 \\ O_2 & F_2 \end{bmatrix} P_4 & O_4 \\ O_4 & M_4 \begin{bmatrix} F_2 & O_2 \\ O_2 & F_2 \end{bmatrix} P_4 \end{bmatrix} P_8\mathbf{X} \\ &= M_8 \begin{bmatrix} M_4 & O_4 \\ O_4 & M_4 \end{bmatrix} \begin{bmatrix} F_2 & O_2 & O_2 & O_2 \\ O_2 & F_2 & O_2 & O_2 \\ O_2 & O_2 & F_2 & O_2 \\ O_2 & O_2 & O_2 & F_2 \end{bmatrix} \begin{bmatrix} P_4 & O_4 \\ O_4 & P_4 \end{bmatrix} P_8\mathbf{X}; \end{aligned}$$

to find $\mathbf{Y}$ we successively 'multiply' $\mathbf{X}$ by $P_8, \ldots, M_8$. The first two products require only permutations of the entries of $\mathbf{X}$; the last two products implement formulae (4.13) to find two 4-transforms and the sought 8-transform.

For other values of $M = 2^\mu$, the algorithm works in the same way. The matrix F_M is thought of as a product of $(\mu - 1) + 1 + (\mu - 1)$ factors. The rightmost $\mu - 1$ factors describe permutations of the entries of $\mathbf{X}$. The central factor is the multiplication-free computation of 2-transforms. The left-most $\mu - 1$ factors successively yield 2^2-transforms, 2^4-transforms, etc.

4.3.3 Practical issues

The FFT also works when M is not a power of 2 (Cooley and Tukey 1965). For instance if M has a prime factor $M_1 \neq 2$, $M = M_1 M_2$, one can work out formulae similar to (4.13) to reduce the computation of an M-transform to that of M_1 transforms of dimension M_2. Iteration of this idea reduces the computation of transforms of arbitrary length to that of transforms whose size is a prime factor

2, 3, 5, ... However, in most implementations, the FFT is most efficient when M is of the form 2^μ and one should try to use numbers of this form.

Note that $F_M\mathbf{X}$ is a complex vector even if, as it is often the case in the applications, $\mathbf{X}$ has real entries. The use of the fully complex FFT algorithm described above is not optimal in terms of efficiency and improvements exist (Press et al. 1989).

All good mathematical software libraries possess implementations of the FFT. These should always be preferred to 'home made' implementations written by users. There is a lack of uniformity in the literature when it comes to defining the matrices F_M and F_M^{-1}; the definition used here has $-i$ in F_M (see (4.1)) and hence $+i$ in F_M^{-1} (see (4.2)), other authors take the signs differently. Here F_M^{-1} carries the factor M^{-1}; other authors attach this factor to the direct transform F_M. These variations should cause no problem.

Our definitions coincide with those in MATLAB. The MATLAB function **fft**(**X**) finds the discrete transform of the vector **X** and **ifft**(**X**) provides the inverse discrete transform. We note that, while mathematical notation has the components of **X** labelled $0, 1, \ldots, M-1$, components in MATLAB run from 1 to M; the mathematical value X_n should then be invoked in a program as $X(n+1)$.

Exercise 22 Write an FFT program. You can check what you write against a coding from a textbook (Press et al. 1989).

Exercise 23 Prove that if the linear system in Exercise 21 is solved by discrete Fourier transforms, the overall number of required multiplications and divisions is less than $M\log_2 M + M$. This should be compared with the $O(M^3)$ cost when using Gaussian elimination (Golub and Van Loan 1989). Since, for realistic values of M, $\log_2 M$ is a small number, the cost of *computing* the solution by Fourier techniques is $O(M)$ for all practical purposes. The cost of *printing* the solution at the end of the computation also grows like $O(M)$!

5. Discrete Fourier Transform vs. Fourier Series

5.1 A first look at aliasing

So far Fourier series, dealing with L-periodic functions of x, and Fourier transforms, dealing with M-vectors, have been presented as unrelated entities. This must now stop.

Let us again consider L-periodic complex-valued functions f as in Sects. 2–3. If $M \geq 1$ is an integer, we discretize the variable x by introducing the grid points $x_n = n\Delta x$, $n = 0, \pm1, \pm2, \ldots$, $\Delta x = L/M$. In applications where the variable x corresponds to physical time, we can think of the grid values $f(x_n)$ as *stroboscopic* samples of f. Due to periodicity, $f(x_n) = f(x_m)$ if $n-m$ differ by an integer multiple of M. Hence only the points $x_0, x_1, \ldots, x_{M-1}$ carry independent information. To each L-periodic function f we associate a vector $\mathbf{X}(f)$ in $\mathbb{C}^M$

defined by $[f(x_0), f(x_1), \ldots, f(x_{M-1})]^T$. Note that $\mathbf{X}(f)$ depends on M, i.e., on the particular grid chosen; for simplicity this dependence is not incorporated into the notation.

Two different functions f_1, f_2 may have $\mathbf{X}(f_1) = \mathbf{X}(f_2)$; this happens if and only if f_1 and f_2 coincide at all grid points. A prime example is given by the Fourier basis functions ϕ_n in (2.2). For these,

$$\mathbf{X}(\phi_n) = \begin{bmatrix} \exp\left(\frac{2\pi 0 n i}{M}\right) \\ \exp\left(\frac{2\pi n i}{M}\right) \\ \exp\left(\frac{2\pi 2 n i}{M}\right) \\ \vdots \\ \exp\left(\frac{2\pi (M-1) n i}{M}\right) \end{bmatrix} \tag{5.1}$$

and it is easy to check that

$$\mathbf{X}(\phi_n) = \mathbf{X}(\phi_m) \quad \Leftrightarrow \quad n \equiv m, \tag{5.2}$$

where $n \equiv m$ means that n and m differ in an integer multiple of M. Hence, *on the grid,* only ϕ_0, ϕ_1, ..., ϕ_{M-1} are different; ϕ_M coincides with ϕ_0, ϕ_{M+1} coincides with ϕ_1, etc. Also, ϕ_{-1} coincides with ϕ_{M-1}, ϕ_{-2} coincides with ϕ_{M-2}, etc. This coincidence is called *aliasing* and plays a crucial role in discrete Fourier analysis; when $\mathbf{X}(\phi_m) = \mathbf{X}(\phi_n)$ we say that ϕ_m is an alias of ϕ_n (see Fig. 7).

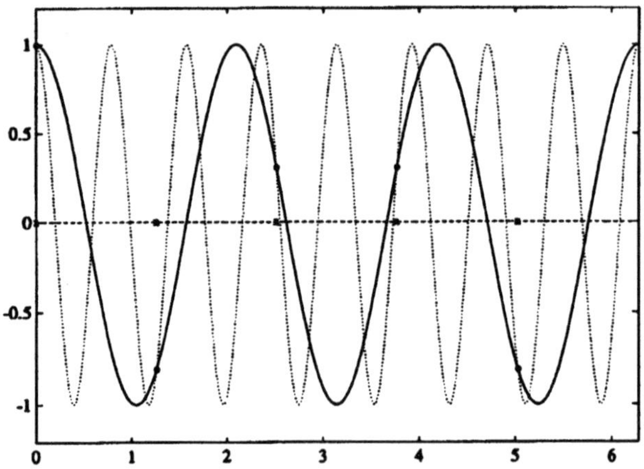

Fig. 7. On the grid resulting from dividing the interval $[0, 2\pi]$ into $M = 5$ equal parts ($\times$ signs) the function $\exp(8ix)$ is an alias of $\exp(3ix)$. This figure corresponds to the real part; a similar figure may be drawn for the imaginary part

Exercise 24 Check (5.1) and (5.2).

5.2 Trigonometric interpolation

5.2.1 The bad way

Let us now compare (4.5) and (5.1):

$$\mathbf{V}_n = (1/M)\mathbf{X}(\phi_n), \quad n = 0, 1, \ldots, M-1. \tag{5.3}$$

This is an important relation between Fourier transforms and Fourier series: the grid values $\mathbf{X}(\phi_n)$ of the Fourier series basis functions ϕ_n coincide, except for the factor M, with the columns $\mathbf{V}_n$ of the matrix F_M^{-1}. We use this coincidence to solve an interpolation problem. Assume that f is a given function and we wish to find coefficients $a_0, a_1, \ldots, a_{M-1}$ such that the trigonometric polynomial

$$\sum_{n=0}^{M-1} a_n \phi_n(x) \tag{5.4}$$

matches f at the grid points, i.e.,

$$f(x_m) = \sum_{n=0}^{M-1} a_n \phi_n(x_m), \quad m = 0, \pm 1, \pm 2, \ldots.$$

This condition may be rewritten as

$$\mathbf{X}(f) = \sum_{n=0}^{M-1} a_n \mathbf{X}(\phi_n)$$

or, in view of (5.2) (see the remark before (4.8)),

$$\mathbf{X}(f) = M \sum_{n=0}^{M-1} a_n \mathbf{V}_n = M F_M^{-1} \mathbf{a}.$$

Therefore

$$\mathbf{a} = (1/M) F_M \mathbf{X}(f), \tag{5.5}$$

the a_n are $1/M$ times the entries of the transform $F_M \mathbf{X}(f)$ of the grid values $\mathbf{X}(f)$ of f.

The trouble with this interpolation is that, if x is not one of the grid points, (5.4) is a very poor approximation to f, see Fig. 8. Why is this? The interpolant (5.4) only combines ϕ_n's with $n \geq 0$, while, according to (2.4), f is, in general, a superposition of multiples of ϕ_n's with positive and *negative* n.

5.2.2 The good way

Having identified the reason for the failure of the interpolant (5.4), it is easy to construct a good interpolant. We certainly need a contribution involving ϕ_{-1}. On the grid, ϕ_{-1} is an alias of ϕ_{M-1}; what we can do is to replace the term $a_{M-1}\phi_{M-1}(x)$ in (5.4) (with $\mathbf{a}$ still given by (5.5)) by $a_{M-1}\phi_{-1}(x)$. This does

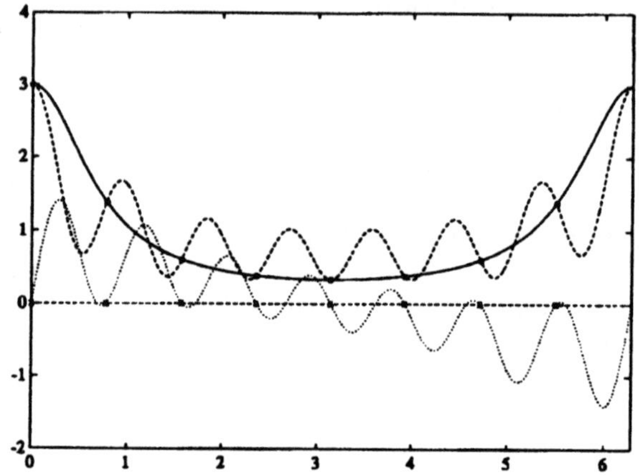

Fig. 8. The real function in Exercise 8 (solid line) and the interpolant (5.4)–(5.5) when $M = 8$. The interpolant is not even real-valued. The real part of the interpolant is the dashed line; the imaginary part the dotted line. At grid points, the real part of the interpolant matches the function and the imaginary part vanishes. The interpolant provides a very poor approximation indeed

not change the value of (5.4) at grid points, so that we are still interpolating f on the grid. Similarly, one replaces $a_{M-1}\phi_{M-2}(x)$ by $a_{M-2}\phi_{-2}(x)$ etc.

To be precise, let us consider separately the cases where M is odd and even. Assume first that M is of the form $M = 2N + 1$. Then we use the interpolant given by (5.5) and

$$a_{N+1}\phi_{-N}(x) + \cdots + a_{M-1}\phi_{-1}(x) + a_0\phi_0(x) + a_1\phi_1(x) + \cdots + a_N\phi_N(x).$$

This coincides with (5.4) (and therefore with f) on the grid, because ϕ_{-n}, $n = 1, \ldots, N$, is an alias of ϕ_{M-n}.

When M is even (in practice this is the commonest situation), one could consider either (Canuto et al. 1988)

$$\begin{aligned} &a_{N+1}\phi_{-N+1}(x) + \cdots + a_{M-1}\phi_{-1}(x) \\ &\quad + a_0\phi_0(x) + a_1\phi_1(x) + \cdots + a_{N-1}\phi_{N-1}(x) + a_N\phi_N(x), \end{aligned}$$

or

$$\begin{aligned} &a_N\phi_{-N}(x) + a_{N+1}\phi_{-N+1}(x) + \cdots + a_{M-1}\phi_{-1}(x) \\ &\quad + a_0\phi_0(x) + a_1\phi_1(x) + \cdots + a_{N-1}\phi_{N-1}(x), \end{aligned}$$

but I prefer to settle for the more symmetric format resulting after averaging these two expressions (Hamming 1973), i.e.,

$$\begin{aligned} &\frac{1}{2}a_N\phi_{-N}(x) + a_{N+1}\phi_{-N+1}(x) + \cdots + a_{M-1}\phi_{-1}(x) \\ &\qquad + a_0\phi_0(x) + a_1\phi_1(x) + \cdots + a_{N-1}\phi_{N-1}(x) + \frac{1}{2}a_N\phi_N(x). \end{aligned}$$

We write this interpolation or *collocation* trigonometric polynomial (Fig. 9) as (a double prime in the summation means that the first and last term should be halved)

$$I_N(f) = \sum_{n=-N}^{N}{}'' \tilde{f}_n \phi_n, \quad \tilde{f}_N = \tilde{f}_{-N}, \tag{5.6}$$

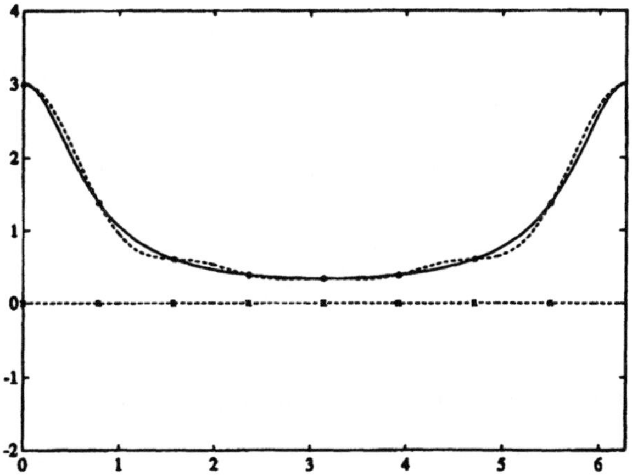

Fig. 9. The real function in Exercise 8 (solid line) and the interpolant (5.6) when $M = 8$. Compare with Fig. 8

with the coefficients given by

$$\tilde{f}_n = a_n, \quad n = 0, 1, \ldots, N-1, \tag{5.7a}$$

$$\tilde{f}_{-n} = a_{M-n}, \quad n = 1, 2, \ldots, N, \tag{5.7b}$$

The a_n are given by (5.5).

5.2.3 Discrete Fourier coefficients

Hereafter we only consider the even M case, $M = 2N$. The coefficients $\tilde{f}_n$ of the interpolant (5.6) are called the *discrete Fourier coefficients* of f. Note that they depend on M, i.e., on the specific grid under consideration; this dependence is however not reflected in the notation. We emphasize that to find the discrete Fourier coefficients of a function is an easy task. It is enough to perform the FFT in (5.5) and then rearrange via (5.7). In MATLAB, the function **fftshift**(**a**) rearranges the output

$$[a_0, a_1, \ldots, a_{N-1}, a_N, a_{N+1}, \ldots, a_{2N-1}]^T$$

of the function **fft** by swapping the left and right halves of the vector:

$$[a_N, a_{N+1}, \ldots, a_{2N-1}, a_0, a_1, \ldots, a_{N-1}]^T. \tag{5.8}$$

This is a handy function because it leaves the coefficients in essentially the same order as they are required in the sum in (5.6).

For vectors $\mathbf{X}(f) = [f(x_0), \dots, f(x_{M-1})]^T$ of grid values of a function f, it is customary to use the norm defined by

$$||\mathbf{X}(f)||^2 = \Delta x \left(|f(x_0)|^2 + \cdots + |f(x_{M-1})|^2\right). \tag{5.9}$$

This differs from the usual norm $|\cdot|$ for vectors (Sect. 1.2) by the presence of the normalizing factor Δx, which is included to ensure that, for Δx small $||\mathbf{X}(f)||$ is an approximation to $||f||$ in (2.1). This normalizing factor affects the value of the norm of the vectors, but not the value of the norm of matrices $||A||$, because $||A||$ is defined (see Sect. 1.2) in terms of quotients of vector lengths. With the definition in (5.9) the following discrete version of Parseval's identity (2.7) holds

$$||\mathbf{X}(f)||^2 = L \sum_{n=-N}^{N}{}'' \, |\tilde{f}_n|^2. \tag{5.10}$$

5.2.4 The trigonometric form of the collocation polynomial

It is possible to write the interpolant $I_N(f)$ in trigonometric form (cf. the derivation of (2.9)). The result is

$$I_N(f) = \tilde{c}_0(f) + \sum_{n=1}^{N-1} \left(\tilde{c}_n(f) \cos \frac{2\pi n}{L} x + \tilde{s}_n(f) \sin \frac{2\pi n}{L} x \right)$$
$$+ \tilde{c}_N(f) \cos \frac{2\pi N}{L} x,$$

with

$$\tilde{c}_0(f) = \tilde{f}_0,$$
$$\tilde{c}_n(f) = \tilde{f}_n + \tilde{f}_{-n}, \quad n = 1, 2, \dots, N,$$
$$\tilde{s}_n(f) = i(\tilde{f}_n - \tilde{f}_{-n}), \quad n = 1, 2 \dots, N-1.$$

The last basis function $\cos(2\pi Nx/L)$ gives rise to the vector of grid values $[1, -1, 1, -1, \dots, -1]^T$ (saw-tooth behavior). Note that $\tilde{f}_N = \tilde{f}_{-N}$ results in $I_N(f)$ not including a term in $\sin((2\pi Nx)/L)$. The absentee $\sin(2\pi Nx/L)$ is, on the grid, an alias of the 0 function, so that it is just as well if it does not feature in $I_N(f)$. There are $N+1$ coefficients $\tilde{c}_n$ and only $N-1$ coefficients $\tilde{s}_n$; this makes in all $M = 2N$ coefficients, which is just right to interpolate at M grid points. If f is real-valued the coefficients $\tilde{c}_n(f)$ and $\tilde{s}_n(f)$ are real.

Exercise 25 Write a program that produces the graph in Fig. 8 with arbitrary values of M. Run your program for $M = 4, 8, 16, \dots$ and try to make sense of the plots you get.

Exercise 26 The saw-tooth vector $[1, -1, 1, -1, \dots, -1]^T$ is one of the columns of the Fourier matrix F_M, $M = 2N$. Why?

Exercise 27 Use (4.3) and (5.5)–(5.7) to prove (5.10).

5.3 More aliasing: discrete Fourier coefficients vs. Fourier coefficients

5.3.1 The relation between discrete Fourier coefficients and Fourier coefficients

There is an alternative way in which the interpolant (5.6) can be constructed. We begin with the Fourier series (2.4), that we rewrite in the form

$$f = \sum_{n=-N+1}^{N} \left(\sum_{m=-\infty}^{\infty} \hat{f}_{n+mM} \phi_{n+mM} \right), \tag{5.11}$$

i.e., we first sum the contributions involving

$$\ldots, \phi_{-N+1-M}, \phi_{-N+1}, \phi_{-N+1+M}, \ldots,$$

then the contributions involving

$$\ldots, \phi_{-N-M}, \phi_{-N}, \phi_{-N+M}, \ldots,$$

etc. The grid values of the right-hand side of (5.11) will not change if we replace ϕ_{n+mM} by its alias ϕ_n. Therefore

$$\sum_{n=-N}^{N}{}'' \left(\sum_{m=-\infty}^{\infty} \hat{f}_{n+mM} \right) \phi_n(x) \tag{5.12}$$

is, provided that the M series $\sum_m \hat{f}_{n+mM}$ converge, an interpolant of f on the grid. (In (5.12) we have achieved symmetry by dividing $\sum_m \hat{f}_{N+mM}$ into two halves and attaching one of the halves to ϕ_N and the other to its alias ϕ_{-N}.) By uniqueness of the interpolant, (5.6) and (5.12) must coincide. This leads to the following formula (Canuto et al. 1988; Hamming 1978) relating the Fourier coefficients $\hat{f}_n$ of f to the discrete Fourier coefficients $\tilde{f}_n$

$$\tilde{f}_n = \sum_{m=-\infty}^{\infty} \hat{f}_{n+mM}, \quad n = 0, \pm 1, \ldots, \pm N. \tag{5.13}$$

5.3.2 Truncation vs. interpolation

It is now expedient to compare the truncation $P_N(f)$ of the Fourier series of f (see (2.6)) with the interpolant $I_N(f)$.

Both $P_N(f)$ and $I_N(f)$ are trigonometric polynomials of the form (2.8) (but $I_N(f)$ has only $2N$ degrees of freedom because $\tilde{f}_N = \tilde{f}_{-N}$). The truncation $P_N(f)$ is characterized by the property that the residual $f - P_N(f)$ is small in the sense that it is orthogonal to the $2N+1$ functions ϕ_n, $n = 0, \pm 1, \ldots, \pm N$. The interpolant $I_N(f)$ is characterized by the property that the residual $f - I_N(f)$ is small in the sense that it vanishes at $2N$ grid points.

The truncation $P_N(f)$ has coefficients $\hat{f}_{-N}, \ldots, \hat{f}_N$ that are independent of the coefficients $\hat{f}_n$ with $|n| > N$, i.e., independent of the modes in f with $|n| > N$. On the other hand, the Fourier coefficients $\hat{f}_n$ with $|n| > N$ do contribute to the interpolant via (5.13). The mode $\hat{f}_n \phi_n(x)$ of the Fourier series 'feeds' the alias mode involving the basis function $\phi_{n'}$ for which $n \equiv n'$.

We noted already that it is easy to find the coefficients $\tilde{f}_n$. On the other hand, the coefficients $\hat{f}_n$ are defined by the integrals (2.5), which in practice should be evaluated numerically (see Exercise 28).

As we discussed in Sect. 2.3, the smoothness of f governs the decay of the $\hat{f}_n$ and hence the velocity of the convergence of $P_N(f)$ to f. The formulae (5.13) may be used (Tadmor 1986) to show that in a like manner, smoother functions have discrete Fourier coefficients that decay faster. Also, the smoother f is the faster the convergence of $I_N(f)$ to f.

5.3.3 The sampling theorem

For functions f that satisfy $\hat{f}_n = 0$ for $|n| \geq N$, it is true that $I_N(f) = f = P_N(f)$. Such functions, being equal to $I_N(f)$ can be reconstructed from its stroboscopic or grid samples $\mathbf{X}(f)$ through (5.5)–(5.7). When written in trigonometric form, such functions only possess harmonics with frequencies n/L below the upper bound N/L. Now, since $\Delta x = L/(2N)$, the frequency upper bound N/L equals $1/(2\Delta x)$; this is called (Hamming 1973) the *Nyquist frequency*. Correspondingly, the periods of the harmonics of f have a *lower* bound $2\Delta x$. The spacing Δx between two consecutive samples must be smaller than half of the smallest period of the harmonics of f, if f is to be reconstructed from its grid values. The function $\sin(2\pi Nx/L)$ whose period is exactly $2\Delta x$ cannot be distinguished on the grid from the 0 function.

Exercise 28 Show that $\tilde{f}_n$, $n = 0, \pm 1, \ldots, \pm N$, is a linear combination of grid values of f and that this linear combination can be seen as a numerical approximation to the integral (2.5) defining $\hat{f}_n$.

Exercise 29 Use (5.13) to compute, for arbitrary $M = 2N$ the discrete Fourier coefficients of the function in Exercise 8. Show that these coefficients decay exponentially as a function of $|n|$, at a rate which does not depend on M.

Exercise 30 Set $M = 4$ and compute the discrete Fourier coefficients of the function in Exercise 8 via (5.5), (5.7). Do you get the same results you found in Exercise 29?

6. Fourier Analysis of Finite-Difference Algorithms: the Time-Continuous Case

6.1 Spatial discretizations of initial value problems

6.1.1 The discretization

We now study the numerical solution of the periodic problem (3.1)–(3.2). For the sake of clarity, it is not advisable to look at the 'general' equation (3.1) and we base our presentation on a model: the advection equation with velocity $c = 1$ (see Sect. 3.3)

$$\partial_t u = -\partial_x u. \tag{6.1}$$

In this section we look at semidiscrete (discrete x, continuous t) approximations to (6.1). In a finite-difference approach, the variable x is discretized as in Section 5.1 and we look for approximations $U_n(t)$ to the solution values $u(x_n, t)$, $n = 0, \pm 1, \pm 2, \ldots, t \geq 0$. By periodicity $U_n(t) = U_m(t)$ whenever $n \equiv m$ and there are really M unknowns $U_n(t)$. These are collected into a vector $\mathbf{U}(t)$. The operator ∂_x is replaced by a suitable finite difference formula, for instance (Mitchell and Griffiths 1980)

$$\partial_x u(x_n, t) \approx \frac{u(x_{n+1}, t) - u(x_n, t)}{\Delta x},$$

(forward differences), or alternatively

$$\partial_x u(x_n, t) \approx \frac{u(x_n, t) - u(x_{n-1}, t)}{\Delta x},$$

(backward differences), or

$$\partial_x u(x_n, t) \approx \frac{u(x_{n+1}, t) - u(x_{n-1}, t)}{2\Delta x}$$

(central differences).

The numerical approximations are then asked to satisfy

$$\frac{d}{dt} U_n(t) = -\frac{U_{n+1}(t) - U_n(t)}{\Delta x}, \quad n = 0, 1, \ldots, M-1,$$

or

$$\frac{d}{dt} U_n(t) = -\frac{U_n(t) - U_{n-1}(t)}{\Delta x}, \quad n = 0, 1, \ldots, M-1,$$

or

$$\frac{d}{dt} U_n(t) = -\frac{U_{n+1}(t) - U_{n-1}(t)}{2\Delta x}, \quad n = 0, 1, \ldots, M-1.$$

Thus $\mathbf{U}(t)$ is asked to satisfy the system of differential equations

$$\frac{d}{dt}\mathbf{U}(t) = A\mathbf{U}(t), \quad t \geq 0, \tag{6.2}$$

where A takes one of the forms

$$A = \begin{bmatrix} 1/\Delta x & -1/\Delta x & 0 & \dots & 0 \\ 0 & 1/\Delta x & -1/\Delta x & \dots & 0 \\ 0 & 0 & 1/\Delta x & \dots & 0 \\ \vdots & \vdots & \vdots & \ddots & \vdots \\ -1/\Delta x & 0 & 0 & \dots & 1/\Delta x \end{bmatrix}, \qquad (6.3a)$$

$$A = \begin{bmatrix} -1/\Delta x & 0 & 0 & \dots & 1/\Delta x \\ 1/\Delta x & -1/\Delta x & 0 & \dots & 0 \\ 0 & 1/\Delta x & -1/\Delta x & \dots & 0 \\ \vdots & \vdots & \vdots & \ddots & \vdots \\ 0 & 0 & 0 & \dots & -1/\Delta x \end{bmatrix}, \qquad (6.3b)$$

$$A = \begin{bmatrix} 0 & -1/(2\Delta x) & 0 & \dots & 1/(2\Delta x) \\ 1/(2\Delta x) & 0 & -1/(2\Delta x) & \dots & 0 \\ 0 & 1/(2\Delta x) & 0 & \dots & 0 \\ \vdots & \vdots & \vdots & \ddots & \vdots \\ -1/(2\Delta x) & 0 & 0 & \dots & 0 \end{bmatrix}. \qquad (6.3c)$$

More sophisticated choices of A are of course possible (Exercise 31).

The matrices in (6.3) and other matrices arising from discretization of (6.1) or more generally other partial differential equations of the form (3.1) have some important features. They are *sparse:* even though they contain $O(M^2)$ entries, the number of nonzero entries is only $O(M)$. This makes a matrix vector product much cheaper than in the general case. Furthermore the matrices are *circulant* (Sect. 4.2). This is a consequence of the periodicity of the boundary conditions and (3.1) having constant coefficients.

The system (6.2) needs an initial condition $\mathbf{U}(0)$, which is usually taken to be the vector $\mathbf{X}(u^0)$ obtained by restriction of the initial condition u^0 in (3.2), i.e.,

$$\mathbf{U}(0) = \mathbf{X}(u^0). \qquad (6.4)$$

In practice one may integrate (6.2), (6.4) with a numerical solver for ordinary differential equations.

The discretization of the spatial variables may alternatively be carried out by finite elements (Strang and Fix 1973). While in cases with several space variables finite elements are more versatile in dealing with the geometry of the problem, in one space dimension there is not much difference between finite elements and finite differences. Finite elements would also lead to a system of the form (6.2); in fact one may think (Mitchell and Griffiths 1980) of the finite element method as of a method to generate good finite difference schemes. In what follows we only deal with finite differences, in the understanding that most of the material can be extended to finite elements.

6.1.2 Consistency

Our aim is now to estimate the error perpetrated by taking $\mathbf{U}(t)$ as an approximation to $u(x,t)$. More precisely we want to measure the size of the vector $\mathbf{E}(t) = \mathbf{U}(t) - \mathbf{X}(u(\cdot,t))$ called the *global error*. This size is defined by (see (5.9))

$$||\mathbf{E}|| = \sqrt{\sum_{n=0}^{M-1} \Delta x |U_n(t) - u(x_n,t)|^2}. \tag{6.5}$$

In order to bound (6.5) an indirect approach is followed, which uses the ideas of *consistency* and *stability* (Richtmyer and Morton 1976; Sanz-Serna 1991).

Consistency refers to the relation between the partial differential equation *problem* being solved and the numerical *problem* (6.2). This is not to be confused with the question we want solved, namely with the relation between the *solutions* of the partial differential equation problem and (6.2).

To study consistency we substitute in (6.2) the vector $\mathbf{X}(u(\cdot,t))$ of the grid values of the true solution u; this gives rise to a residual

$$\mathbf{L}(t) = \frac{d}{dt}\mathbf{X}(u(\cdot,t)) - A\mathbf{X}(u(\cdot,t)) \tag{6.6}$$

called the *truncation* or *local* error.

With the choice (6.3a), the n-th component of $\mathbf{L}(t)$ is given by

$$L_n(t) = \partial_t u(x_n,t) + \frac{u(x_{n+1},t) - u(x_n,t)}{\Delta x}, \tag{6.7}$$

or, after Taylor expanding,

$$L_n(t) = \partial_t u(x_n,t) + \frac{1}{\Delta x}\left[\Delta x \partial_x u(x_n,t) + \frac{\Delta x^2}{2}\partial_{xx} u(x_n,t) + \cdots\right].$$

Since u satisfies (6.1),

$$L_n(t) = \frac{\Delta x}{2}\partial_{xx} u(x_n,t) + \cdots = O(\Delta x), \quad \Delta x \to 0.$$

The definition of $||\cdot||$ (see Exercise 27 and (6.5)) involves summation of $M = L/\Delta x$ terms, but also includes a Δx factor. Therefore $O(\Delta x)$ entries imply $O(\Delta x)$ norms and

$$||\mathbf{L}(t)|| = O(\Delta x), \quad \Delta x \to 0. \tag{6.8}$$

The conclusion is that, with the choice (6.3a), the problem (6.2) is an approximation to the true problem (6.1), in the sense that solutions u of (6.1) satisfy (6.2) except for an $O(\Delta x)$ remainder. The finite-difference approximation (6.2) is then said to be consistent. Since only the first power of Δx appears in (6.8), we say that (6.2)–(6.3a) is consistent *of the first order.*

A similar argument reveals that (6.2) with (6.3b) is also consistent of the first order. For (6.3c),

$$||\mathbf{L}(t)|| = O(\Delta x^2), \quad \Delta x \to 0,$$

this is consistency of the second order.

6.1.3 Stability

We say that (6.2) is *stable,* if, for each finite interval $0 \leq t \leq T$, a constant $C(T)$ can be found such that, for all solutions $\mathbf{U}(t)$ of (6.2),

$$||\mathbf{U}(t)|| \leq C(T)||\mathbf{U}(0)||, \quad 0 \leq t \leq T; \tag{6.9}$$

small initial conditions lead to small solutions. What is important here is that $C(T)$ has to be independent, not only of the initial condition $\mathbf{U}(0)$, but also of the parameter Δx. Independence of Δx is a delicate business: as $\Delta x \to 0$, the entries of the matrix A in (6.2) blow up and one would expect that $\mathbf{U}(t)$ may also grow.

There are two things I would like to point out:

(i) The requirement (6.9) is a discrete analogue of the well-posedness estimate (3.8). In fact (3.8) implies

$$||u(\cdot,t)|| \leq C(T)||u(\cdot,0)||, \quad 0 \leq t \leq T,$$

with $C(T) = \exp(\alpha(P(\partial_x))T)$.

(ii) Since $\mathbf{U}(t) = \exp(tA)\mathbf{U}(0)$, (6.9) is equivalent to

$$||e^{tA}|| \leq C(T), \quad 0 \leq t \leq T. \tag{6.10}$$

For the choice (6.3c), (6.2) is stable. In fact, since A is skew-symmetric, $\exp(tA)$ is unitary,

$$(e^{tA})^* = e^{tA^*} = e^{-tA} = (e^{tA})^{-1},$$

(Sect. 1.2) so that (6.10) holds with $C(T) = 1$. The stability for the choices (6.3a), (6.3b) is discussed later.

6.1.4 Convergence

Subtraction of (6.6) from (6.2) leads to the following system of differential equations for the global error we wish to estimate

$$\frac{d}{dt}\mathbf{E}(t) = A\mathbf{E}(t) - \mathbf{L}(t). \tag{6.11}$$

This is similar to the system (6.2) satisfied by the numerical solution itself; the difference is that (6.2) is homogeneous while (6.11) has the truncation error $\mathbf{L}(t)$ as a forcing term.

The solution of (6.11) may be written via the variation of constants formula (Strang 1986)

$$\mathbf{E}(t) = e^{tA}\mathbf{E}(0) - \int_0^t e^{(t-s)A}\mathbf{L}(s)\,ds. \tag{6.12}$$

Thus $\mathbf{E}(t)$ is the sum of two contributions. The first $\exp(tA)\mathbf{E}(0)$ represents the time evolution of the initial global error $\mathbf{E}(0)$; if $\mathbf{U}(0)$ is taken as in (6.4), this first contribution vanishes. The second contribution is a superposition of terms $-\exp((t-s)A)\mathbf{L}(s)$; each of the terms being superposed is the effect at time t of the forcing term $-\mathbf{L}(s)$ acting as an initial condition at time s (Duhamel's principle).

Assume that the numerical method (6.2) is stable and consistent of order p, then, using (6.10) in (6.12),

$$\begin{aligned}||\mathbf{E}|| &\leq ||e^{tA}||||\mathbf{E}(0)|| + \int_0^t ||e^{(t-s)A}||||\mathbf{L}(s)||\, ds \\ &\leq C(T)||\mathbf{E}(0)|| + TC(T)O(\Delta x^p), \quad 0 \leq t \leq T,\end{aligned}$$

and, with the standard choice (6.4) for $\mathbf{U}(0)$,

$$||\mathbf{E}|| \leq TC(T)O(\Delta x^p) = O(\Delta x^p), \quad 0 \leq t \leq T;$$

the global errors decay as $O(\Delta x^p)$ as the grid is refined. This is called *convergence of order p*. To sum up, we have just proved that *stability and consistency of order p yield convergence of order p*. Conversely, both stability and consistency are necessary for convergence; this is what the celebrated Lax equivalence theorem says (Richtmyer and Morton 1967; Sanz-Serna and Palencia 1985).

There is a subtle point we should not avoid (Richtmyer and Morton 1967; Sanz-Serna 1985). The investigation of consistency (see Sect. 6.1.2) is based on Taylor expansions which may only be carried out if the true solution u is smooth enough. (More precisely if u has continuous derivatives up to a given order depending on the specific finite-difference scheme being investigated; for instance the derivation of (6.8) required the existence and continuity of the functions $\partial_t u$, $\partial_{xx} u$.) Hence we have really proved convergence of order p only for smooth solutions. What is the situation for nonsmooth solutions? With the square-wave function in Exercise 3 as an initial condition and the forward difference matrix A in (6.3a), the truncation error $L_n(t)$ in (6.7) equals $-2/\Delta x$ when $x_n = \pi$ and $t = 0$; the finer the grid the worse the approximation! Nevertheless it can be shown that, for a scheme that is stable and consistent of order p, convergence holds for *all* solutions, regardless of their smoothness. Smooth solutions have $||\mathbf{E}(t)|| = O(\Delta x^p)$, other solutions have $||\mathbf{E}(t)|| = O(\Delta x^q)$ with $q < p$; the exact value of q depends on the exact smoothness of u. In the square-wave example convergence is $O(\Delta x^{1/4})$ (Sanz-Serna, 1985).

Exercise 31 For the problem (6.1) write a finite-difference formula of the form

$$\begin{aligned}&\partial_x u(x_n, t) \approx \\ &\quad \alpha_2 u(x_{n+2}, t) + \alpha_1 u(x_{n+1}, t) + \alpha_0 u(x_n, t) + \alpha_{-1} u(x_{n-1}, t) + \alpha_{-2} u(x_{n-2}, t),\end{aligned}$$

with the α_m chosen to achieve the highest possible order of consistency. Analyze the stability and convergence of the resulting scheme.

Exercise 32 For the heat equation construct difference schemes of the form (6.2) with orders of consistency 2, 4 or 6.

6.2 The Von Neumann stability analysis

We have not yet discussed the stability of (6.2) when A is given by (6.3a) or (6.3b). This stability will now be investigated by Fourier analysis.

With the choice (6.3a) or (6.3b), the matrix A in (6.2) is, as any other circulant matrix (see Sect. 4.2), normal (Section 1.2). Therefore $\|\exp(tA)\| = \exp(t\alpha(A))$, with $\alpha(A)$ the spectral abscissa of A (Section 1.2). Then the stability condition (6.10) holds if and only if

$$\sup_{\Delta x} \alpha(A) < \infty. \tag{6.13}$$

This is called the Von Neumann condition and is a direct analogue of the well posedness condition $\alpha < \infty$ in Sect. 3.2 but here there is an extra parameter Δx.

The eigenvalues of the most general circulant matrix were found in (4.6) by Fourier analysis. For (6.3b) (backward differences) the eigenvalues are

$$\sigma_n = -\frac{1}{\Delta x} + \frac{1}{\Delta x}\exp\left(\frac{2\pi(M-1)ni}{M}\right), \quad n = 0, 1, \ldots, M-1;$$

or, by periodicity,

$$\sigma_n = -\frac{1}{\Delta x} + \frac{1}{\Delta x}\exp\left(-\frac{2\pi ni}{M}\right), \quad n = 0, 1, \ldots, M-1; \tag{6.14}$$

this leads to $\alpha(A) \leq 0$ and therefore to stability with stability constant $C(T) = 1$.

On the other hand, for (6.3a) (forward differences) the eigenvalues are

$$\sigma_n = \frac{1}{\Delta x} - \frac{1}{\Delta x}\exp\left(\frac{2\pi ni}{M}\right), \quad n = 0, 1, \ldots, M-1; \tag{6.15}$$

thus $2/\Delta x$ is an eigenvalue ($n = N = M/2$) and $\alpha(A) \geq 2/\Delta x$. As the grid is refined for fixed t, $\|\exp(tA)\| \geq \exp(2/\Delta x)$ grows exponentially and we have instability. By the Lax equivalence theorem the scheme (6.2)–(6.3a) is not convergent, in spite of the fact that it is a 'reasonable' discretization of (6.1). I would like to emphasize that it is the growth of $\exp(tA)$ for fixed t as $\Delta x \to 0$ which prevents stability and convergence. This growth is not to be confused with growth for fixed Δx and $t \to \infty$, see Exercise 34.

Let us summarize. We use Fourier analysis to find the eigenvalues σ_n of the system (6.2) we are investigating. The eigenvalues of $\exp(tA)$ are then $\exp(t\sigma_n)$ and, *due to the normality of A*, the norm of $\exp(tA)$, which we want bounded, coincides with the eigenvalue $\exp(t\sigma_n)$ with maximum modulus. This arises from the σ with maximum real part.

Exercise 33 Use the Von Neumann method to investigate the stability of the difference schemes constructed in Exercises 31 and 32.

Exercise 34 Consider the equation $\partial_t u = \partial_x u + u$. Study the well-posedness of the corresponding periodic initial value problem (cf. Exercise 16). Discretize this problem by central differences and analyze the stability and convergence of the resulting discretization. Note that $\|\exp(tA)\|$ grows with t and, nevertheless, the numerical method is stable.

Exercise 35 Discretize the system (3.7) by the finite difference scheme

$$\frac{d}{dt}V_n(t) = c\frac{W_n(t) - W_{n-1}(t)}{\Delta x}, \quad n = 0, 1, \ldots, M-1,$$
$$\frac{d}{dt}W_n(t) = c\frac{V_{n+1}(t) - V_n(t)}{\Delta x}, \quad n = 0, 1, \ldots, M-1.$$

Introducing the vector of unknowns $\mathbf{U} = [V_0, W_0, \ldots, V_{M-1}, W_{M-1}]^T$, write the difference equations in the format (6.2), with A of dimension $2M$ (M is, as always, the number of grid points). Is the matrix A you obtain a circulant matrix? The answer should be no. Use Fourier analysis to find the eigenvalues of the matrix A you have found. The result should be $\pm(2ci/\Delta x)\sin(\pi n\Delta x/L)$, $n = 0, 1, \ldots, M-1$ (Hint: Assume that the eigenvectors have $V_n = A\exp((2\pi ni\Delta x)/L)$ and $W_n = B\exp((2\pi ni\Delta x)/L)$ and substitute in $A\mathbf{U} = \sigma\mathbf{U}$.) In general, for systems of partial differential equations the matrix is *not* circulant, but is such that its eigenvalues can be explicitly found by Fourier analysis. The condition that the real part of the eigenvalues of A should be bounded above is still necessary for stability, because for any matrix $\|\exp(tA)\| \geq \rho(\exp(tA)) = \exp(t\alpha(A))$, $t > 0$. However the study of the eigenvalues of A is not in general sufficient for stability, because if A is not normal the norm $\|\exp(tA)\|$ may be strictly larger than the spectral radius $\rho(\exp(tA)) = \exp(t\alpha(A))$. For this reason for linear, constant coefficient *systems* of partial differential equations with periodic boundary conditions, the Von Neumann condition (6.13) is necessary but not sufficient for stability (Richtmyer and Morton 1967).

6.3 The roles of stability and consistency from a Fourier viewpoint

The Von Neumann stability test provides the easiest application of Fourier methods to the numerical analysis of initial value problems. There are other applications certainly worth studying. For instance, it is useful to gain insight, via Fourier analysis, into the equivalence between convergence and consistency plus stability.

We still look at the advection equation (6.1) solved by any of the methods in (6.2)–(6.3). The theoretical solution was found in (3.6) and is given by

$$u(x,t) = \sum_{n=-\infty}^{\infty} \exp(\mu_n t)\hat{u}_n^0\phi_n(x), \quad \mu_n = -2\pi ni/L.$$

We evaluate u at grid points in order to make it possible a comparison with the numerical solution $\mathbf{U}(t)$. The result is

$$\mathbf{X}(u(\cdot,t)) = \sum_{n=-\infty}^{\infty} \exp(\mu_n t)\hat{u}_n^0 \mathbf{X}(\phi_n), \quad \mu_n = -2\pi n i/L. \tag{6.16}$$

We could now use the aliasing relations (5.2) to replace each $\mathbf{X}(\phi_n)$ by an $\mathbf{X}(\phi_m)$, with $-N \le m \le N$; we performed such a replacement when we wrote formula (5.11). However we prefer to leave (6.16) as it stands.

We next write the numerical solution $\mathbf{U}(t)$ in a format similar to (6.16). Before we do this, let me go back to the system (4.4) with *general* circulant matrix. One easily checks that, for $n = 0, \pm 1, \pm 2, \ldots$, the vector $\mathbf{X}(\phi_n)$ in (5.1) is an eigenvector of the matrix in (4.4); the corresponding eigenvalue is

$$\sigma_n = a_0 + a_1 \exp\left(\frac{2\pi n i}{M}\right) + \cdots + a_{M-1} \exp\left(\frac{2\pi(M-1)ni}{M}\right). \tag{6.17}$$

Does this mean that the $M \times M$ matrix A possesses infinitely many eigenvectors/values? Certainly it does not. If $n \equiv m$, then $\mathbf{X}(\phi_n) = \mathbf{X}(\phi_m)$, so that we have only found M distinct eigenvectors. Indeed, upon recalling (5.3), we see that the $\mathbf{X}(\phi_n)$ are scaled versions of the columns $\mathbf{V}_n$ of F_M^{-1} we resorted to in Sect. 4.2. Correspondingly, in (6.17) $\sigma_n = \sigma_m$ whenever $n \equiv m$, because $\exp(2\pi \ell n i/M)$ is an M-periodic function of n. Comparison with (4.6) gives us the reassuring conclusion that we have again found the M eigenvectors/functions we had in Sect. 3.2. The difference is that now we let n to be arbitrary in (6.17), while in (4.6) n was between 0 and $M-1$. The solution of (4.4) with initial condition $\mathbf{X}(u^0)$ is then

$$\sum_{n=-\infty}^{\infty} \exp(\sigma_n t)\hat{u}_n^0 \mathbf{X}(\phi_n). \tag{6.18}$$

This certainly has the correct value at time $t = 0$; furthermore, it satisfies (4.4) as one easily checks by substitution in the differential system. We could use of the aliasing relations (5.2) to rewrite (6.18) as a sum with only M terms; the result would be the solution (4.11) we found in Sect. 4.2.

It is time to leave the general problem (4.4) and return to our finite-difference method (6.2)–(6.3). The numerical solution with initial condition (6.4) is given by (6.18) with the σ_n equal to the eigenvalues of the matrix A corresponding to the specific choice of finite-difference method; for instance for the central-difference method (6.3c), the eigenvalues (6.17) are readily found to be

$$\sigma_n = -\frac{i}{\Delta x} \sin\left(\frac{2\pi n}{M}\right) = -\frac{i}{\Delta x} \sin\left(\frac{2\pi n \Delta x}{L}\right). \tag{6.19}$$

For backward and forward differences the eigenvalues were found in (6.14) and (6.15).

Subtraction of (6.16) from (6.18) provides the expression for the global error

$$\mathbf{E}(t) = \sum_{n=-\infty}^{\infty} \left[\exp(\sigma_n t) - \exp(\mu_n t)\right] \hat{u}_n^0 \mathbf{X}(\phi_n). \tag{6.20}$$

This is the representation we wish to discuss. Let us first fix a value of n and look at the corresponding term in the series (6.20). We have an exponent μ_n coming from the theoretical solution and an exponent σ_n from the numerical solution. To be specific, assume that we use central differences with σ_n given by (6.19). As the grid is refined ($\Delta x \to 0$), Taylor expansion in (6.19) yields

$$\begin{aligned}\sigma_n &= -\frac{i}{\Delta x}\left(\frac{2\pi n\Delta x}{L}\right) + \frac{1}{6}\frac{i}{\Delta x}\left(\frac{2\pi n \Delta x}{L}\right)^3 - \cdots \\ &= \mu_n + \frac{4\pi^3 n^3 i}{3L^3}\Delta x^2 + \cdots,\end{aligned}$$

so that σ_n approaches μ_n. A fixed mode becomes better and better approximated as $\Delta x \to 0$. This is the Fourier analysis expression of the consistency of the method. In fact, above, $\sigma_n = \mu_n + O(\Delta x^2)$ because (6.3c) leads to consistency of the second order. For forward or backward differences $\sigma_n = \mu_n + O(\Delta x)$. This is the good news: consistency guarantees that all is well as $\Delta x \to 0$ with n fixed.

The bad news is that, on any fixed grid you may be using, there are numbers n for which σ_n and μ_n are grossly different. This is made clear in Fig. 10, that should be studied carefully. While the figure refers to (6.2), (6.3c), a similar discussion holds for the choices (6.3a) or (6.3b).

If, for each Δx, there are terms in the series (6.20) for which $\exp(\sigma_n t) - \exp(\mu_n t)$ is large, how is it possible to get convergence, i.e. small $\mathbf{E}(t)$? It is the $\hat{u}_n^0$ that come to the rescue: they must decay as $|n| \to \infty$, because the series in (2.7) converges under the only assumption that u^0 is in $\mathcal{L}^2[0, L]$. Furthermore, we noticed in Sect. 2.3, that, the smoother the initial datum u^0, the faster the $\hat{u}_n^0$ decay. It is this decay that is implicit at the heart of convergence and explains why the rate of convergence decreases for nonsmooth solutions. The mathematical details are as follows (Richtmyer and Morton 1967). Assume, that, at a given time $t > 0$, we want to make the norm of $\mathbf{E}(t)$ less than a given small quantity via a suitable choice of Δx. One begins by finding an index ν for which

$$\sum_{|n|>\nu} [\hat{u}_n^0]^2 \tag{6.21}$$

is small. This is possible because of the convergence of the series in (2.7). Once this ν is known, we take Δx small enough to ensure that

$$\sum_{|n|\leq\nu} [\exp(\sigma_n t) - \exp(\mu_n t)]\, \hat{u}_n^0 \mathbf{X}(\phi_n)$$

is small; this is possible for a consistent scheme because the sum involves a finite number of modes, and for each mode σ_n approaches μ_n as the grid is refined. This leaves us with the remaining terms

$$\sum_{|n|>\nu} \left[e^{\sigma_n t} - e^{\mu_n t}\right] \hat{u}_n^0 \mathbf{X}(\phi_n). \tag{6.22}$$

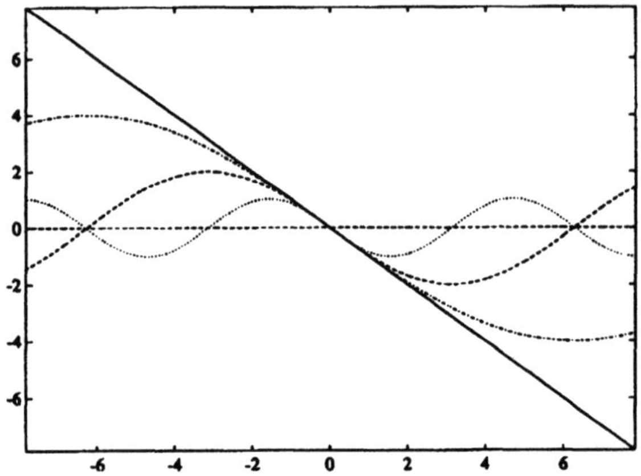

Fig. 10. The horizontal axis corresponds to the wave number $2\pi n/L$. The solid line gives the imaginary part of the exponent μ_n in the theoretical solution (6.16). The sinusoids give the imaginary part of of the exponent σ_n in the central difference numerical solution (6.18)–(6.19). The dotted line corresponds to $\Delta x = 1$, the dashed line has $\Delta x = 1/2$ and the dash-dotted line has $\Delta x = 1/4$. Let us first look at a fixed location in the horizontal axis (i.e., a fixed wave number): the numerical σ_n approach, as $\Delta x \to 0$, the theoretical μ_n. This is a reflection of the consistency of the scheme. Let us then look, for fixed Δx, at σ_n as a function of the wave number. The numerical σ_n is close to μ_n only for those wave numbers $2\pi|n|/L$ that are small relatively to $(\Delta x)^{-1}$. In other words, only wave lengths that are large relatively to the grid spacing Δx are well approximated on any given grid. For instance, when the wave length L/n is $12\Delta x$, one has $\sigma_n = -i/(2\Delta x)$ and $\mu_n = -i\pi/(6\Delta x)$; since $\pi/6 = 0.52$, this means a relative error of less than 5%. When the wave length is only $4\Delta x$, $\mu_n = -i\pi/(2\Delta x)$ and $\sigma_n = -i\Delta x$ with relative error of about 50%. For the Nyquist wave number given by $\pi/\Delta x$ (saw tooth mode) the approximation is completely wrong because $\sigma_n = 0$. Note (see the dotted line) that σ_n as a function of the wave number is a periodic function that repeats itself after the Nyquist value. This periodicity is a consequence of aliasing. Smaller Δx lead to larger Nyquist wave numbers; the grid supports more essentially different modes

Here the $\hat{u}_n^0$ are small thanks to (6.21), and the term $\exp(\mu_n t)$ is bounded by the well-posedness of the theoretical problem. Therefore (6.22) will be small, provided that $\exp(\sigma_n t)$ is under control. Now the control of the size of $\exp(\sigma_n t)$ coincides, as we discussed in the previous section, with the issue of the stability of the method.

Let me summarize. Consistent finite difference schemes approximate badly the small wave length modes. Any reasonable initial condition is relatively poor in small wave length modes and the numerical method will perform satisfactorily if the small wave lengths, that are being misrepresented, are controlled and remain small. This control corresponds to stability.

Exercise 36 Draw a figure similar to Fig. 10 for the method constructed in Exercise 31. Compare with Fig. 10. Observe the advantages of the high-order method (Kreiss and Oliger 1973; Fornberg 1975, 1987, 1990).

6.4 Numerical dissipation and numerical dispersion

Let us again consider the central difference solution (6.18)–(6.19) that we write in trigonometric form as (see Sect. 3.3)

$$A_0\mathbf{X}(1) + \sum_{n=1}^{\infty} A_n\mathbf{X}\left(\cos\left[\frac{2\pi n}{L}x - \frac{1}{\Delta x}\sin\left(\frac{2\pi n\Delta x}{L}\right)t - \psi_n\right]\right).$$

With the terminology of Sect. 3.3, the numerical solution to the dispersionless problem (6.1) thus turns out to be a dispersive wave with dispersion relation

$$\omega_n = \frac{1}{\Delta x}\sin(\kappa_n \Delta x), \quad \kappa_n = \frac{2\pi n}{L};$$

in the limit $\Delta x \to 0$ one has $\omega_n = \kappa_n$, which is the correct relation for the partial differential equation (6.1). For any (nonzero) Δx, no matter how small, there is a dependence on κ_n of the phase speed ω_n/κ_n. For instance, small wave numbers κ_n travel with velocities close to the correct value 1, while the wave number $\kappa_n = \pi/\Delta x$ (the Nyquist value corresponding to the sawtooth mode) stands still at 0 phase velocity. As a result of the spurious dispersion introduced by the process of discretization, the shape of the numerical solution will change with t (cf. Fig. 5).

Spurious, numerically induced dispersion is not the only problem discretization may bring. Spurious numerical dissipation is also a common occurrence, see Exercise 37. The study of the properties of dissipation and dispersion of numerical schemes is a key ingredient in the analysis of the schemes and is related to several theories from physics (Vichnevetsky 1987a, 1987b, 1989, 1990, 1992; Trefethen 1982, 1983).

Exercise 37 Write in trigonometric form the solution of the backward difference method (6.2), (6.3b) and notice the spurious dissipative behavior.

7. Fourier Analysis of Finite-Difference Algorithms: the Fully-Discrete Case

7.1 Full discretization of initial value problems

7.1.1 Time discretization

In Sect. 6 the variable t remained continuous. We now study fully-discrete numerical methods where t is also discretized (the term 'fully' indicates that all independent variables become discrete). We denote by Δt the time increment and consider the grid values $t^m = m\Delta t$, $m = 0, 1, 2, \dots$

In order not to blur the exposition, I shall still concentrate on the model equation (6.1); however the material has wider applicability. A fully-discrete finite-difference scheme for (6.1) may easily be obtained by replacing $\partial_t u$ and $\partial_x u$ by suitable increment quotients. It is however more advisable not to discretize t and x simultaneously. We proceed in two stages. First we discretize x and obtain, say, one of the methods in (6.2)–(6.3). Once the system of differential equations (6.2) is available, we discretize its independent variable t. The discretization of t then takes place in the context of a system of ordinary differential system rather than of a partial differential equation (Sanz-Serna and Verwer 1989). Any of the standard numerical methods (Lambert 1991; Hairer and Wanner 1991, Hairer et al. 1993, Sanz-Serna and Calvo 1994) for systems of the form

$$\frac{d}{dt}\mathbf{Y}(t) = \mathbf{F}(t, \mathbf{Y}(t)), \tag{7.1}$$

is in principle eligible. Some well-known one-step possibilities are the *Euler rule*

$$\mathbf{Y}^{m+1} = \mathbf{Y}^m + \Delta t\mathbf{F}(t^m, \mathbf{Y}^m), \quad m = 0, 1, 2, \dots, \tag{7.2}$$

($\mathbf{Y}^m$ is the numerical approximation to $\mathbf{Y}(t^m)$), the *backward Euler rule*

$$\mathbf{Y}^{m+1} = \mathbf{Y}^m + \Delta t\mathbf{F}(t^{m+1}, \mathbf{Y}^{m+1}), \quad m = 0, 1, 2, \dots, \tag{7.3}$$

and the *trapezoidal rule*

$$\mathbf{Y}^{m+1} = \mathbf{Y}^m + \frac{\Delta t}{2}\mathbf{F}(t^{m+1}, \mathbf{Y}^{m+1}) + \frac{\Delta t}{2}\mathbf{F}(t^m, \mathbf{Y}^m), \quad m = 0, 1, 2, \dots. \tag{7.4}$$

Higher-order methods include the celebrated (but obsolete) classical Runge-Kutta fourth-order method.

The Euler method is *explicit;* it provides a formula for finding $\mathbf{Y}^{m+1}$ once the preceding $\mathbf{Y}^m$ is available. The backward Euler and trapezoidal rules are *implicit:* to find each $\mathbf{Y}^{m+1}$ one has to solve a system of algebraic equations, which may be a costly task. *Multistep methods* where $\mathbf{Y}^{m+1}$ is linked to $\mathbf{Y}^m$, $\mathbf{Y}^{m-1}$, etc. are possible and may be very efficient; I am sorry they cannot be considered here.

All reasonable one-step methods (including (7.2)–(7.4)) have the property that, when the system (7.1) takes the simple linear form

$$\frac{d}{dt}\mathbf{Y}(t) = B\mathbf{Y}(t), \tag{7.5}$$

B a constant matrix, the approximation $\mathbf{Y}^{m+1}$ may be obtained from $\mathbf{Y}^m$ by matrix multiplication. More precisely

$$\mathbf{Y}^{m+1} = R(\Delta t B)\mathbf{Y}^m, \quad m = 0, 1, 2, \ldots, \tag{7.6}$$

where $R(z)$ is a rational function (see Sect. 1.2) that depends on the method but not on the specific problem of the form (7.5). For instance, $R(z) = 1 + z$ for the Euler rule, while the backward Euler rule has $R(z) = 1/(1 - z)$ and for the trapezoidal rule $R(z) = (1 + z/2)/(1 - z/2)$. Thus with each one-step method one associates a rational function; it turns out that many theoretical properties of the method may be studied by looking at the corresponding $R(z)$. In practice, (7.6) is implemented by solving the linear system of equations $P(\Delta t A)\mathbf{Y}^{m+1} = Q(\Delta t A)\mathbf{Y}^m$, where P and Q denote the numerator and denominator of R. When solving this system of equations the sparsity of A is important.

The formula (7.6) should be compared with the corresponding expression for the true solution

$$\mathbf{Y}(t^{m+1}) = \exp(\Delta t B)\mathbf{Y}(t^m), \quad m = 0, 1, 2, \ldots.$$

Such a comparison reveals that $R(z)$ should approximate $\exp(z)$; for instance with the Euler rule $R(z) = 1 + z$ consists of the first two terms of the expansion of $\exp(z)$ in powers of z. In general for a method of order p, $R(z)$ differs from $\exp(z)$ in terms of order $O(z^{p+1})$.

7.1.2 Fully discrete methods

As described above, the fully discrete scheme is obtained by time discretization of the time continuous problem (6.2). The fully discrete solution $\mathbf{U}^m$ at time t^m is a vector $[U_0^m, U_1{}^m, \ldots, U_{M-1}^m]^T$, where U_n^m is an approximation to $u(x_n, t^m)$. According to (7.6), the vectors $\mathbf{U}^m$ are recursively found from the formula

$$\mathbf{U}^{m+1} = R(\Delta t A)\mathbf{U}^m, \quad m = 0, 1, 2, \ldots, \tag{7.7}$$

where A is the matrix in (6.2) and $R(z)$ the rational function of the specific time-stepping method being employed. The initial $\mathbf{U}^0$ is usually taken as in (6.4).

As an illustration we present the method corresponding to using the Euler rule (7.2) along with backward differences in space (see(6.3b)). When written componentwise, the formulae (7.7) become

$$U_n^{m+1} = U_n^m - \Delta t \frac{U_n^m - U_{n-1}^m}{\Delta x}, \quad n = 0, 1, \ldots, M-1, \quad m = 0, 1, 2, \ldots. \tag{7.8}$$

7.1.3 Consistency, stability and convergence in the fully discrete case

The analysis of methods of the form (7.7) is also based on the ideas of consistency, stability and convergence. The *truncation errors* $\mathbf{L}^m$ are again defined by substituting the grid values of the theoretical solution in (7.7)

$$\Delta t \mathbf{L}_{m+1} = \mathbf{X}(u(\cdot, t^{m+1})) - R(\Delta t A)\mathbf{X}(u(\cdot, t^m)). \tag{7.9}$$

Note the normalizing factor Δt in the left-hand side. This is introduced because the format (7.7) does not directly approximate the partial differential equation (6.1), it rather approximates Δt times (6.1). (Look at the example in (7.8): one would have to divide by Δt before having a discrete analogue to (6.1).) Consistency of order p in space and q in time means that $||\mathbf{L}^m||$ behaves as $O(\Delta x^p + \Delta t^q)$ upon grid refinement. The discretization (7.8) has $p = q = 1$. Some authors define the truncation error to be the right-hand side of (7.9); for those authors the truncation error of (6.1) is $O(\Delta x \Delta t + \Delta t^2)$.

The discretization (7.7) is said to be *stable* (cf. (6.9)) if, for each finite time interval $0 \leq t \leq T$, a constant $C(T)$ can be found such that for all solutions of (7.7)

$$||\mathbf{U}^m|| \leq C(T)||\mathbf{U}^0||, \quad 0 \leq t^m \leq T.$$

Here it is crucial that $C(T)$ should not depend on Δx and Δt. From (7.7), $\mathbf{U}^m = R(\Delta t A)^m \mathbf{U}^0$, so that the scheme is stable if and only if (cf. (6.10))

$$||R(\Delta t A)^m|| \leq C(T), \quad 0 \leq t^m \leq T. \tag{7.10}$$

The global errors are now given by $\mathbf{U}^m - \mathbf{X}(u(\cdot, t^m))$ and a have a representation similar to (6.12), namely

$$\mathbf{E}^m = R(\Delta t A)^m \mathbf{E}^0 - \Delta t \sum_{\ell=1}^{m} R(\Delta t A)^{m-\ell} \mathbf{L}^\ell. \tag{7.11}$$

From this formula it is easily concluded, as in Sect. 6.1.4, that a stable scheme with consistency of order p in space and order q in time has, for smooth solutions, global errors $||\mathbf{E}^m||$ that behave as $O(\Delta x^p + \Delta t^q)$.

Exercise 38 Write componentwise the nine schemes resulting from combining (6.3a)–(6.3b) with (7.2)–(7.4). Study the consistency.

Exercise 39 Prove (7.11).

7.2 The Von Neumann stability analysis

The stability condition (7.10) may be investigated by Fourier analysis. If A is normal, then $R(\Delta t A)^m$ is also normal and (Sect. 1.2)

$$\|R(\Delta t A)^m\| = \rho(R(\Delta t A)^m) = \rho(R(\Delta t A))^m;$$

for stability this should be bounded by $C(T)$ whenever $0 \leq t^m \leq T$. It is not difficult to show (Richtmyer and Morton 1967) that this is equivalent to the existence of a constant $C'(T)$, independent of Δx and Δt such that for all Δx and Δt

$$\rho(R(\Delta t A)) \leq 1 + C'(T)\Delta t.$$

Upon recalling that the eigenvalues of $R(\Delta t A)$ are given by $R(\Delta t \sigma)$ with σ eigenvalue of A, we conclude that stability is equivalent to

$$|R(\Delta t \sigma_n)| \leq 1 + C'(T)\Delta t, \tag{7.12}$$

as σ_n runs through all the eigenvalues of A. These were found by Fourier analysis in Sect. 6.2. The condition (7.12) is the Von Neumann condition for fully discrete schemes.

Take (7.8) as an example and assume that the space and time grids are refined so as to keep the mesh ratio $r = \Delta t/\Delta x$ a constant. Here $R(z) = 1 + z$ and the eigenvalues of A are given in (6.14). Therefore the stability requirement reads

$$\left|1 + r\left(\exp\left(-\frac{2\pi n i}{M}\right) - 1\right)\right| \leq 1 + C'(T)\Delta t, \quad n = 0, 1, \ldots, M-1;$$

since the left-hand side is independent of Δt this condition can only hold if

$$\left|1 + r\left(\exp\left(-\frac{2\pi i}{M}\right) - 1\right)\right| \leq 1, \quad n = 0, 1, \ldots, M-1. \tag{7.13}$$

It is easy to check that (7.13) is fulfilled if and only if $r \leq 1$. Therefore the scheme is stable and convergent only if the ratio r is kept below 1; this behavior is called *conditional stability*.

Exercise 40 Study the stability of the nine schemes introduced in Exercise 38.

7.3 The roles of stability and consistency from a Fourier viewpoint

The solution $\mathbf{U}^m$ given by the method (7.7) can be written as

$$\mathbf{U}^m = \sum_{n=-\infty}^{\infty} R(\Delta t \sigma_n)^m \hat{u}_n^0 \mathbf{X}(\phi_n), \quad m = 0, 1, 2, \ldots \tag{7.14}$$

This is a fully discrete analogue of (6.18). Subtraction from (6.16) (evaluated at $t = t^m$) leads to the following expression for the global error

$$\mathbf{E}^m = \sum_{n=-\infty}^{\infty} [R(\Delta t\sigma_n)^m - \exp(\Delta t\mu_n)^m]\hat{u}_n^0 \mathbf{X}(\phi_n), \quad m = 0, 1, 2, \ldots.$$

This is a fully discrete version of (6.20) and may be analyzed as in Sect. 6.3.

For a consistent method, any fixed mode of the theoretical solution can be well approximated by choosing Δx and Δt suitably small. If Δx is small, then, as we know, σ_n is close to μ_n. If Δt is also small (relatively to $1/|\mu_n|$) then $R(\Delta t\sigma_n)$ is close to $\exp(\Delta t\mu_n)$ because $R(z)$ approximates $\exp(z)$ for $|z|$ small. Once more, the bad news is that for any given values of Δx and Δt there are values of n for which $R(\Delta t\sigma_n)$ is a very poor approximation to $\exp(\Delta t\mu_n)$. For this reason consistency is not sufficient to guarantee convergence: one needs some control on the short wave length components (stability). As illustrated by the example in Sect. 7.2 stability may impose an upper bound on the value of Δt that can be used on a given spatial mesh. It is often the case that this upper bound forces Δt to be smaller (or even much smaller) than the value one would like to use for consistency reasons, i.e., the value that would make $\Delta t|\mu_n|$ small enough for the modes that are significantly present in the theoretical solution. This situation where stability rather than the natural time-scale of the theoretical solution dictates the choice of time step is called stiffness and unfortunately is a common occurrence in numerical differential equations (Hairer and Wanner 1991; Dekker and Verwer 1984). It is stiffness that makes implicit time stepping methods appealing for partial differential equations, explicit schemes at best lead to conditional stability (Sanz-Serna and Verwer 1989).

Exercise 41 Prove that (7.14) is indeed the solution of (7.7) subject to (6.4).

7.4 Numerical dissipation and numerical dispersion

Let us particularize (7.14) to the scheme considered in Sect. 7.2 (backward differences in space and Euler time-stepping, $r = \Delta t/\Delta x$ a constant). The result is

$$\mathbf{U}^m = \sum_{n=-\infty}^{n} \{1 + r\,[\exp(-\kappa_n i\Delta x) - 1]\}^m\, \hat{u}_m^0 \mathbf{X}(\phi_n), \quad \kappa_n = 2\pi n/L. \tag{7.15}$$

We now wish to write the quantity in curly brackets as the exponential of its logarithm. As $\Delta x \to 0$ the Taylor expansion of this logarithm is

$$\begin{aligned}\log\{1 + r\,[\exp(-\kappa_n i\Delta x) - 1]\} &= \log\left\{1 - r\kappa_n i\Delta x - \frac{1}{2}r\kappa_n^2\Delta x^2 + \cdots\right\}\\ &= -r\kappa_n i\Delta x - \frac{1}{2}r(1-r)\kappa_n^2\Delta x^2 + \cdots\\ &= \Delta t\left(-\kappa_n i - \frac{1}{2}(1-r)\kappa_n^2\Delta x + \cdots\right).\end{aligned}$$

Thus, (7.15) may be rewritten as

$$\mathbf{U}^m = \sum_{n=-\infty}^{n} \exp(-\kappa_n i t^m) \exp\left(-\frac{1}{2}(1-r)\kappa_n^2 \Delta x t^m + \cdots\right) \hat{u}_m^0 \mathbf{X}(\phi_n).$$

Here the first exponential provides the correct advection with velocity 1; the second exponential arises from the error in the numerical method. For $r < 1$, the behavior of the leading, $O(\Delta x)$, term is *dissipative* (see Sect. 3.3). The dissipation coefficient is proportional to Δx and, as expected, finer spatial grids induce less spurious dissipation. Note also that the dissipation coefficient decreases as r approaches 1: unexpectedly, for this numerical scheme, longer time steps on a fixed spatial grid create less dissipation, see Figs. 11–12. For $r > 1$ we know from Sect. 7.2 that the scheme is unstable and hence useless; here we see that the numerical solution behaves as the solution to an ill posed backward heat equation. The case $r = 1$ is exceptional: in the formula above the $O(\Delta x)$ terms in the global error disappear. Indeed, for $r = 1$ all error terms disappear and the scheme provides the exact theoretical solution. This is possible because we are dealing with a very simple partial differential equation: the theoretical solution is constant along the characteristic lines $x = t + \xi$, and for $r = 1$ so is the numerical solution because, then from (7.8), $U_n^{m+1} = U_{n-1}^m$.

As we may see numerical methods have a dynamics of their own, which may be quite different from that of the equation being integrated (Sanz-Serna 1992). Modified equations are a popular way of analyzing spurious dissipation and dispersion (Warming and Hyett 1974; Griffiths and Sanz-Serna 1986).

Exercise 42 Repeat the analysis in this section for the scheme for (6.1) based on central differences and trapezoidal time-stepping. Are the leading terms in the error dissipative or dispersive? Check your analysis by running a program.

8. The Practical Relevance of Fourier Analysis of Difference Methods

In Sects. 6–7 we have seen how Fourier techniques provide a powerful means for analyzing the stability and other properties of difference methods. The insights derived by Fourier analysis are essential when understanding difference methods. This insight, by itself, justifies the study of Fourier analysis.

The fly in the ointment is that, *strictly speaking,* Fourier analysis is only applicable under very restrictive conditions: periodic boundary conditions, linear equations, constant coefficients. This limitation in scope is particularly unfortunate because the difference numerical methods being analyzed have themselves no limitation in their range of applicability. Indeed one of the prime advantages of difference methods is their versatility: it is easy to write down a difference scheme for virtually any problem one may encounter.

In spite of these comments, Fourier techniques *are used in practice* in the analysis of linear problems with variable coefficients, of nonlinear problems and

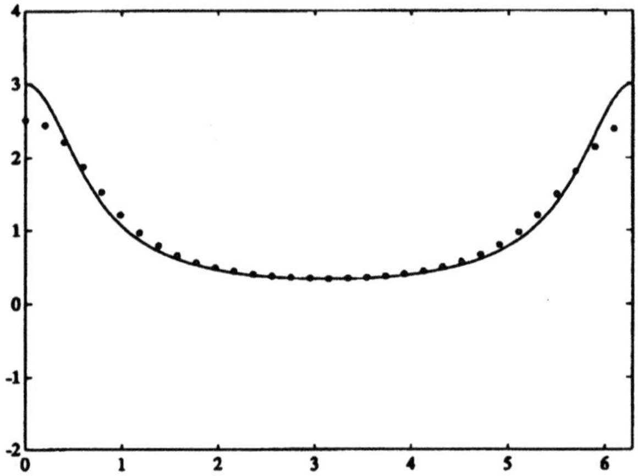

Fig. 11. The solution at time $t = 2\pi$ of the advection equation $\partial_t u = -\partial_x u$, on $0 \leq x \leq L = 2\pi$ for the initial datum in Exercise 8. The true solution is the solid line; this coincides with the initial condition because in 2π units of time the $u(\cdot, t)$ has moved 2π units to the right. The numerical method (7.8) is applied with $M = 32$ (so that $\Delta x = \pi/16$) and $\Delta t = 8\Delta x/9$ (so that 36 time steps have been needed). The dots represent the numerical solution; clearly the scheme is dissipative

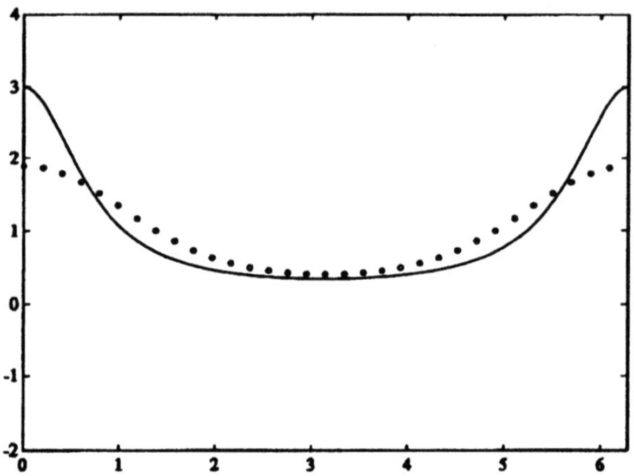

Fig. 12. The experiment in Fig. 11 has been repeated with the only change that now $\Delta t = \Delta x/4$. With this smaller time increment 128 time steps have to be taken to reach the final time $t = 2\pi$. In spite of the extra work, the result is more dissipative than that in Fig. 11

of problems whose boundary conditions are not periodic. For a linear problem with variable coefficients (say $\partial_t u = a(x,t)\partial_x u + b(x,t)u$) one 'freezes' the coefficients at a fixed, representative value ($a_0 = a(x_0, t_0)$, $b_0 = b(x_0, t_0)$) and analizes the resulting constant coefficient problem ($\partial_t u = a_0 \partial_x u + b_0 u$). For a nonlinear problem one linearizes around a representative particular solution (for instance, for $\partial_t u = u\partial_x + \partial_{xx} u$, linearizing around $u = 0$ implies discarding the quadratic term $u\partial_x u$ to get $\partial_t u = \partial_{xx} u$). When the boundary conditions are not periodic, one changes during the analysis the boundary conditions into periodic conditions. In general (but there are some exceptions), such analyses are not mathematically justified. They nevertheless provide useful rule-of-thumb indications of the behavior of the scheme being employed. As expected, the predictions based on Fourier analysis tend to be optimistic. A Fourier analysis stability limit, such as $\Delta t \leq (1/2)\Delta x^2$, is likely to be too generous; the real scheme will have a smaller stability limit due to nonlinear effects, boundary conditions etc. that are ignored in the analysis.

9. Spectral Methods

9.1 Spectral methods for periodic linear constant coefficient problems

9.1.1 The Galerkin approach

For periodic, constant-coefficient, linear problems (3.1), the solution is available in closed form (3.6). Why do we then need difference methods for these problems? Well, we do not really need them; it is better to use (3.6) directly. It is then ironic that Fourier analysis of difference methods is mathematically justified precisely in those cases where the difference method is not really useful!

The only difficulty with (3.6) is that it comprises infinitely many terms; too many for a practical solution. We have to truncate somewhere and then the numerical solution is defined to be

$$u_G^N(x,t) = \sum_{n=-N}^{N} \exp(\mu_n t)\hat{u}_n^0 \phi_n(x). \tag{9.1}$$

In u_G^N the superscript N indicates how many modes are being kept and G means Galerkin. Galerkin methods are projection methods and (9.1) is a projection method because

$$u_G^N(\cdot,t) = P_N(u(\cdot,t)); \tag{9.2}$$

at each time t the numerical solution is the orthogonal projection (see Exercise 2) of the theoretical solution onto the space of trigonometric polynomials of degree $\leq N$. Hence the error equals $u(\cdot,t) - P_N(u(\cdot,t))$, which we know (Sect. 2.3) converges quickly to 0 if u is smooth. How quickly depends on the exact smoothness; *the error may even be exponentially small as* $N \to \infty$.

To effectively construct (9.1) we need the Fourier coefficients $\hat{u}_n^0$, with $|n| \leq N$; these would have to be calculated by numerical evaluation of the integrals (2.5).

The method (9.1) is said to be a *spectral Galerkin method.* It uses the explicit knowledge of the spectrum of eigenvalues of the operator $P(\partial_x)$ in the equation (3.1) being solved.

9.1.2 The collocation approach

One could avoid the computation of the $\hat{u}_n^0$ in (9.1) if these were replaced by the discrete Fourier coefficients $\tilde{u}_n^0$ of u^0 relative to a grid with $M = 2N$ points. The result would be an *alternative* numerical *spectral* approximation given by

$$u_\psi^N(x,t) = \sum_{n=-N}^{N}{}'' \exp(\mu_n t)\tilde{u}_n^0 \phi_n(x). \tag{9.3}$$

The subscript ψ means *pseudospectral* and indeed (9.3) is called a pseudospectral solution or, alternatively, a spectral collocation solution. Now the $\tilde{u}_n^0$ are found by discrete Fourier transform as explained in Sect. 5.2.2. The cost of forming (9.3) is then $O(M \log_2 M)$ operations, which is competitive with finite difference methods. Without the fast transform, (9.3) would require $O(M^2)$ operations, which is not really so appealing.

As we discussed, (9.1) is based on projections: one projects u^0 to find the required coefficients and the method gives back the projection of the theoretical solution, see (9.2). In the pseudospectral solution (9.3) one interpolates u^0 on the grid. However the method does not yield at time t the interpolant of $u(\cdot, t)$. That would be too much; it would mean that the method was actually exact at the grid points. To clarify this, recall that, by (5.13),

$$\tilde{u}_n^0 = \sum_{m=-\infty}^{\infty} \hat{u}_{n+mM}^0$$

and hence

$$u_\psi^N(x,t) = \sum_{n=-\infty}^{\infty} E_n(t)\hat{u}_n^0 \phi(x)$$

where

$$E_n(t) = \exp(\mu_m t),$$

if $n \equiv m$, $|m| < N$ (i.e., m is the mode below the Nyquist limit into which the mode n becomes aliased), and

$$E_n(t) = \frac{1}{2}\exp(\mu_N t) + \frac{1}{2}\exp(\mu_{-N} t),$$

if $n \equiv N$ (i.e., n is aliased into the Nyquist limit).

Comparing with the theoretical solution (3.6), we see that the source of error in (9.3) is the following. High wave numbers are aliased when interpolating u^0 and later are evolved with exponentials $\exp(\mu_m t)$ that really correspond to the modes into which they have been aliased. High wave numbers are kept in the numerical solution but falsified. On the Galerkin solution they are just suppressed, which is in a way another form of falsification.

In any case, both the Galerkin and pseudospectral solution approximate without error all the modes representable in the grid (cf. Fig. 10). With these methods the errors depend on the high Fourier coefficients $\hat{u}_n^0$, $|n| > N$. For this reason, the accuracy of u_G or u_ψ depends only on the decay of the $\hat{u}_n^0$ and therefore is only limited by the smoothness of u^0. The smoother u^0 the faster the convergence. If u^0 is very regular, the errors may be exponentially small as $N \to \infty$. For a stable time-continuous difference scheme of order of consistency $O(\Delta x^p)$, errors are never better than $O(\Delta x^p)$ for very smooth solutions and may be worse than that if u is not so smooth.

Exercise 43 Use (9.3) to solve the heat equation. Write a program that, for a given initial u^0, plots the numerical solution at any given time.

9.2 Pseudospectral difference matrices

9.2.1 First derivatives

The spectral methods introduced in Sect. 9.1 for periodic, constant coefficient, linear problems can be extended to periodic, variable coefficient, linear problems and even to periodic, nonlinear problems. This extension is presented in Sect. 9.3. We now make a small detour. We wish to study the following problem. We are given the grid values $\mathbf{X}(f)$ of an L-periodic function $f(x)$ and are asked to find approximations to the grid values $\mathbf{X}(f')$ of the derivative. We could of course use standard finite-difference formulae, such as $(f(x_{n+1}) - f(x_n))/\Delta x$, but such standard formulae do not take into account that we are dealing with periodic functions. A better recipe is: (i) interpolate the given grid values as in (5.6), (ii) differentiate the interpolant I_N to find, in view of (2.12),

$$I_N(f)' = \sum_{n=-N}^{N}{}'' \lambda_n \tilde{f}_n \phi_n(x),$$

and (iii) evaluate the derivative at the grid points. The end result is a vector $\mathbf{X}'(f)$ of approximations to the values of f' at grid points (not to be confused with the vector of exact derivative values $\mathbf{X}(f')$)

$$\mathbf{X}'(f) = \sum_{n=-(N-1)}^{N-1} \lambda_n \tilde{f}_n \mathbf{X}(\phi_n). \tag{9.4}$$

Note that the terms with $|n| = N$ do not feature in (9.4). Why? Recall from (5.6) that $\tilde{f}_{-N} = \tilde{f}_N$. On the other hand, $\lambda_{-N} = -\lambda_N$ and then the $|n| = N$ terms in $I_N(f)'$ equal $\lambda_N \tilde{f}_N[\phi_N(x) - \phi_{-N}(x)]$, i.e., $2i\lambda_N \tilde{f}_N \sin(2\pi Nx/L)$; the function $\sin(2\pi Nx/L)$ has zero grid values and therefore the $|n| = N$ terms do not contribute to (9.4).

The process of finding $\mathbf{X}'(f)$ as a function of the data $\mathbf{X}(f)$ is linear and therefore there exists an $M \times M$ matrix D such that

$$\mathbf{X}'(f) = D\mathbf{X}(f).$$

This matrix is called the *pseudospectral differentiation matrix*. It is a real matrix: for real $\mathbf{X}(f)$, (9.4) contains the complex conjugate of each of its terms.

The following representation is most important

$$D = \left(P\frac{1}{M}F_M\right)^{-1} \Lambda \left(P\frac{1}{M}F_M\right). \tag{9.5}$$

Here P is the $M \times M$ permutation matrix

$$P = \begin{bmatrix} O_N & I_N \\ I_N & O_N \end{bmatrix}$$

and Λ is an $M \times M$ diagonal matrix

$$\Lambda = \mathrm{Diag}\,(0, \lambda_{-N+1}, \lambda_{-N+2}, \ldots, \lambda_{-1}, \lambda_0, \lambda_1, \ldots, \lambda_{N-2}, \lambda_{N-1}).$$

In (9.5), D is written as the product of three factors. The rightmost factor comprises the matrix $(1/M)F_M$ that finds the discrete Fourier coefficients as in (5.5) and the permutation matrix P that rearranges them as in (5.8). The central matrix Λ multiplies each discrete Fourier coefficient by the corresponding eigenvalue λ_n. Note that Λ includes a 0 entry that suppresses the $|n| = N$ contribution. The leftmost factor in the right-hand side of (9.5) undoes the action of the rightmost factor, i.e., computes grid values given discrete Fourier coefficients.

The Fourier matrix F_M has all its entries $\neq 0$ and, as a consequence, the matrix D is not a sparse matrix: D is full. When using finite differences, differentiation is performed through a sparse matrix, see e.g. (6.3). For standard central differences (6.3c), the differentiation matrix has two nonzero entries per row. For fourth-order differencing (Exercise 31), sixth-order differencing, ... the number of nonzero entries per row is four, six, ...; higher order implies less sparsity. The pseudospectral matrix can be seen (Fornberg 1975, 1987, 1990) as the full limit of difference matrices of increasing order of accuracy and decreasing sparsity.

Since P is its own inverse, (9.5) may be rewritten as

$$D = F_M^{-1} P \Lambda P F_M. \tag{9.6}$$

To find $D\mathbf{X}$ for a given vector $\mathbf{X}$ requires (i) an FFT, (ii) a permutation (in MATLAB, this is achieved by the function **fftshift**, see Sect. 5.2.2), (iii) M complex multiplications by the diagonal entries of Λ, (iiii) another permutation and (v) an inverse FFT transform. This involves a computational cost that is essentially $O(M)$. I emphasize that it is not advisable to find explicitly D by multiplying the five matrices in (9.6): if one uses the explicit form of D and carries out the multiplication $D\mathbf{X}$ by the standard matrix-times-vector recipe, then the work is $O(M^2)$.

9.2.2 Higher derivatives

The idea in Sect. 9.2.1 is readily generalized to higher derivatives. To obtain approximations to the grid values of f'', we again differentiate the interpolant to get

$$I_N(f)'' = \sum_{n=-N}^{N}{}'' \lambda_n^2 \tilde{f}_n \phi_n(x),$$

and then evaluate at grid points

$$\mathbf{X}''(f) = \sum_{n=-N}^{N}{}'' \lambda_n^2 \tilde{f}_n \mathbf{X}(\phi_n). \tag{9.7}$$

Now the $|n| = N$ terms do not disappear because $\lambda_{-N}^2 = \lambda_N^2$. In matrix form

$$\mathbf{X}''(f) = D^{(2)}\mathbf{X}(f),$$

with

$$D^{(2)} = F_M^{-1} P \Lambda^{(2)} P F_M,$$

where, in turn,

$$\Lambda^{(2)} = \operatorname{Diag}\left(\lambda_{-N}^2, \lambda_{-N+1}^2, \lambda_{-N+2}, \dots, \lambda_{-1}^2, \lambda_0^2, \lambda_1^2, \dots, \lambda_{N-2}^2, \lambda_{N-1}^2\right).$$

A small point: $\Lambda^{(2)}$ is not the square of Λ; the $(1,1)$ entry of Λ^2 is zero. For this reason

$$D^2 = \left(F_M^{-1} P \Lambda P F\right)\left(F_M^{-1} P \Lambda P F_M\right) = F_M^{-1} P \Lambda^2 P F_M$$

does not coincide with $D^{(2)}$. If you first interpolate and then differentiate twice you get a contribution with $|n| = N$ (in the trigonometric representation of the interpolant as in Sect. 5.2.4, the second derivative of $\cos(2\pi N x/L)$ is a multiple of $\cos(2\pi N x/L)$ and stays). If you interpolate, differentiate once, evaluate, and again interpolate and differentiate, the contribution with $|n| = N$ perishes in the first differentiation and never raises from the dead. Nevertheless, the difference between D^2 and $D^{(2)}$ is in practice very small: these two matrices only differ in the way they treat the highest discrete Fourier mode.

Higher derivative matrices $D^{(k)}$ can be constructed in a similar way. Alternatively these may be replaced by powers D^k of the first-derivative matrix D. Independently of the value of k the multiplication $D^{(k)}\mathbf{X}$ (or $D^k\mathbf{X}$), requires a direct and an inverse FFT plus M multiplications.

Exercise 44 Prove that D is a skew-symmetric matrix. Prove that D is a circulant matrix.

Exercise 45 Use (9.6) to find explicitly D when $M = 4$. Check that, for the matrix D_4 you have found, $D_4\mathbf{X}(f)$ provides the correct nodal values of the derivative of f when f is one of the functions 1, $\cos(2\pi x/L)$, $\sin(2\pi x/L)$,

$\cos(4\pi x/L)$. What is the situation when $f(x) = \sin(4\pi x/L)$? Find $D_4\mathbf{X}(f)$ for $f(x) = \cos(2\pi n x/L)$ or $f(x) = \sin(2\pi n x/L)$ and explain the results in terms of aliasing. Compare the entries in D_4 with those of the 4×4 matrix that arises from standard central differences (this is the negative of the matrix in (6.3c)).

9.3 The pseudospectral method for periodic nonlinear problems

Both the Galerkin and pseudospectral methods presented in Sect. 9.1 can be extended outside the class of periodic, constant coefficient, linear problems (Canuto et al. 1988; Gottlieb and Orszag 1977). The extension is much easier for the pseudospectral case and we therefore only consider pseudospectral methods.

To see how this extension, first suggested by Kreiss and Oliger in 1972, works, let us first look at the example of the heat equation $\partial_t u = \partial_{xx} u$. The pseudospectral solution (9.3) is

$$u_\psi^N(x,t) = \sum_{n=-N}^{N}{}'' \exp(\lambda_n^2 t)\tilde{u}_n^0 \phi_n(x).$$

Denote by $\mathbf{U}(t)$ the M-vector of grid values of u_ψ^N at time t. Obviously

$$\mathbf{U}(t) = \sum_{n=-N}^{N}{}'' \exp(\lambda_n^2 t)\tilde{u}_n^0 \mathbf{X}(\phi_n) \tag{9.8}$$

and

$$\frac{d}{dt}\mathbf{U}(t) = \sum_{n=-N}^{N}{}'' \lambda_n^2 \exp(\lambda_n^2 t)\tilde{u}_n^0 \mathbf{X}(\phi_n). \tag{9.9}$$

Now, from (9.8), $\mathbf{U}(t)$ has discrete Fourier coefficients $\exp(\lambda_n^2 t)\tilde{u}_n^0$, so that comparing the right-hand side of (9.9) with (9.7) we see that this right-hand side is none other than $D^{(2)}\mathbf{U}(t)$. Therefore, (9.9) may be written in a simple form:

$$\frac{d}{dt}\mathbf{U}(t) = D^{(2)}\mathbf{U}(t).$$

This is very similar to a time-continuous finite difference scheme of the format (6.2); the only difference is that instead of a (sparse) finite-difference matrix we now have a (full) pseudospectral difference matrix. The pseudospectral matrix differentiates exactly all Fourier modes below the Nyquist limit. As the grid is refined more modes are differentiated without error.

This example gives the key to writing pseudospectral methods for problems that, *while being periodic,* have variable coefficients or are nonlinear. One considers as unknown a vector $\mathbf{U}(t)$ of grid values and one writes a system of differential equations

$$(d/dt\mathbf{U}(t)) = \mathbf{F}(t, \mathbf{U}(t)), \tag{9.10}$$

for $\mathbf{U}(t)$. The right-hand side function $\mathbf{F}$ is constructed from the partial differential equation being solved by replacing the derivative operators ∂_x^k by the matrices $D^{(k)}$ (or D^k). For instance, for the Korteweg-de Vries equation,

$$\partial_t u = -3\partial_x u^2 - \partial_{xxx} u,$$

a pseudospectral scheme (Frutos and Sanz-Serna 1992) would have $\mathbf{F}$ given by the nonlinear function

$$\mathbf{F}(\mathbf{V}) = -3D\mathbf{V}^2 - D^3\mathbf{V},$$

where the vector $\mathbf{V}^2$ is obtained by squaring the entries of $\mathbf{V}$.

In (9.10) the variable t is still continuous so that one has to integrate in time by some numerical method for ordinary differential equations (Sect. 7.1.1). An automatic package from a mathematical software library may be a sensible choice, but one also may consider some home-made algorithm. In any case one should be careful because (9.10) is both stiff and full. With any time integration method one may chose, one needs to be able to write a subroutine that evaluates the right-hand side function $\mathbf{F}$ at any given vector $\mathbf{V}$. In the Korteweg-de Vries example, the multiplications by D^3 and D are carried out by FFT as discussed in Sect. 9.2.1. The right-most F_M^{-1} implicit in D (see (9.6)) and in D^3 can be taken as a common factor. Then the evaluation of $\mathbf{F}$ requires two transforms and an inverse transform, plus M multiplications to find the entries of $\mathbf{V}^2$ and $2M$ additional multiplications by the diagonal entries of Λ and Λ^3.

In practice it may be better not to use the system (9.10) directly. One rather performs the time integration in terms of the transformed vector $\tilde{\mathbf{U}}(t) = PF_M\mathbf{U}(t)$, whose entries are, except for a normalizing factor M, the discrete Fourier coefficients of the solution. From (9.10), the transformed vector satisfies the differential system

$$\frac{d}{dt}\tilde{\mathbf{U}}(t) = \tilde{\mathbf{F}}(t, \tilde{\mathbf{U}}(t)),$$

with

$$\tilde{\mathbf{F}}(\tilde{\mathbf{V}}) = PF_M\mathbf{F}(F_M^{-1}P\tilde{\mathbf{V}}).$$

In the Korteweg-de Vries example, the new right-hand side function $\tilde{\mathbf{F}}$ is

$$\begin{aligned} PF_M\mathbf{F}(F_M^{-1}P\tilde{\mathbf{V}}) &= PF_M\left(-3D\mathbf{V}^2 - D^3\mathbf{V}\right), \\ &= -3\Lambda PF_M\mathbf{V}^2 - \Lambda^3\tilde{\mathbf{V}} \end{aligned}$$

where

$$\mathbf{V} = F_M^{-1}P\tilde{\mathbf{V}}. \tag{9.11}$$

Therefore when the system is written in terms of the transformed vector, an evaluation of the right-hand side function $\tilde{\mathbf{F}}$ demands an inverse Fourier transform (9.11) to find the nodal values $\mathbf{V}$ from discrete Fourier coefficients, M multiplications to find the entries of $\mathbf{V}^2$, a further discrete Fourier transform to find $F_M\mathbf{V}^2$ and the $2M$ multiplications by the diagonal entries of Λ and Λ^3. This saves a transform when compared with the evaluation of $\mathbf{F}$.

The main limitation of the Fourier pseudospectral approach is the restriction to periodic boundary conditions. Finite differences may cope successfully with any boundary condition. Pseudospectral methods based on polynomial rather than trigonometric basis functions of course exist (Canuto et al. 1988; Gottlieb and Orszag 1977) and can deal with nonperiodic boundary conditions.

The good news is that the rate of convergence of Fourier pseudospectral methods is only restricted by the smoothness of the solution u. For very smooth solutions errors are exponentially small as $\Delta x \to 0$. This is to be compared with the situation for finite differences where the error is never better than $O(\Delta x^p)$, with p determined by the specific scheme being used.

In my experience (Abia and Sanz-Serna 1990; Frutos et at. 1990, 1991; Frutos and Sanz-Serna 1989, 1992) the superiority of spectral methods over finite differences or finite elements is dramatic. Let me report an experiment (Abia and Sanz-Serna 1990). The equation being integrated is nonlinear and describes waves in a fluidized bed. A pseudospectral method, with accurate time integration, yielded an error of 6×10^{-3} when $M = 4$. The error is reduced down to 1×10^{-5} when M is doubled ($M = 8$). This is an error reduction by a factor 600! A second order finite-difference scheme had with $M = 32, 64, 128$ errors of 3×10^{-2}, 7×10^{-3}, 1×10^{-3}. Here the error is only divided by a factor of about 4 when M is doubled. Note also that finite differences with $M = 128$ (CPU time 13 seconds) are 100 times less accurate than the pseudospectral method with $M = 8$ (CPU time 7 seconds).

Exercise 46 Write a program for the pseudospectral method for the Korteweg-de Vries equation. Use the classical Runge-Kutta formula for the integration in time (Lambert 1991). Write a program based on finite differences and compare with the pseudospectral method.

I am thankful to J. de Frutos who helped me with the preparation of the manuscript. I have been supported by grant DGICYT PB92-254.

References

Abia, L. and Sanz-Serna, J.M. (1990): Computing **44** 187

Canuto, C., Hussaini, M.Y., Quarteroni, A. and Zang, T.A. (1988): *Spectral Methods in Fluid Dynamics* (Springer, New York)

Cooley, J. W. and Tukey, J. W. (1965): Math. Comput. **19** 297

Dekker, K. and Verwer, J.G. (1984): *Stability of Runge-Kutta Methods for Stiff Nonlinear Differential Equations* (North Holland, Amsterdam)

Fornberg, B. (1975): SIAM J. Numer. Anal. **12** 509

Fornberg, B. (1987): Geophysics **52** 483

Fornberg, B. (1990): SIAM J. Numer. Anal. **27** 904

Frutos, J. de, Ortega, T. and Sanz-Serna, J.M. (1990): Comput. Methods Appl. Mech. Eng. **80** 417

Frutos, J. de, Ortega, T. and Sanz-Serna, J.M. (1991): Math. Comput. **57** 109

Frutos, J. de and Sanz-Serna, J.M. (1989): J. Comput. Phys. **83** 407

Frutos, J. de and Sanz-Serna, J.M. (1992): J. Comput. Phys. **103** 160

Golub, G.H. and Van Loan, C.F. (1989): *Matrix Computations* (John Hopkins University Press, Baltimore)

Gottlieb, D. and Orszag, S.A. (1977): *Numerical Analysis of Spectral Methods: Theory and Applications* (SIAM-CBMS, Philadelphia)

Griffiths, D.F. and Sanz-Serna, J.M. (1986): SIAM J. Sci. Stat. Comput. **7** 994

Hairer, E., Nørsett, S.P. and Wanner, G. (1993): *Solving Ordinary Differential Equations I, Nonstiff Problems* (Springer, Berlin)

Hairer, E. and Wanner, G. (1991): *Solving Ordinary Differential Equations II, Stiff and Differential-Algebraic Problems* (Springer, Berlin)

Hamming, R.W. (1973): *Numerical Methods for Scientists and Engineers* (McGraw-Hill, New York)

Horn, R.A. and Johnson, C.R. (1985): *Matrix Analysis* (Cambridge University Press, Cambridge)

Kreiss, H. and Oliger, J. (1972): Tellus **24** 199

Kreiss, H. and Oliger, J. (1973): *Methods for the Approximate Solution of Time Dependent Problems* Global Atmospheric Research Program Publications Series No, 10 (World Meteorological Organization, Geneva)

Lambert, J.D. (1991): *Numerical Methods for Ordinary Differential Equations, The Initial Value Problem* (Wiley, Chichester)

Mitchell, A.R. and Griffiths, D.F. (1980): *The Finite Difference Method in Partial Differential Equations* (Wiley, Chichester)

Press W.H., Flannery, B.P., Teukolsky, S.A. and Vetterling, W.T. (1989): *Numerical Recipes, The Art of Scientific Computing* (Cambridge University Press, Cambridge)

Richtmyer R.D. and Morton, K.W. (1967): *Difference Methods for Initial-Value Problems* (Wiley-Interscience, New York)

Sanz-Serna, J.M. (1985): "Stability and Convergence in Numerical Analysis, a Simple, Comprehensive Account", in Nonlinear Differential Equations ed. by Hale, J.K. and Martínez-Amores, P. (Pitman, Boston) pp. 64–113

Sanz-Serna, J.M. (1991): "Two Topics in Nonlinear Stability", in Advances in Numerical Analysis, ed. by W. Light (Clarendon Press, Oxford), pp. 147-174.

Sanz-Serna, J.M. (1992): "Numerical Ordinary Differential Equations vs. Dynamical Systems", in The Dynamics of Numerics and the Numerics of Dynamics, ed. by Broomhead, D.S. and Iserles, A. (Clarendon Press, Oxford) pp. 81–106.

Sanz-Serna, J.M. and Calvo, M.P. (1994): *Numerical Hamiltonian Problems* (Chapman & Hall, London)

Sanz-Serna, J.M. and Palencia, C. (1985): Math. Comput. **45** 143

Sanz-Serna, J.M. and Verwer, J.G. (1989): Appl. Numer. Math. **5** 117

Strang, G. (1986): *Introduction to Applied Mathematics* (Wellesley-Cambridge Press, Wellesley)

Strang, G. and Fix, G.J. (1973): *An Analysis of the Finite Element Method* (Prentice-Hall, Englewood Cliffs)

Tadmor E. (1986): SIAM J. Numer. Anal. **23** 1

Trefethen, LL. N. (1982): SIAM Review **24** 113

Trefethen, LL. N. (1983): J. Comput. Phys. **49** 199

Vichnevetsky, R. (1987a): Appl. Numer. Math. **3** 133

Vichnevetsky, R. (1987b): Int. J. Numer. Methods Fluids **7** 409

Vichnevetsky, R. (1989): Int. J. Numer. Methods Fluids **9** 623

Vichnevetsky, R. (1990): Computers Math. Applics. **19** 59

Vichnevetsky, R. (1992): Appl. Numer. Math. **10** 195

Warming, R.F. and Hyett, B.J. (1974): J. Comput. Phys. **14** 159

Whitham, G.B. (1974): *Linear and Nonlinear Waves* (Wiley, New York)

A Fast Algorithm for the Generation of Random Numbers with Exponential and Normal Distributions

Julio F. Fernández[1,2] and Juan Rivero[2]

[1] Instituto Carlos I de Física Teórica y Computacional and Departamento de Física Aplicada, Universidad de Granada, 18071-Granada, Spain.
[2]Instituto Venezolano de Investigaciones Científicas, Apartado 21827, Caracas 1020A, Venezuela

Abstract: Algorithms for the generation of pseudorandom numbers with normal and exponential distributions are described here. No transcendental functions need to be evaluated; furthermore, only two uniform deviates per generation are required; no tables are used. These algorithms are much faster than other exponential and normal random number generators.

1. Introduction

Many simulations in computational physics require random numbers with a given distribution. Exponential and normal distributions are often needed. Existing random number generators are often inaccurate and they are time consuming. It has been shown recently that the inaccuracy issue can sometimes be reasonably serious [1]. Exponential and normally distributed random numbers are generated particularly slowly because one or more transcendental functions and/or several uniform deviates must be evaluated for each random number generated [2,3]. New algorithms for the generation of pseudorandom numbers with normal and exponential distributions are described in this lecture. No transcendental functions need to be evaluated; furthermore, only two uniform deviates per generation are required. Its accuracy is easy to control. No tables are used. It is much faster than other generators.

As part of an introduction for students who are unfamiliar with this subject, one of the simplest methods to generate exponentially distributed random numbers is explained next. Take a random number x (supplied, for instance by a built in generator in your computer), distributed uniformly in the interval $(0, 1)$, that is, $P(x) = 1$, for $0 < x < 1$, and $P(x) = 0$, for $x < 0$ and $x > 1$. There is a function $y(x)$ that transforms the *uniform deviate* x into the desired *exponential deviate* y; it must fulfill

$$P(x)|dx| = P(y)|dy| \tag{1.1}$$

where $P(y)$ is the desired exponential distribution (e^{-y}). Substitution of $P(x)$ and $P(y)$ into the above equation gives

$$\left| \frac{dx}{dy} \right| = e^{-y} \tag{1.2}$$

The solution to this equation is $x = c \pm e^{-y}$, where c is an arbitrary constant. Since we know that x is within $(0, 1)$, it follows that either

$$x = e^{-y} \tag{1.3}$$

or

$$x = 1 - e^{-y}. \tag{1.4}$$

The first of these equations gives

$$y = -\ln x \tag{1.5}$$

which is the desired transformation. (The other solution, $y = \ln(1 - x)$ entails an additional subtraction, and is therefore of little interest.) Unfortunately, this is a slow algorithm. It takes most computers about 10 times longer to compute a logarithm than it does to do simple arithmetic operations, such as sums, subtractions and multiplications.

Implementation of this idea for the generation of normal deviates turns out to be somewhat more complicated. Equation (1.2) gives

$$\left| \frac{dx}{dy} \right| = \frac{1}{\sqrt{2\pi}} e^{-y^2/2} \tag{1.6}$$

Unfortunately this expression integrates to an error function, which is not included in the list of functions given in most computer languages. This problem is avoided by the Box-Muller procedure, which is a generalization of the above method [3]. It requires the evaluation of a logarithm and a square root (it takes nearly as much computer time to compute a square root as a logarithm does).

There are other ways to generate exponential and normal deviates. We will not describe them here. We next list a few of them and state their main shortcomings. A method often used to generate exponentially distributed random numbers is *the comparison method* [4]. Its main drawback is that it requires, on the average, that one generate 4.3 uniform random deviates per exponential random number computed (it is therefore slow). One can also index a table of the desired distribution with a uniform random number. This method is not accurate, due to the necessarily coarse resolution of the table with respect to machine precision [2]. In Ref. [5], a method to generate random numbers by means of memory lookup is proposed and implemented in hardware; but this implementation is obviously not universally available.

In addition to the Box-Muller method (mentioned already), there exist several algorithms for the generation of normal deviates. The comparison method may be used [4]. As in the exponential case, it demands several uniform deviates per normal deviate. The sum method [3], which is based on the central limit theorem of probability theory, requires too many uniform deviates to achieve reasonable accuracy. Consequently, it is either inaccurate or very slow.

We now turn to the algorithms that we have designed for the generation of pseudorandom numbers with normal and exponential distributions. We first give the physical insight that motivated them. No attempt whatsoever to be general is made. The algorithms are then given. Their derivation is given in the section that follows. The performance of the algorithms (including CPU times) is examined in the last section.

2. Physical Motivation

There is a physical idea behind these algorithms (non-physicists will probably do best skipping right on to the next section). We take up the exponential distribution first. Consider a closed assembly of N simple classical harmonic oscillators that interact weakly among themselves. After the assembly reaches equilibrium (assuming that it does), the probability density that any one given system have energy ε fulfills $P(\varepsilon) \propto e^{-\beta\varepsilon}$, where β is a parameter clearly related to the total energy of the assembly. Simulation of such an assembly is the method we propose to generate random exponential deviates. It is up to us to postulate the interaction. We are guided by the following considerations: (a) the interaction must be simple to implement on a computer; (b) it need not have anything to do with any real physical system; (c) in order to achieve the Gibbs distribution, the interaction must however not violate the following rules: conservation of energy, reversibility, and ergodicity [6]. Accordingly, we use the following (non-deterministic) interacting scheme: select any two systems i and j; let their energies be ϵ_i and ϵ_j; we let $\epsilon_i \rightarrow \alpha E$ and $\epsilon_j \rightarrow (1-\alpha)E$, where α is a uniform random number within the (0,1) interval, and $E = \epsilon_i + \epsilon_j$. This process is repeated indefinitely. For simplicity, we may set $\epsilon_j = 1$ initially for all j. After each system has exchanged energy with some other system a few times, we expect the desired exponential distribution to ensue. After that, each time we let two systems interact, we may use either their ingoing or outgoing energies as a pair of random numbers that are exponentially distributed.

For the generation of random numbers that are distributed normally, consider a system of molecules in (for simplicity) one dimension. Everything is as above, except that we are not interested now in how the *energies* are distributed; we are interested in how the molecular *velocities* are distributed. However, energy is the quantity that must be conserved. We will be using the following interaction scheme: select any two systems i and j; think of their velocities, v_i and v_j, as the two components of a two dimensional vector, and rotate it by an random angle θ that is uniformly distributed within $(0, 2\pi)$. Obviously, $v_i^2 + v_j^2$ is conserved. In order to save computer time, a few random angles (say, θ_n for $n = 1, 2, \ldots, K$) should be selected at the outset and the corresponding sines and cosines should also be evaluated once and for all at the outset ($c_n = \cos\theta_n$ and $s_n = \sin\theta_n$); thereafter, a pair c_m, s_l should be chosen at random from the K such pairs available each time a rotation is to be performed. Again, this process is repeated indefinitely. For simplicity, we may set $v_j = 1$ initially for all j. After each system has exchanged energy with som other system a few times, we expect the desired normal (Maxwellian) velocity distribution to ensue. After that, each time we let two molecules interact, we may use either their ingoing or outgoing velocities as a pair of random numbers that are distributed normally.

The algorithms are derived with no appeal to physical ideas in the last section. We will speak of the value of a "register" instead of the energy of a "system" or the velocity of a "molecule". The algorithms follow next.

3. The Algorithms

3.1 Exponential random deviates

We first describe the algorithm for the generation of random numbers u drawn from the exponential distribution

$$f(u) = \lambda e^{-\lambda u}.$$

Let $IU(k,j)$ and $RU(x_1,x_2)$ be uniformly distributed random variables; furthermore let $IU(k,j)$ be an integer such that $k \le IU(k,j) \le j$, and let $RU(x_1,x_2)$ be a real number such that $x_1 < RU(x_1,x_2) < x_2$. IU and RU can be generated by any convenient method. (For fast performance, however, see below.) We consider an array of N elements, or registers, $r_1, r_2 \ldots r_N$; their initial values must fulfill $r_n \ge 0$ for all n. Let $\lambda = \sum_n r_n/N$ (we have chosen $r_n = \lambda$ for all n in all our computer runs). The algorithm conserves the *sum* of these registers.

The algorithm is as follows. The registers are updated according to:

$$\begin{array}{rrl}
 & m = & (m+1) \bmod N \\
1 & n = & IU(0, N-1) \\
if & n = & m \quad goto \quad 1 \\
 & x = & RU(0,1) \\
 & S = & r_m + r_n \\
 & r_m = & xS \\
 & r_n = & (1-x)S
\end{array} \tag{3.2}$$

r_n and r_m is the desired output: they are exponentially distributed random numbers.

A few remarks of some practical importance follow:

(a) It is convenient to let $N = 2^p$, because modulo operations can then be performed with a much faster logical *and*. Registers should then be numbered $0, 1, \ldots, 2^p - 1$. $L = 2^p - 1$ would be defined at the outset, and the modulo operation would be replaced by $m = (m+1).and.L$ (the corresponding instruction would read, $m = iand((m+1), L)$, in Fortran in some computers).

(b) For fast performance it is important to be selective about the method to be used for the generation of $IU(0, N-1)$ and of $RU(0,1)$. Generators that come with computers are often poor choices. They are usually several times slower than, for instance, the shift register algorithm [7,8], which is the one we prefer. Such slow generators bring the performance of our algorithm down to the level of much slower ones.

(c) One may either start out with a set of N *exponential* deviates for $r_1, r_2, \ldots, r_N$, which may take up a bit of work to set up, or one may initially set all $r_i = \lambda$. In the latter case, the generator must be first "warmed" up by going through it $q \times N$ times, where $q \sim 5$. (This point, as well as other subtleties of the algorithm, are is discussed in [9].)

(d) For reasons given in the last section, there are errors of about 1 part in N in the random numbers generated. 1024 registers seems to be a reasonable choice for most purposes.

3.2 Normal random deviates

We now give the algorithm for the generation of random numbers u with a probability density function given by

$$f(u) = \frac{1}{\sqrt{2\pi\sigma}} e^{-\frac{u^2}{2\sigma^2}}. \tag{3.3}$$

As in the exponential case, consider an array of N registers $r_1, r_2, \ldots, r_N$; instead of conserving the sum of all the registers, however, the method conserves the sum of their squares. This is accomplished as follows: think of two registers r_m and r_n as components of a two-dimensional vector, on which we perform rotations by randomly chosen angles. What makes the method fast is that the corresponding rotation matrices may be precomputed for a finite set of angles.

We initially assign uniform random numbers in the interval $(0, 2\pi)$ to θ_k and we compute $c_k \equiv \cos\theta_k$ and $s_k \equiv \sin\theta_k$, which remain fixed for the whole run, for all $k = 0, 1, \ldots K-1$. The algorithm follows,

$$\begin{array}{rrl} & m = & (m+1) \bmod N \\ 1 & n = & IU(0, N-1) \\ if & n = & m \quad goto \quad 1 \\ & k = & IU(0, K-1) \\ & x = & r_m c_k + r_n s_k \\ & r_n = & -r_m s_k + r_n c_k \\ & r_m = & x \end{array} \tag{3.4}$$

r_n and r_m is the desired output: they are two normally distributed random numbers. All the remarks made above for the generator of random exponential deviates hold here as well. Furthermore, it turns out that a few values of θ suffice (for an explanation, see Ref. [9]); $K = 2^4$ is certainly sufficiently large.

4. Derivation

One may think of our algorithms as random walks in $(N-1)$-dimensional bounded hypersurfaces. They are bounded because no register is allowed to have negative values. Our algorithms (which are ergodic Markov processes and fulfill "microscopic reversibility" [10]) fulfill the necessary conditions for a constant probability density, $P(r_1, r_2, \ldots, r_N) = c$, for $\sum_n r_n = N\lambda$ for the first algorithm and for $\sum_n r_n^2 = N\lambda^2$ for the second algorithm, ensuing after infinitely many iterations [6]. Therefore, $p(x)$ (the probability density that $r_1 = x$) is proportional to the $N-2$ dimensional volume Ω of the region specified by $r_2 + r_3 + \ldots + r_N = N\lambda - x$ (for the algorithm for exponential deviates). It follows simply from either dimensional or similarity considerations (or from straightforward integration) that

$$\Omega \propto (N\lambda - x)^{N-2},$$

and, since $p(x) \propto \Omega$, it follows that $p(x) = c(1 - x/N\lambda)^{N-2}$, or, more conveniently for our purposes,

$$p(x) = ce^{-\frac{x}{\lambda} g_N(x)},$$

where c is a normalizing constant, and

$$g_N(x) = 1 - \frac{2}{N}(1 - \frac{x}{4\lambda}) + O(N^-2).$$

It follows that if, for instance, a 1% accuracy in the distribution is desired, as many as $N = 2^7$ registers should be used, and so on.

For the algorithm for normal deviates, $p(x)$ is proportional to the surface area Ω of a $N-2$ dimensional sphere of radius $(N\lambda^2 - x^2)^{1/2}$. Clearly, $\Omega = c_N(N\lambda^2 - x^2)^{N/2-1}$, and c_N is a function of N whose value is irrelevant for our purposes. It follows that, $p(x) = c[1 - (x^2/(N\lambda)]^{N/2-1}$, or, more conveniently,

$$p(x) = ce^{-\frac{x^2}{\lambda^2} q_N(x)},$$

where c is a normalizing constant, and

$$q_N(x) = 1 - \frac{2}{N}(1 - \frac{x^2}{4\lambda^2}) + O(N^{-2}).$$

Once again, errors in the distribution of the random numbers generated are about 1 part in N (the number of registers used).

5. Performance

We first make a few brief comments on some of the most relevant questions about the performance of our algorithms. They are dealt with in more detail in Ref. [9].

It is easier to initialize all registers to the same value. Then, the algorithms must be run long enough to achieve "equilibrium". Intuitive considerations, as well as the numerical tests we have performed, show that about 5 "passes" per register are enough.

Some issues, such as recurrence times, that are relevant for generators (that are self contained) of uniform random deviates, do not arise here. This is because our generators (as most exponential and normal generators) only *process* uniform random deviates that are fed into them.

Some errors and correlations do arise because the number N of registers used can only be finite. As shown in the previous section, errors in the distributions are only about 1 part in N, however. It is easily shown [9] that correlations also vanish as $1/N$.

We have checked numerically the above relations for systematic deviations of random number distributions from the ones desired; chi-square values were obtained for histograms of random number frequencies. Analysis of these results show that fractional systematic errors in the frequency of random numbers generated are indeed given approximately by $1/N$ both for the exponential and for the normal distributions. For instance, we find that 1024 registers are sufficient for deviations within about 0.1% in the $0 \leq x/\lambda \leq 10$ range and in the $0 \leq x/\lambda \leq 5$ exponential and normal distributions, respectively. Smaller deviations from the desired distribution can obviously be obtained by increasing the number of registers. Execution times depend weakly on N (e.g., as N varies from $N = 32$ to $N = 512$, execution times increase only by about 20% –at least on IBM and SUN workstations).

Distributions obtained for 10^8 exponential deviates generated are shown in Fig. 2 for 1024 registers. The outcome is, except at the tails, quite satisfactory. The performance at the tails is, of course, due to poor statistics caused by insufficient sampling.

Plots of distributions obtained for 10^8 normal deviates generated are shown in Fig. 1. Once again, the only significant deviations from a gaussian distribution occur only at the "tails". Therein, fluctuations are too large because the number of recorded events is too small.

We have tested for speed our exponential and normal random number generators against a few other generators. In order to understand the outcome, it helps to realize that arithmetic operations are performed in most computers in less than 1/10 of the time it takes them to evaluate the transcendental functions that are a part of other generators. Even the *fast version* of the Box-Muller generator [11] requires (in addition to three uniform deviates) the evaluation of a transcendental function and a square root for each pair of random numbers generated. We have used FORTRAN code on some computer workstations. Our

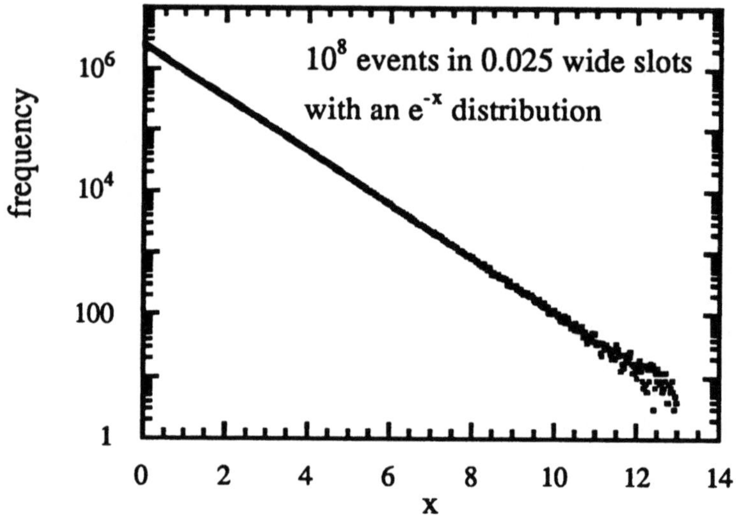

Fig. 1. This histogram shows data from a computer run that generated 10^8 random numbers with an e^{-x} distribution. The vertical axis [labeled *log(frequency)*] shows the decimal logarithm of how many random numbers generated fall into 0.025 wide bins.

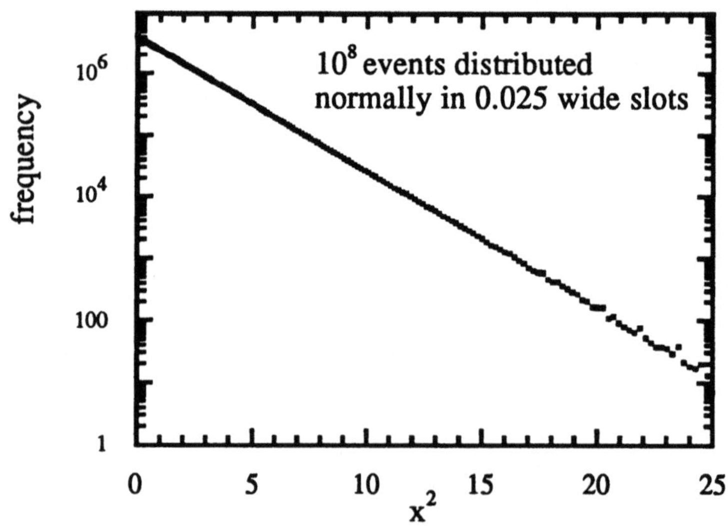

Fig. 2. This histogram shows data from a computer run that generated 2×10^8 random numbers with an $e^{-x^2/2}$ distribution. The vertical axis [labeled *log(frequency)*] shows the decimal logarithm of how many random numbers generated fall into 0.025 wide

normal random number generator is 5 – 10 times faster than the fast version of the Box-Muller algorithm (about 10 times faster on IBM-6000 RISC workstations, and 5 – 7 times faster on SUN SPARCS and IPX workstations). Our exponential random number generator turns out to be about 3 times faster than the algorithm that takes the logarithm of a uniform random deviate, and about 2.5 times faster than the von Newmann algorithm [12].

Financial support from Junta de Andalucia, and from DGCIT, through grant # PB91-0709, of Spain, is gratefully acknowledged.

References

1. For a recent account and further references, see, for instance, I. Vattulainen, T. Ala-Nissila, and K. Kankaala *Phys. Rev. Lett.* **73**, 2513 (1994).
2. See, for instance, R. Toral and A. Chakrabarti, *Comp. Phys. Comm.* **74**, 327 (1993) and references therein.
3. W.H. Press, B.P. Flannery, S.A. Teukolsky, and W.T. Vetterling, *Numerical Recipes*, (Cambridge University Press, Cambridge, UK, 1989); P. Bratley, B.L. Fox and L.E. Schrage, *A guide to Simulation* (Springer-Verlag, New York, 1983).
4. G.E. Forsythe, *Math. of Comp.* **26**, 817, (1972).
5. T. Ishikawa, P.Y. Wang, K. Wakui, and T. Yabe, *Comp. Phys. Commun.* **70**, 501, (1992).
6. O. Penrose, *Foundations of Statistical Mechanics* (Plenum Press, New York, 1970).
7. G. Marsaglia, B. Narasimhan, and A. Zaman, *Comp. Phys. Comm.* **60**, 345, (1990).
8. S. Kirkpatrick and E.P. Stoll, *J. Comp. Phys* . **40**, 517 (1981).
9. J. F. Fernández and J. Rivero (To be published).
10. By *microscopic reversibility* we mean that the probability that any two given registers i and j go from values r_i, r_j to r_i', r_j' equals the probability that they go from r_i', r_j' to r_i, r_j. See also Ref. [6].
11. See Press et al, Ref. [3], p. 202.
12. See P. Bratley et al, Ref. [3].

Numerical Simulations of a Nonlinear Klein-Gordon Model. Applications

P.J. Pascual[1], S. Jiménez[2] and L. Vázquez[3]

[1]Instituto de Ingeniería del Conocimiento
Universidad Autónoma de Madrid
Mod. C-XVI, Cantoblanco, 28049-Madrid, Spain.
[2]Departamento de Ingeniería
Universidad Alfonso X El Sabio,
P.O.B. 016, 28691-Villanueva de la Cañada, Madrid, Spain.
[3]Departamento de Física Teórica, Facultad de Ciencias Físicas,
Universidad Complutense, 28040-Madrid, Spain

1. Introduction

The purpose of these notes is to present an application of computational physics. We will study the stochastic perturbations of the solitary wave solutions (soliton and breather) associated with the integrable Klein-Gordon system

$$\phi_{tt} - \phi_{xx} + \sin\phi = 0 \tag{1.1}$$

which is the well-known Sine-Gordon (SG) equation.

In a more general context, we are going to study a nonlinear system with disorder that can be formulated in terms of a nonlinear wave equation. Such kind of equation appears in different contexts (Bass et al., 1988; Dodd et al., 1982; Kivshar et al., 1989; Konotop and Vázquez, 1994; Sánchez and Vázquez, 1991) as the propagation of waves in continuous media, particle physics, condensed matter physics,..etc.

The contribution is organized as follows. In Section 2, we present the main features of the solitary waves while in Sections 3 and 4, we discuss the features of finite difference schemes for the Klein-Gordon model and a Lagrangian system associated with the collective coordinates study. In Sections 5 and 6 it is analyzed the effect of stochastic perturbations on the soliton and breather solutions of the SG model. It is also included the effect of the dissipation and a periodic driving force. Finally, in Section 7 some applications to the modelization of the charge-density waves is presented.

2. Solitary Waves

The solitary waves are solutions of the nonlinear wave equations characterized by the following properties:

1. Constance in time of their wave form and velocity.
2. They represent structures which do not spread: the effect of the dispersion is compensated by the effect of the nonlinearity.
3. In the framework of the classical extended particles, the solitary waves are localizations of energy, momentum, charge, magnetic moment as well as other physical quantities.

To concrete the above properties, let us consider the simplest nonlinear relativistic equation of the mathematical physics described by the equation

$$\phi_{tt} - \Delta\phi + m^2\phi + F(\phi\phi^*)\phi = 0 \tag{2.1}$$

the equation (1.1) is a particular case of (2.1), being ϕ real. The solitary waves of (2.1) are the stationary solutions (without radiation) of the form

$$\phi(\boldsymbol{x}, t) = e^{-i\omega t}\phi(\boldsymbol{x}) \tag{2.2}$$

and localized, in the sense that the associated physical quantities, charge and energy, must be finite

$$Q = \omega \int \phi(\boldsymbol{x})^2 d^3\boldsymbol{x} \tag{2.3}$$

$$E = \int [(\omega^2 + m^2)\phi^2 + |\nabla\phi|^2 + G(\phi)] d^3\boldsymbol{x} \tag{2.4}$$

where $F(\phi) = dG/d\phi$. The charge Q and the energy E define the natural functional spaces to study the associated mathematical problems. In this context, we can summarize the following topics to be studied:

1. The existence of solitary waves (Berestycki and Lions, 1983; Brezis and Lieb, 1984; Cazenave and Vázquez, 1986).
2. The general evolution problem associated to the nonlinear wave equation (Haraux, 1981).
3. The stability problem (Vázquez, 1982; Stubbe and Vázquez, 1988).
4. The integrability (Dodd et al., 1982).
5. Computational problems (Dodd et al., 1982; Konotop and Vázquez, 1994).

A solitary wave is stable in a general sense if it is not destroyed under the influence of a general perturbation. Thus, we have different definitions of stability, that can be classified into two large groups:

1. *Stability with respect to a perturbation of the initial data*
   a) Dynamical stability: A solitary wave is dynamically stable when small perturbations of its amplitude do not destroy it.
   b) Lyapunov stability: It happens when the initially perturbed solitary wave remains in a neighborhood (defined in an appropriate functional space) of the unperturbed solitary wave.
   c) Stability with constraints.

d) Energetic stability.
e) Soliton stability. A solitary wave is called a **soliton** if preserves its shape after collision with other solitary waves. The solitons are associated with the integrable systems, where the dynamics is well understood. In the non-integrable systems the dynamics of the solitary waves is more complicated and their stability is an important tool to study the non-integrable dynamics.

2. *Structural stability: Stability with respect to a perturbation of the evolution equation*
   a) Static stability: Behavior under external static potentials.
   b) Stability under time-dependent potentials.
   c) Stability under stochastic perturbations and disorder.

3. Numerical Treatment of the Sine-Gordon Equation

3.1. Numerical schemes

Let us consider the plain SG equation without any perturbation. We want to solve numerically the Cauchy problem

$$\begin{aligned} &\phi_{tt} - \phi_{xx} + \sin\phi = 0, \\ &\phi(x,0) = f_0(x), \quad \phi_t(x,0) = f_1(x). \end{aligned} \tag{3.1}$$

Besides general schemes to integrate numerically such a P.D.E. (Ritchmyer and Morton, 1967; Potter, 1977), the numerical schemes proposed to solve specifically (3.1) are generally based on the properties of the solutions: the first one is due to (Perring and Skyrme, 1962) and is based on characteristics; other schemes can be found in (Ablowitz et al.; Dodd et al., 1982). Our approach is a conservative numerical scheme (Pascual and Vázquez, 1985; Pascual, 1984) based on the technique of (Strauss and V'azquez, 1978).

The Strauss-Vázquez schemes have been successfully applied to other nonlinear wave equations such as ϕ-4 models, nonlinear Schrödinger equations, etc. (Jiménez and Vázquez, 1990; Jiménez, 1994) and also with stochastic perturbations as the ones we consider in this work (Sánchez and Vázquez, 1989; Rodríguez, 1988; Rodríguez and Vázquez, 1990). Applications to finite-dimension equations (O.D.E.) have been also widely considered (Pascual and Vázquez, 1986; Vázquez, 1984; Jiménez, 1994; Jiménez and Vázquez, 1989, 1991).

The main feature of the scheme is to present a discrete counterpart of the conservation law for the energy. The numerical scheme is

$$\frac{\phi_l^{n+1} - 2\phi_l^n + \phi_l^{n-1}}{\Delta t^2} - \frac{\phi_{l+1}^n - 2\phi_l^n + \phi_{l-1}^n}{\Delta x^2} - \frac{\cos\phi_l^{n+1} - \cos\phi_l^{n-1}}{\phi_l^{n+1} - \phi_l^{n-1}} = 0 \tag{3.2}$$

where Δt and Δx are the temporal and spatial mesh steps, and $\phi_l^n = \phi(l\Delta x, n\Delta t)$. The nonlinear term comes from the discrete version of the analogous term in the Euler-Lagrange equations:

$$\frac{\partial}{\partial \phi}(1 - \cos \phi).$$

Scheme (3.2) is a second order, centered, finite difference scheme. It can be considered explicit since we obtain ϕ_l^{n+1} directly from ϕ_l^{n-1}, ϕ_{l-1}^n, ϕ_l^n and ϕ_{l+1}^n, although the corresponding algebraic equation is a nonlinear one that must be solved, for instance, with a Newton method. The convergence and stability for this scheme have been studied in (Guo et al., 1986), and they are guaranteed provided the Courant-Friedrichs-Lewy condition holds (Potter, 1977)

$$\frac{\Delta t^2}{\Delta x^2} < 1$$

(which is the same as for the linear equation).

The discrete energy and its conservation law are obtained in a similar way as with the continuous equation: we multiply (3.2) by a discrete counterpart of ϕ_t

$$\frac{\phi_l^{n+1} - \phi_l^{n-1}}{2\Delta t}$$

and we obtain, after rearranging terms and performing a summation over the spatial index l,

$$\frac{E^{n+1} - E^n}{\Delta t} = 0,$$

where

$$E^n = \sum_{l=-\infty}^{\infty} \Delta x \left[\frac{1}{2} \left(\frac{\phi_l^{n+1} - \phi_l^n}{\Delta t} \right)^2 + \frac{1}{2} \left(\frac{\phi_{l+1}^{n+1} - \phi_l^{n+1}}{\Delta x} \right) \left(\frac{\phi_{l+1}^n - \phi_l^n}{\Delta x} \right) \right.$$
$$\left. + 1 - \frac{\cos \phi_l^{n+1} + \cos \phi_l^n}{2} \right]. \tag{3.3}$$

In practice we integrate numerically the problem over a finite region, say $[-L, L]$, and thus have a finite range for the index l. This implies, on one hand, a small error on the conservation law (Pascual, 1984) and, on the other hand, some boundary conditions at the discrete level. We have chosen a region $[-L, L]$ such that the soliton is confined in an inner and much smaller interval, and fixed boundary conditions on the solution: $\phi(\pm L, t) = \phi(\pm L, 0) =$constant. In the case of breathers, we have chosen free boundary conditions such that $\phi_x(\pm L, t) = 0$.

Besides the energy, equation (3.1) has the momentum as a conserved integral. If we multiply now (3.2) by a discrete representation of ϕ_x:

$$\frac{\phi_{l+1}^n - \phi_{l-1}^n}{2\Delta x}$$

we obtain, after rearranging terms and performing the summation over l

$$\frac{P^{n+1} - P^n}{\Delta t} = - \sum_{l=-\infty}^{\infty} \Delta x \left(\frac{\cos \phi_l^{n+1} - \cos \phi_l^{n-1}}{\phi_l^{n+1} - \phi_l^{n-1}} \right) \left(\frac{\phi_{l+1}^n - \phi_{l-1}^n}{2} \right),$$

where

$$P^n = -\frac{1}{2}\sum_{l=-\infty}^{\infty} \Delta x \left(\frac{\phi_l^{n+1}-\phi_l^n}{\Delta t}\right)\left(\frac{\phi_{l+1}^{n+1}-\phi_{l-1}^{n+1}}{2\Delta x}\right).$$

As we see, there is not an exact conservation law for the discrete momentum, even before introducing the finite range of summation, but anyhow the error is small (Pascual, 1984). It is possible to have a scheme, similar to (3.2), that preserves the momentum instead of the energy, but its general behavior is bad compared to the one we use (Jiménez and Vázquez, 1990).

The position of the center of mass (center of energy) is defined in the continuous case as

$$x_{\rm cm}(t) = \frac{\displaystyle\int x e(x,t)\,dx}{\displaystyle\int e(x,t)\,dx}$$

where $e(x,t)$ is the energy density

$$e(x,t) = \frac{1}{2}(\phi_t^2 - \phi_x^2) + 1 - \cos\phi.$$

In the discrete level, we define from (3.3) the energy density as

$$\begin{aligned} e_l^n =& \frac{1}{2}\left(\frac{\phi_l^{n+1}-\phi_l^n}{\Delta t}\right)^2 + \frac{1}{2}\left(\frac{\phi_{l+1}^{n+1}-\phi_l^{n+1}}{\Delta x}\right)\left(\frac{\phi_{l+1}^n-\phi_l^n}{\Delta x}\right) \\ &+ 1 - \frac{\cos\phi_l^{n+1}+\cos\phi_l^n}{2}. \end{aligned}$$

and the discrete position of the center of mass is taken as

$$\begin{aligned} x_{\rm cm}^n =& \frac{1}{E^n}\sum_l \Delta x x_l \left[\frac{1}{2}\left(\frac{\phi_l^{n+1}-\phi_l^n}{\Delta t}\right)^2 + 1 - \frac{\cos\phi_l^{n+1}+\cos\phi_l^n}{2}\right] \\ &+ \frac{1}{E^n}\sum_l \Delta x \frac{x_{l+1}+x_l}{2}\frac{1}{2}\left(\frac{\phi_{l+1}^{n+1}-\phi_l^{n+1}}{\Delta x}\right)\left(\frac{\phi_{l+1}^n-\phi_l^n}{\Delta x}\right) \end{aligned}$$

where $x_l = l\Delta x$. The last term has been multiplied by $(x_{l+1}+x_l)/2$ instead of x_l to maintain the symmetry.

From the definition of the continuous velocity, we may define the discrete one:

$$v = \frac{dx_{\rm cm}}{dt}; \quad v^n = \frac{x_{\rm cm}^n - x_{\rm cm}^{n-1}}{\Delta t}$$

There is an alternative definition from the momentum:

$$v' = \frac{P}{E}; \quad v'^n = \frac{P^n}{E^n}$$

but both definitions appear to give the same results in practice (Pascual, 1984).

Remark: If the SG equation is perturbed by a linear dissipation as follows:

$$\phi_{tt} - \phi_{xx} + \sin\phi + \alpha\phi_t = 0 \tag{3.4}$$

where α is a positive constant. The total energy variation is given by the equation

$$\frac{dE}{dt} = -\alpha \int dx\phi_t^2 \tag{3.5}$$

One possible scheme that satisfies the above property is a natural generalization of the scheme (3.2):

$$\frac{\phi_l^{n+1} - 2\phi_l^n + \phi_l^{n-1}}{\Delta t^2} - \frac{\phi_{l+1}^n - 2\phi_l^n + \phi_{l-1}^n}{\Delta x^2} - \frac{\cos\phi_l^{n+1} - \cos\phi_l^{n-1}}{\phi_l^{n+1} - \phi_l^{n-1}} + \alpha\frac{\phi_l^{n+1} - \phi_l^{n-1}}{2\Delta t} = 0$$

being the variation of the discrete energy (3.3):

$$\frac{E^{n+1} - E^n}{\Delta t} = -\alpha \sum_{l=-\infty}^{\infty} \left(\frac{\phi_l^{n+1} - \phi_l^{n-1}}{2\Delta t}\right)^2$$

3.2 A stochastic Sine-Gordon model

We will study the two kinds of perturbed SG equations (Pascual, 1992):

1 *Multiplicative Noise with/without Dissipation*

$$\phi_{tt} - \phi_{xx} + (1 + \nu(t))\sin\phi + \alpha\phi_t = 0 \tag{3.6}$$

2 *Additive Noise with a (Periodic) Driven Force and with/without Dissipation*

$$\phi_{tt} - \phi_{xx} + \sin\phi + \alpha\phi_t = \nu(t) + F(t) \tag{3.7}$$

where $\alpha\phi_t$ represents a linear dissipation term, while where $\nu(t)$ is a white gaussian noise:

$$\langle\nu(t)\rangle = 0; \quad \langle\nu(t)\nu(t+\tau)\rangle = 2D\delta(t). \tag{3.8}$$

If we understand the perturbed SG equations (3.6) and (3.7) in the Stratonovich sense (van Kampen, 1981), we may introduce the white noise $\nu(t)$ in the numerical scheme at the first order in Δt (Helfand, 1979), using a random number generator with a gaussian distribution of zero mean value. The variance σ^2 has to be taken as follows. When discretizing, we pass from the *continuous* stochastic variable $\nu(t)$ to a discrete one by means of an integral:

$$\nu^n = \frac{1}{\Delta t}\int_{n\Delta t}^{(n+1)\Delta t} \nu(\tau)\, d\tau$$

if we use now (3.8) we have that

$$\langle\nu^n\nu^m\rangle = \frac{1}{\Delta t^2}\int_{n\Delta t}^{(n+1)\Delta t}\int_{m\Delta t}^{(m+1)\Delta t} \langle\nu(\tau)\nu(\tau')\rangle\, d\tau\, d\tau' = \frac{2D}{\Delta t}\delta_{nm}$$

and

$$\sigma^2 = \langle (\nu^n)^2 \rangle = \frac{2D}{\Delta t}.$$

As a consequence, ν^n is a good approximation to the white noise $\nu(t)$ in the region of the frequency spectrum that corresponds to the correlation function (3.8) that lays below $(\Delta t)^{-1}$, which in our case is much bigger than the inverse of any time scale involved.

For the equations (3.6)-(3.7), a consistent stochastic finite-difference scheme is obtained from (3.2) as follows

$$\frac{\phi_l^{n+1} - 2\phi_l^n + \phi_l^{n-1}}{\Delta t^2} - \frac{\phi_{l+1}^n - 2\phi_l^n + \phi_{l-1}^n}{\Delta x^2} - \left(1 + \frac{\nu^{n+1} + \nu^{n-1}}{2}\right) \frac{\cos\phi_l^{n+1} - \cos\phi_l^{n-1}}{\phi_l^{n+1} - \phi_l^{n-1}} + \alpha \frac{\phi_l^{n+1} - \phi_l^{n-1}}{2\Delta t} = 0$$

$$\frac{\phi_l^{n+1} - 2\phi_l^n + \phi_l^{n-1}}{\Delta t^2} - \frac{\phi_{l+1}^n - 2\phi_l^n + \phi_{l-1}^n}{\Delta x^2} - \frac{\cos\phi_l^{n+1} - \cos\phi_l^{n-1}}{\phi_l^{n+1} - \phi_l^{n-1}} + \alpha \frac{\phi_l^{n+1} - \phi_l^{n-1}}{2\Delta t} = \frac{\nu^{n+1} + \nu^{n-1}}{2}$$

The discretization of the time dependent terms is fixed on the basis of the energy variation for the underlying continuous equation (Pascual and Vázquez, 1985; Rodríguez and Vázquez, 1990; Konotop and Vázquez, 1994).

4. Numerical Schemes for Euler-Lagrange Equations

4.1 Schemes for the collective coordinates study

Let us consider a Lagrangian of the form

$$L(x, v) = T_{xx}(x)v^2 - V(x),$$

where x is the position and $v = \dot{x}$ is the velocity. Such a Lagrangian may arise in the collective coordinates study of the SG equation. The canonical momentum associated to x and the Euler-Lagrange equation are

$$p = \frac{\partial L}{\partial v}, \quad \frac{dp}{dt} = \frac{\partial L}{\partial x}. \tag{4.1}$$

Let us consider a numerical scheme (Pascual, 1992) to simulate (4.1) as follows

$$\begin{aligned}
x^{n+1} &= x^n + v^n \Delta t \\
p^{n+1} &= \frac{L(x^n, v^{n+1}) - L(x^n, v^n)}{v^{n+1} - v^n} = T_{xx}(x^n)(v^{n+1} + v^n) \\
\frac{p^{n+1} - p^n}{\Delta t} &= \frac{L(x^{n+1}, v^{n+1}) - L(x^n, v^{n+1})}{x^{n+1} - x^n} = \frac{V(x^{n+1}) - V(x^n)}{x^{n+1} - x^n}.
\end{aligned}$$

Starting from x^n, v^n and p^n we obtain x^{n+1} from the first equation, substituting the second equation in the third we obtain v^{n+1} and finally from the second equation we have p^{n+1}. It is easily seen that there is a discrete energy exactly preserved

$$E^n = p^n v^n - L(x^n, v^n) = T_{xx}(x^n)\, v^n v^{n-1} + V(x^n).$$

The scheme can be easily extended to a two-dimensional case with Lagrangian of the form

$$L(x, \dot{x}; y, \dot{y}) = T_{xx}(x,y)\, \dot{x}^2 + T_{xy}(x,y)\, \dot{x}\dot{y} + T_{yy}(x,y)\, \dot{y}^2 - V(x,y). \tag{4.2}$$

The numerical scheme is (Pascual, 1992)

$$\begin{aligned}
x^{n+1} &= x^n + \dot{x}^n \Delta t \\
y^{n+1} &= y^n + \dot{y}^n \Delta t \\
{p_x}^{n+1} &= \frac{L(x^n, \dot{x}^{n+1}; y^n, \dot{y}^n) - L(x^n, \dot{x}^n; y^n, \dot{y}^n)}{\dot{x}^{n+1} - \dot{x}^n} \\
{p_y}^{n+1} &= \frac{L(x^{n+1}, \dot{x}^{n+1}n; y^n, \dot{y}^{n+1}) - L(x^{n+1}, \dot{x}^{n+1}; y^n, \dot{y}^n)}{\dot{y}^{n+1} - \dot{y}^n} \\
\frac{{p_x}^{n+1} - {p_x}^n}{\Delta t} &= \frac{L(x^{n+1}, \dot{x}^{n+1}; y^n, \dot{y}^n) - L(x^n, \dot{x}^{n+1}; y^n, \dot{y}^n)}{x^{n+1} - x^n} \\
\frac{{p_y}^{n+1} - {p_y}^n}{\Delta t} &= \frac{L(x^{n+1}, \dot{x}^{n+1}; y^{n+1}, \dot{y}^{n+1}) - L(x^{n+1}, \dot{x}^{n+1}; y^n, \dot{y}^{n+1})}{y^{n+1} - y^n}
\end{aligned}$$

and the conserved energy is now

$$E^n = {p_x}^n \dot{x}^n + {p_y}^n \dot{y}^n - L(x^n, v^n).$$

4.2 Extension to a general Lagrangian

Although fairly general, the previous schemes are based on Lagrangians separable into a kinetic and a potential parts. If we apply the schemes, as they are defined, to a non-separable Lagrangian, we may obtain meaningless results (Jiménez and Malavasi, 1994). It is possible nevertheless to extend the previous ideas to a general Lagrangian, in one, two or any dimensions, with or without explicit time dependence (Jiménez and Malavasi, 1994).

Let us consider, for simplicity, a one dimensional Lagrangian with no explicit time dependence $L(x, \dot{x})$. The Euler-Lagrange equation is in this case

$$\frac{d}{dt}\frac{\partial L}{\partial \dot{q}} - \frac{\partial L}{\partial q} = 0,$$

and we can define a momentum and an energy

$$p(t) = \frac{\partial L}{\partial \dot{q}},$$

$$E(t) = \dot{q}p - L(q,\dot{q}).$$

From the Euler-Lagrange equation and the definition of p we deduce the conservation of the energy:

$$\frac{dE}{dt} = \ddot{q}p + \dot{q}\dot{p} - \frac{\partial L}{\partial q}\dot{q} - \frac{\partial L}{\partial \dot{q}}\ddot{q} = \ddot{q}\left(p - \frac{\partial L}{\partial \dot{q}}\right) + \dot{q}\left(\dot{p} - \frac{\partial L}{\partial q}\right) = 0. \tag{4.3}$$

The idea is to start from the conservation law for the energy and reproduce expression (4.3) at discrete level, deducing from there a suitable discrete representation for both the Euler-Lagrange equation and the definition of the momentum. For this, we need to represent at discrete level the total time derivative of L in terms of the partial derivatives with respect to q and $\dot{q}$. Since there is not an unique way to do it, we end with several schemes, but choosing the simpler representations and imposing the schemes to be symmetric in the variables we obtain the following scheme

$$\begin{aligned} \dot{x}^n &= \frac{x^{n+1} - x^n}{\Delta t} \\ p^{n+1} &= \frac{L(x^{n+1}, \dot{x}^{n+1}) - L(x^{n+1}, \dot{x}^n)}{\dot{x}^{n+1} - \dot{x}^n} \\ \frac{p^{n+1} - p^n}{\Delta t} &= \frac{L(x^{n+1}, \dot{x}^n) - L(x^n, \dot{x}^n)}{x^{n+1} - x^n} \end{aligned}$$

and the energy is

$$E^n = p^n \dot{x}^n - L(x^n, \dot{x}^n).$$

In the case of a Lagrangian with an explicit time-dependence of the form $L(x, \dot{x}; t)$, the energy is not conserved but has a variation law. The process to build the scheme is nevertheless the same, and the discrete energy is such that its variation is similar to the continuous one. We start from

$$\frac{dE}{dt} = \ddot{q}\left(p - \frac{\partial L}{\partial \dot{q}}\right) + \dot{q}\left(\dot{p} - \frac{\partial L}{\partial q}\right) - \frac{\partial L}{\partial t} = -\frac{\partial L}{\partial t}.$$

At the discrete level, considering now the energy as

$$E^n = p^n \dot{x}^n - L(x^n, \dot{x}^n; t^n),$$

where $t^n = n\Delta t$, we obtain the following scheme:

$$\begin{aligned} \dot{x}^n &= \frac{x^{n+1} - x^n}{\Delta t} \\ p^{n+1} &= \frac{1}{2}\Bigg(\frac{L(x^{n+1}, \dot{x}^{n+1}; t^{n+1}) - L(x^{n+1}, \dot{x}^n; t^{n+1})}{\dot{x}^{n+1} - \dot{x}^n} \\ &\quad + \frac{L(x^{n+1}, \dot{x}^{n+1}; t^n) - L(x^{n+1}, \dot{x}^n; t^n)}{\dot{x}^{n+1} - \dot{x}^n}\Bigg) \\ \frac{p^{n+1} - p^n}{\Delta t} &= \frac{L(x^{n+1}, \dot{x}^n; t^n) - L(x^n, \dot{x}^n; t^n)}{x^{n+1} - x^n} \end{aligned}$$

and the variation of the discrete energy is

$$\frac{E^{n+1}-E^n}{\Delta t} = \frac{1}{2}\left(\frac{L(x^{n+1},\dot{x}^n;t^{n+1})-L(x^{n+1},\dot{x}^n;t^n)}{\Delta t} + \frac{L(x^n,\dot{x}^n;t^{n+1})-L(x^n,\dot{x}^n;t^n)}{\Delta t}\right).$$

Due to the expression of $\dot{x}^n$, this is an $\mathcal{O}(2)$ expression centered around $(x^{n+1/2};t^{n+1/2})$.

The scheme in two dimensions with no explicit time-dependence is constructed in a similar way, giving:

$$\begin{aligned}
x^{n+1} &= x^n + \dot{x}^n \Delta t\\
y^{n+1} &= y^n + \dot{y}^n \Delta t\\
{p_x}^{n+1} &= \frac{1}{2}\left(\frac{L(x^{n+1},\dot{x}^{n+1};y^{n+1},\dot{y}^{n+1})-L(x^{n+1},\dot{x}^n;y^{n+1},\dot{y}^{n+1})}{\dot{x}^{n+1}-\dot{x}^n}\right.\\
&\quad\left.+\frac{L(x^{n+1},\dot{x}^{n+1};y^{n+1},\dot{y}^n)-L(x^{n+1},\dot{x}^n;y^{n+1},\dot{y}^n)}{\dot{x}^{n+1}-\dot{x}^n}\right)\\
{p_y}^{n+1} &= \frac{1}{2}\left(\frac{L(x^{n+1},\dot{x}^{n+1};y^{n+1},\dot{y}^{n+1})-L(x^{n+1},\dot{x}^{n+1};y^{n+1},\dot{y}^n)}{\dot{y}^{n+1}-\dot{y}^n}\right.\\
&\quad\left.+\frac{L(x^{n+1},\dot{x}^n;y^{n+1},\dot{y}^{n+1})-L(x^{n+1},\dot{x}^n;y^{n+1},\dot{y}^n)}{\dot{y}^{n+1}-\dot{y}^n}\right)\\
\frac{{p_x}^{n+1}-{p_x}^n}{\Delta t} &= \frac{1}{2}\left(\frac{L(x^{n+1},\dot{x}^n;y^{n+1},\dot{y}^n)-L(x^n,\dot{x}^n;y^{n+1},\dot{y}^n)}{x^{n+1}-x^n}\right.\\
&\quad\left.+\frac{L(x^{n+1},\dot{x}^n;y^n,\dot{y}^n)-L(x^n,\dot{x}^n;y^n,\dot{y}^n)}{x^{n+1}-x^n}\right)\\
\frac{{p_y}^{n+1}-{p_y}^n}{\Delta t} &= \frac{1}{2}\left(\frac{L(x^{n+1},\dot{x}^n;y^{n+1},\dot{y}^n)-L(x^{n+1},\dot{x}^n;y^n,\dot{y}^n)}{y^{n+1}-y^n}\right.\\
&\quad\left.+\frac{L(x^n,\dot{x}^n;y^{n+1},\dot{y}^n)-L(x^n,\dot{x}^n;y^n,\dot{y}^n)}{y^{n+1}-y^n}\right)
\end{aligned}$$

and the conserved energy is now, as in the previous two-dimensional scheme,

$$E^n = {p_x}^n\dot{x}^n + {p_y}^n\dot{y}^n - L(x^n,\dot{x}^n;y^n,\dot{y}^n).$$

This process can be extended to more degrees of freedom and / or explicit time-dependence. The process starts with the discrete energy. From its variation law we identify the terms according to the continuous equation. The different ways of expressing the discrete variation of the Lagrangian give rise to different numerical schemes. Finally, by asking the schemes to represent the continuous symmetries at the discrete level, we determine the ones best suited to treat the problem.

5. Sine-Gordon Equation Under Multiplicative Noise

In this section we will study the SG equation perturbed with a multiplicative noise with/without dissipation:

$$\phi_{tt} - \phi_{xx} + (1 + \nu(t)) \sin \phi + \alpha \phi_t = 0 \tag{5.1}$$

where $\nu(t)$ is the gaussian white noise (3.8). First, we will discuss the analytical approximations that we will use to face the problem, and then we will study the particular behavior of a soliton and a breather.

5.1 The soliton

As it can be seen, for the perturbed SG equation:

$$\begin{aligned} \phi_{tt} - \phi_{xx} + \sin \phi &= f(\phi, \phi_t; t) \\ f(\phi, \phi_t; t) &= -\alpha \phi_t + \epsilon R(\phi; t) \end{aligned} \tag{5.2}$$

the energy and momentum corresponding to the SG equation vary with time as:

$$\begin{aligned} \frac{dE}{dt} &= \int_{-\infty}^{\infty} f(\phi, \phi_t; t) \phi_t \, dx \\ \frac{dP}{dt} &= -\int_{-\infty}^{\infty} f(\phi, \phi_t; t) \, \phi_x \, dx \end{aligned} \tag{5.3}$$

On the other hand, from the definition of center of mass x_{cm} we have that:

$$\frac{dx_{\text{cm}}}{dt} = -v_{\text{cm}} + \frac{1}{E} \int_{-\infty}^{\infty} (x - x_{\text{cm}}) \, f(\phi, \phi_t; t) \, \phi_t \, dx \tag{5.4}$$

where by definition:

$$v_{\text{cm}} = \frac{P}{E} \tag{5.5}$$

In the adiabatic approximation, $\phi = \phi_0$, for a soliton as an initial state, we take the ansatz:

$$\begin{aligned} \phi(x,t) &= 4 \tan^{-1} \left[\exp \left(\frac{x - X(t)}{\sqrt{1 - U^2(t)}} \right) \right] \\ \phi_t(x,t) &= \frac{-2U(t)}{\sqrt{1 - U^2(t)}} \quad \operatorname{sech} \left[\frac{x - X(t)}{\sqrt{1 - U^2(t)}} \right] \end{aligned} \tag{5.6}$$

so we obtain:

$$\begin{aligned} E &= \frac{8}{\sqrt{1 - U^2(t)}} \\ P &= \frac{8U(t)}{\sqrt{1 - U^2(t)}} \\ x_{\text{cm}} &= X(t) \\ v_{\text{cm}} &= U(t) \end{aligned} \tag{5.7}$$

which are the same results as for the non perturbed case.

If we now insert the ansatz in (5.3) and (5.4), we obtain the dynamic equations of the soliton, considered as a particle:

$$\begin{aligned} \frac{dU}{dt} &= -\frac{1}{4}(1-U^2)\int_{-\infty}^{+\infty} f(\phi,\phi_t;t)\,\mathrm{sech}\theta\,dx \\ \frac{dX}{dt} &= U - \frac{1}{4}U\sqrt{1-U^2}\int_{-\infty}^{+\infty} f(\phi,\phi_t;t)\,\theta\,\mathrm{sech}\theta\,dx \\ \frac{dP}{dt} &= -2\sqrt{1-U^2}\int_{-\infty}^{+\infty} f(\phi,\phi_t;t)\,\mathrm{sech}\theta\,dx \end{aligned} \tag{5.8}$$

where

$$\theta = \frac{x - X(t)}{\sqrt{1-U^2(t)}}$$

These are the basic equations of the evolution of the parameters (position, velocity and momentum) of a SG perturbed soliton in the adiabatic approximation. We will make reference to them later on when considering the different kinds of perturbations that we will study.

For the multiplicative noise without dissipation we have:

$$f(\phi,\phi_t;t) = \epsilon R(\phi;t) = -\nu(t)\sin\phi$$

and substituting in (5.8) we obtain:

$$\begin{aligned} \dot{U} &= 0 \\ \dot{X} &= U - \frac{1}{2}U(1-U^2)\nu(t) \\ \dot{P} &= 0 \end{aligned} \tag{5.9}$$

So we have that the evolution of the velocity and momentum are deterministic, and they remain constant: $U = U_0$, $P = P_0$, which is equivalent to say that their probability distribution is of the form (for the velocity, for example) $W(U,t) = \delta(U-U_0)$.

Regarding the position of the soliton, by integrating (5.9) we obtain:

$$X = X_0 + U_0 t - \frac{1}{2}U_0(1-U_0^2)\int_0^t dt\,\nu(t)$$

and from this equation we can obtain the evolution of the mean values. In this particular case, we can solve the Fokker-Planck equation associated with the probability distribution function W(X,t):

$$\frac{\partial W}{\partial t} = -U_0\frac{\partial W}{\partial X} + D_X\frac{\partial^2 W}{\partial X^2}\ ;\ D_X = D\left[\frac{1}{2}U_0(1-U_0^2)\right]^2 \tag{5.10}$$

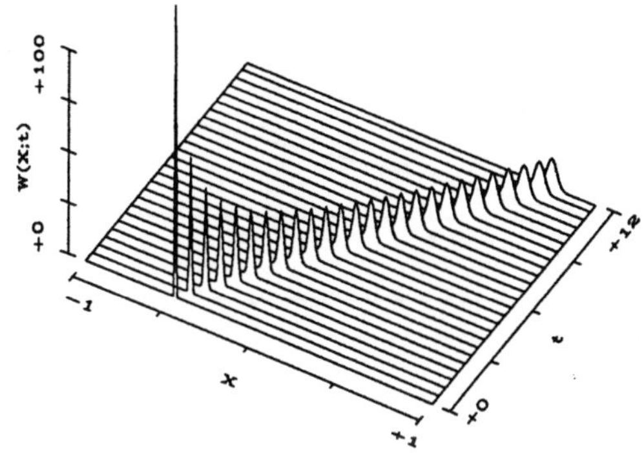

Fig. 1. Evolution of the probability distribution function $W(X,t)$ corresponding to a soliton under multiplicative noise, starting at $X_0 = -0.5$ with velocity $U_0 = 0.1$ and noise $\sigma = 1$.

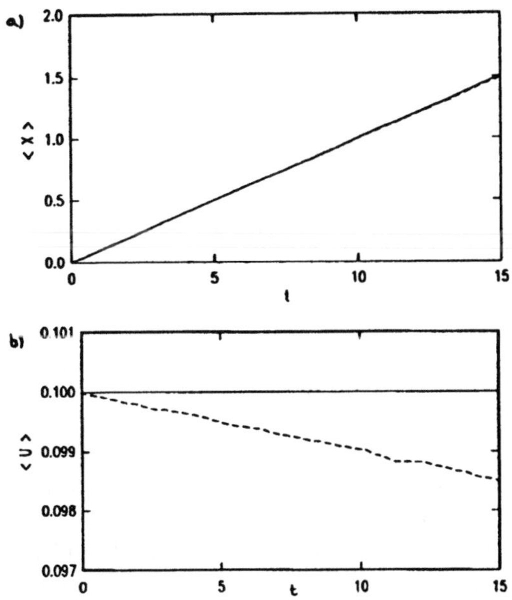

Fig. 2. Evolution of the mean values of the velocity and position of a soliton under multiplicative noise. Comparison of a) $\langle X \rangle$, b) $\langle U \rangle$ obtained through numerical simulation (dotted line) with the ones given by the equation (5.12) (solid line) for a noise $\sigma = 0.5$; $X_0 = -0.$, $U_0 = 0.1$. Results are shown for 100 realizations.

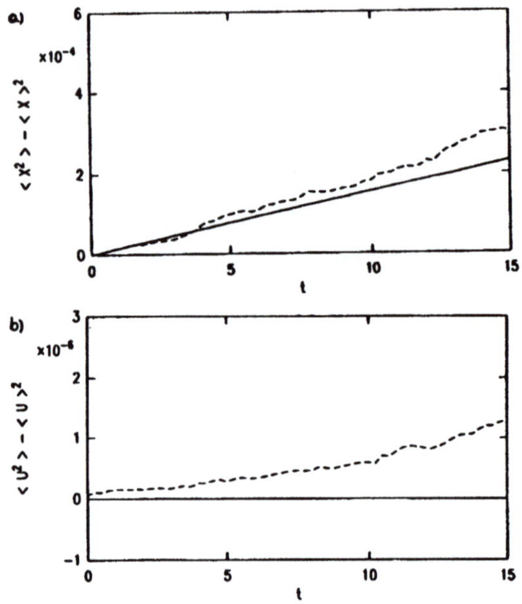

Fig. 3. Evolution of the mean deviations of the velocity and position of a soliton under multiplicative noise. Comparison of a) $\langle X^2\rangle - \langle X\rangle^2$, b) $\langle U^2\rangle - \langle U\rangle^2$ obtained through numerical simulation (dotted line) with the ones given by equation (58) $X_0 = -0.$, $U_0 = 0.1$. Results are shown for 100 realizations.

and for $W(X;0) = \delta(X - X_0)$, we obtain:

$$W(X,t) = \frac{1}{\sqrt{4\pi D_X t}} \exp\left[-\frac{(X - X_0 - U_0 t)^2}{4D_X t}\right] \tag{5.11}$$

In Fig. 1, the evolution of $W(X,t)$ for a soliton starting at $X_0 = -0.5$ with velocity $U_0 = 0.1$ is shown; as we can see, the distribution function moves at constant velocity U_0 while it spreads out, with the diffusion constant D of the noise being corrected by a factor $[U_0(1 - U_0^2)/2]^2$ due to the multiplicative character of the noise, so slower and faster solitons have less diffusion.

When comparing with numerical simulations, it is easier to study the mean values of the system variables. If we always start with a soliton located at the origin ($X_0 = 0$) and with initial velocity U_0 (i.e., $W(X,U;0) = \delta(X - X_0)$ $\delta(U - U_0)$), we can obtain the evolution of the most significant mean values either from equation (5.9) or from the associated Fokker-Planck equation (5.10), arriving at:

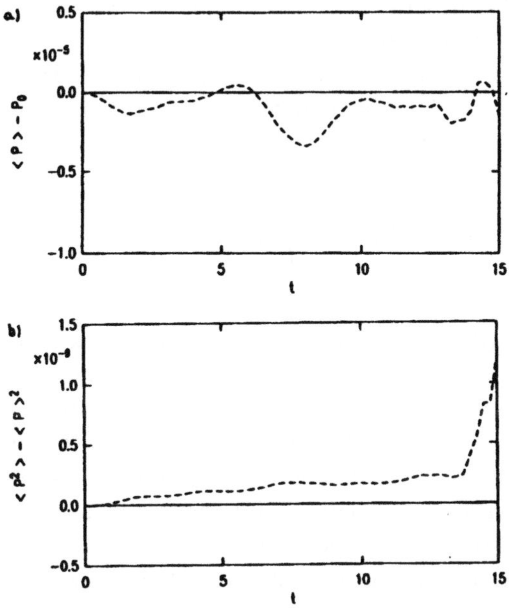

Fig. 4. Evolution of the mean values of the momentum and its deviation of a soliton under multiplicative noise. Comparison of a) $\langle P\rangle$, b) $\langle P^2\rangle - \langle P\rangle^2$ obtained through numerical simulation (dotted line) with the ones given by equation (5.12) (solid line) for a noise $\sigma = 0.5$; $X_0 = -0.$, $U_0 = 0.1$. Results are shown for 100 realizations.

$$\begin{aligned}
\langle P\rangle &= P_0 \\
\langle P^2\rangle - \langle P\rangle^2 &= 0 \\
\langle U\rangle &= U_0 \\
\langle U^2\rangle - \langle U\rangle^2 &= 0 \\
\langle X\rangle &= U_0 t \\
\langle X^2\rangle - \langle X\rangle^2 &= 2D_X t = \frac{1}{2}U_0^2(1-U_0^2)^2 D t
\end{aligned} \tag{5.12}$$

In figures 2, 3 and 4, comparison of the numerical simulation results with the ones given by the equation (5.12) are shown; mean values are obtained for 100 realizations of the noise. It can be seen that, while the momentum fits very well and remains constant with an almost zero deviation, the velocity decreases clearly with a seemingly linear decay. This behavior can be explained keeping in mind that the conservation of momentum is more general, and comes from the fact that we have a temporal perturbation, and so the system is spatially invariant and conserves the momentum. On the other hand, the constant character of the velocity comes from the ansatz: we assume that the effect of the perturbation is to vary the parameters that characterize a soliton (its position and velocity), discarding other effects. We see that this approximation has its limit, and that the discarded effects, the shape change of the soliton and the emission of radiation, are revealed in this case.

In (Parrondo et al., 1990), an expression is given for the evolution of the mean value of the energy of the SG equation under any stochastic perturbation; for our case this expression is:

$$\frac{d\langle E_{SG}\rangle}{dt} = D \int dx \, \langle [\sin(\phi)]^2 \rangle \tag{5.13}$$

In figure 5 a comparison of $\langle E \rangle$ obtained numerically with the above expression is shown, giving a good fit. We observe that $\langle E \rangle$ increases, this implies a decreasing of the velocity since $U = P/E$ and the momentum is constant. Thus we have that in mean value the system absorbs energy continuously from the stochastic perturbation, transforming this energy into radiation.

Regarding the position X, small discrepancies are observed, that may be attributed to the variation of U with time.

Figure 6 shows the evolution of a soliton and its energy density for one of the realizations of the noise considered in the mean values calculations. We can see that the effect of radiation is small enough to not be noticeable in this representation.

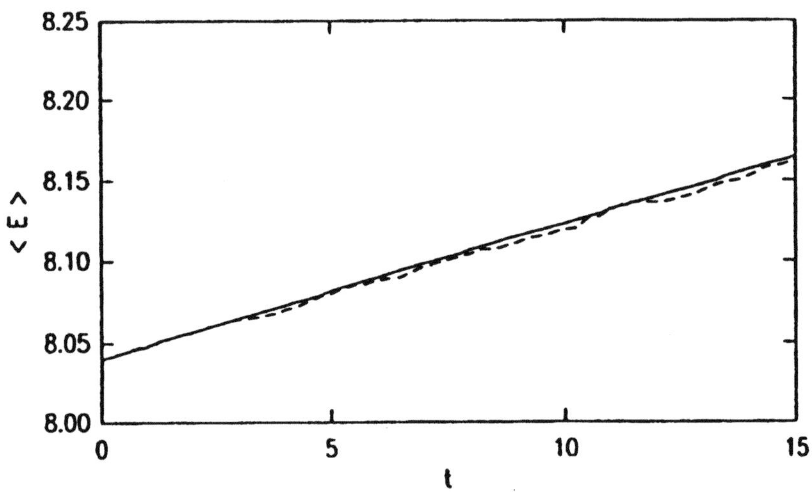

Fig. 5. Evolution of the mean value of the energy of a soliton under multiplicative noise. Comparison of $\langle E \rangle$ obtained through numerical simulation (dotted line) with the equation (5.13) (solid line) for a noise $\sigma = 0.5$; $X_0 = -0.$, $U_0 = 0.1$. Results are shown for 100 realizations.

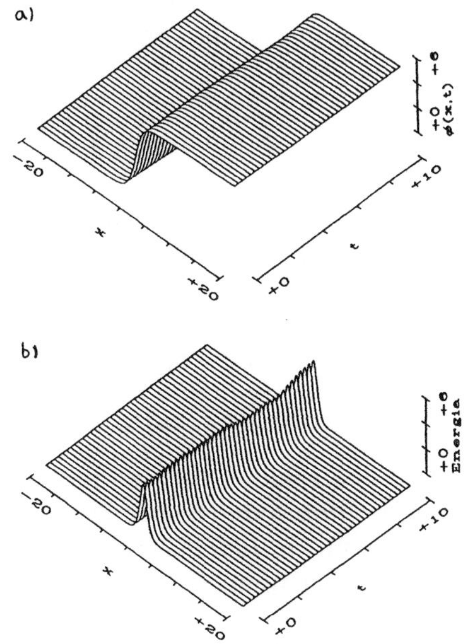

Fig. 6. Evolution of a soliton and its energy density under the multiplicative noise. The parameters are the following $\sigma = 0.5$; $X_0 = 0.$, $U_0 = 0.1$.

Comparing to the additive noise case described in (Pascual and Vázquez, 1985), the evolution of the velocity, discarding radiation effects, is deterministic, remaining constant with a zero mean deviation, while for the additive perturbation case, the mean value of the velocity is constant and its mean deviation is non zero. Regarding the position, for the multiplicative noise the stochastic perturbation appears explicitly in its evolution equation, and for the additive case the noise enters through the velocity. This affects the behavior of the mean deviation, linear in t for the multiplicative noise, while proportional to t^3 for the additive noise.

5.2 The breather

The adiabatic approximation for a breather is a bit more complicated because, besides the initial velocity and position, there are two new parameters: an internal frequency, $\cos\mu$, and its initial phase, $(\cos\mu)t_0$, which may vary with time:

$$\phi_B = 4\tan^{-1}\left(\frac{\tan\mu\sin[(\cos\mu)\gamma(t-ux-t_0)]}{\cosh[(\sin\mu)\gamma(x-ut-x_0)]}\right)\ ;\ \gamma = \frac{1}{\sqrt{1-u^2}} \tag{5.14}$$

More interesting than the evolution of the velocity and position of the center of mass, as in the case of the soliton, it is the evolution of the internal oscillation parameters, and specially the crossing through threshold values ($|\sin\mu| \leq 1$), which indicates the passing from a breather bound regime to a soliton-antisoliton pair free one.

So, let us concentrate in the study of the internal structure of the breather, and let us assume that it is centered at the origin with zero center of mass velocity, which corresponds to the equation:

$$\phi_B = 4\tan^{-1}\left(\frac{\tan\mu\sin[(\cos\mu)(t-t_0)]}{\cosh[(\sin\mu)x]}\right) = 4\tan^{-1}\left(\frac{k\sin(\omega_B t-\varphi)}{\omega_B\cosh(kx)}\right) \quad (5.15)$$

In this equation, we can see that when $|k| = |\sin\mu|$ becomes greater than 1, if we change $k = \sin\mu$ by $1/\sqrt{1-u^2}$, and consequently $\omega_B = \cos\mu$ by $\pm\imath u/\sqrt{1-u^2}$, we pass from a bound to a free regime, given by:

$$\phi_B = -4\tan^{-1}\left(\frac{\sinh\left[u(t-t_0)/\sqrt{1-u^2}\right]}{u\cosh\left[(x-x_0)/\sqrt{1-u^2}\right]}\right) \quad (5.16)$$

(as a matter of fact, the breather solution may be regarded as an analytical continuation of the soliton-antisoliton state (Bishop et al., 1989)).

In references (Legrand and Reinisch, 1987; Legrand, 1987), collective coordinates are given for a constant perturbation:

$$\phi_{tt} - \phi_{xx} + \sin\phi = \epsilon\ ;\ (\epsilon > 0)$$

in order to obtain the threshold value of the perturbation associated with the transformation of the breather into a soliton-antisoliton pair. Starting from the equality:

$$4\tan^{-1}\exp[k(x+z)] + 4\tan^{-1}\exp[-k(x-z)] - 2\pi = 4\tan^{-1}\left[\frac{\sinh(kz)}{\cosh(kx)}\right]$$

where $z = z(t)$, which shows the sum of a soliton and an antisoliton of wave number k as a breather profile, the right hand side of the equation is chosen as the ansatz for the breather, i.e.:

$$\sinh[kz(t)] = \frac{k}{\omega_B}\sin(\omega_B t+\varphi)\ ;\ k\le 1$$

In (Legrand and Reinisch, 1987) it is assumed that $k = 0$, so that only one collective coordinate is used; defining $X = 2kz$ as the collective coordinate which gives the internal separation between the two pulses of the breather, the Hamiltonian for the perturbed SG equation:

$$H = \int_{-\infty}^{\infty} dx\left(\frac{1}{2}\phi_t^2 + \frac{1}{2}\phi_x^2 + 1 - \cos\phi - \epsilon\phi\right)$$

becomes

$$H = \frac{2}{k}\left[1+\frac{X}{\sinh X}\right]\left[\dot{X}^2 + 4\tanh^2\left(\frac{X}{2}\right)\right] + 8k\left[1-\frac{X}{\sinh X}\right] - \frac{2\pi}{k}\epsilon X \quad (5.17)$$

By studying this Hamiltonian, we can obtain the behavior of the threshold value $\epsilon_{cr}(\omega_B, \varphi)$, obtaining a good qualitative fit with direct numerical simulations of the SG equation, except when φ approaches $\pi/2$. Around $\varphi = \pi/2$ oscillations of the breather are rapidly damped and at the same time, large amplitude nonlinear waves are generated, so that the transition from a breather to a soliton-antisoliton pair is not so clear, as it is for other values of the phase where these large amplitude waves do not appear.

In (Legrand, 1987), this ansatz is refined, with two collective coordinates, $k(t)$ and $Y(t) = kz(t)$. The Hamiltonian is now a bit more complex:

$$\begin{aligned} H = & \, 8\frac{\dot{Y}^2}{k}\left[1+\frac{2Y}{\sinh 2Y}\right] - 16Y\dot{Y}\frac{\dot{k}}{k^2} \\ & + \frac{2}{3}\frac{\dot{k}^2}{k^3}\left[(\pi^2+4Y^2)\left[1-\frac{2Y}{\sinh 2Y}\right]+8Y^2\right] \\ & - 8k\left[1-\frac{2Y}{\sinh 2Y}\right] - \frac{8}{k}\tanh^2 Y\left[1+2Y\sinh 2Y\right] + \frac{4\pi}{k}Y\epsilon \end{aligned} \tag{5.18}$$

where the ansatz for the breather is:

$$\phi(x, Y(t), k(t)) = 4\tan^{-1}\left[\frac{\sinh Y(t)}{\cosh[k(t)x]}\right] \tag{5.19}$$

Dissipation is also considered in this work, obtaining better results for $\varphi = \pi/2$, though stability is not clear when the breather becomes a soliton-antisoliton pair, because as the pulses get closer ($Y \to 0$), the numerical simulation becomes unstable [11]. This can be seen as an indication of the lack of validity of the ansatz (5.19) to obtain the Hamiltonian (5.18) when pulses get enough close so that they interact strongly. As a matter of fact, when $Y = 0$, only the kinetic energy term ($8\dot{Y}^2\left[1+2Y/\sinh 2Y\right]/k$), is not zero; this may mean troubles in determining the evolution of k in this point; it can also be shown that the equations for the momenta ($P = \partial H/\partial\dot{Y}$, $Q = \partial H/\partial\dot{k}$) and $\dot{Y}, \dot{k}$ are a non invertible system when $Y = 0$.

The results in (Legrand and Reinisch, 1987; Legrand, 1987) have been repeated applying the new conservative schemes, described in Section 4, to the Hamiltonian systems (5.17) and (5.18).

In the case of the evolution associated with the Hamiltonian (5.18), we observed that the system becomes unstable when the separation between the pulses, given by coordinate $Y(t)$, decreases to zero. Figure 7 shows the results when the perturbation is $\epsilon = 0.01$, and we can see how as $Y(t) \to 0$, $k(t)$ becomes unstable and decreases abruptly.

[11] The Hamiltonian in (Legrand, 1987) has a mistake in one sign, which is the cause of instability. That is why instability does not show up in (Legrand, 1987).

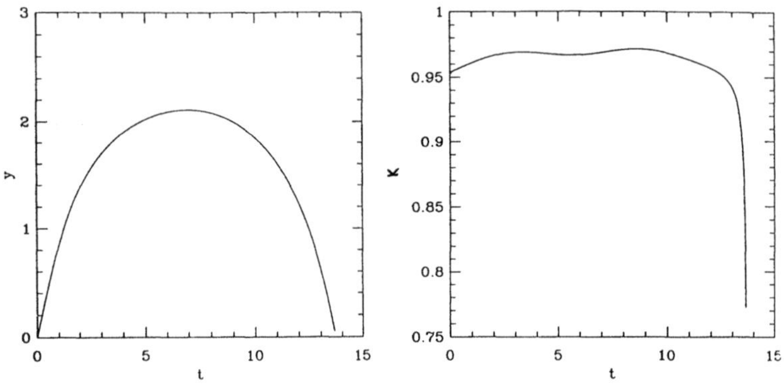

Fig. 7. Breather under a constant perturbation. Evolution of the collective coordinates $Y(t)$ and $k(t)$ from the Hamiltonian (5.18) with $\epsilon = 0.01, \omega_B = 0.3$ and $\varphi = 0$.

We tried to avoid the problem at $Y = 0$ by using the symmetries of the system, but the different terms in the equation has different kind of symmetries, and simulations couldn't overcome the problem. We also tried to expand the Hamiltonian around $Y = 0$, and keep only the first terms, using this approximated Hamiltonian when Y was near zero, but we did not succeed. Reference (Caputo and Flytzanis, 1990) mentions also more classical approaches (fourth order Runge-Kutta), and more elaborated ones (adaptive Runge-Kutta, with variable time step) without success in traversing $Y = 0$; in fact, it is mentioned that our approach seems more reliable as it conserves energy, a key point when k becomes unstable, because it is no longer constant, which is an indication that the problem is in the starting equation and not in the particular numerical scheme used.

It is worthwhile to try to clarify the causes of this instability, as discussed in (Caputo and Flytzanis, 1990). If

$$L(a_i, \dot{a}_i, t) = \int \mathcal{L}(\phi_t, \phi_x, \phi, t) dx$$

is the Lagrangian of the system as a function of the collective coordinates $\{a_i\}$ (Y, k in our case), and $\mathcal{L}(\phi_t, \phi_x, \phi, t)$ is the Lagrangian density corresponding to the SG equation, where ϕ depends on the chosen ansatz, $\phi = \phi(x, a_i(t))$, then we can obtain the evolution of the collective coordinates:

$$\frac{\partial L}{\partial a_i} - \frac{d}{dt}\left(\frac{\partial L}{\partial \dot{a}_i}\right) = 0$$

that is to say

$$\int_{-\infty}^{\infty} dx \left[\frac{\partial l}{\partial \phi}\frac{\partial \phi}{\partial a_i} + \frac{\partial l}{\partial \phi_x}\frac{\partial \phi_x}{\partial a_i} + \frac{\partial l}{\partial \phi_t}\frac{\partial \phi_t}{\partial a_i} + \frac{\partial}{\partial t}\left(\frac{\partial l}{\partial \phi_t}\frac{\partial \phi_t}{\partial \dot{a}_i}\right)\right] = 0$$

Using the equality $\frac{\partial \phi_t}{\partial a_i} = \frac{\partial \phi}{\partial a_i}$, integrating and assuming that $\frac{\partial \phi}{\partial a_i} \to 0$ when $x \to \pm\infty$, then:

$$\int_{-\infty}^{\infty} dx \left[\frac{\partial l}{\partial \phi} - \frac{\partial}{\partial t}\left(\frac{\partial l}{\partial \phi_t}\right) - \frac{\partial}{\partial x}\left(\frac{\partial l}{\partial \phi_x}\right)\right] \frac{\partial \phi}{\partial a_i} = 0$$

The expression in brackets corresponds to the left hand side of the SG equation, obtained imposing a space-time stationary variation of $\mathcal{L}(\phi_t, \phi_x, \phi, t)$; the above equation may then be seen as if the projection of the SG operator over the mode $\partial\phi/\partial a_i$ is zero, and this is the interpretation used in (Legrand, 1987) to obtain the evolution for the dissipative case, projecting the perturbed SG equation over $\partial\phi/\partial a_i$. For the (Y, k) coordinates we have:

$$\frac{\partial \phi}{\partial Y} = 4\frac{\cosh Y \cosh^2 kx}{1 + \sinh^2 Y} \; ; \; \frac{\partial \phi}{\partial k} = -4\frac{x \sinh Y \sinh kx}{\cosh^2 kx + \sinh^2 Y}$$

We observe then that while the first mode, $\partial\phi/\partial Y$, never becomes zero, the second one, $\partial\phi/\partial k$, is zero when $Y = 0$, so that the evolution equation for k holds for any evolution of Y and k. We are projecting the SG operator over a zero mode, which does not provide enough information, and the system may change to another trajectory different of the one where the ansatz fulfills the SG equation.

This is true for every ansatz and every equation derived from a Lagrangian, and even, by using the interpretation in (Legrand and Reinisch, 1987), for any perturbed equation, like the above mentioned dissipative case. However, the ansatz that we have used for the soliton is not of the same type, but imposes conditions not only to $\phi(x,t)$ but to $\phi_t(x,t)$ also, and this is the base for obtaining the evolution equations of the collective coordinates of the soliton.

We think that is more interesting the approach suggested in (Bishop et al., 1989). The parameters used there to describe the evolution of a soliton-antisoliton pair and a breather are the values of the wave function and its temporal derivative at the origin:

$$\begin{aligned} u(t) &\equiv \phi(x = 0, t) \\ \tilde{u}(t) &\equiv \phi_t(x = 0, t) \end{aligned} \tag{5.20}$$

For the case we are studying, where the center of mass remains at the origin, there are some equations that relate the parameters of a soliton-antisoliton pair (velocity and time to collision) or a breather (internal frequency and initial phase) with these new parameters. Moreover, for pure SG equation, the evolution equations for u and $\tilde{u}$:

$$\begin{aligned} u_t &= \tilde{u} \\ \tilde{u}_t &= -\sin u - \tan\frac{u}{4}\left[\frac{\tilde{u}^2}{4} + \sin^2\frac{u}{2}\right] \end{aligned} \tag{5.21}$$

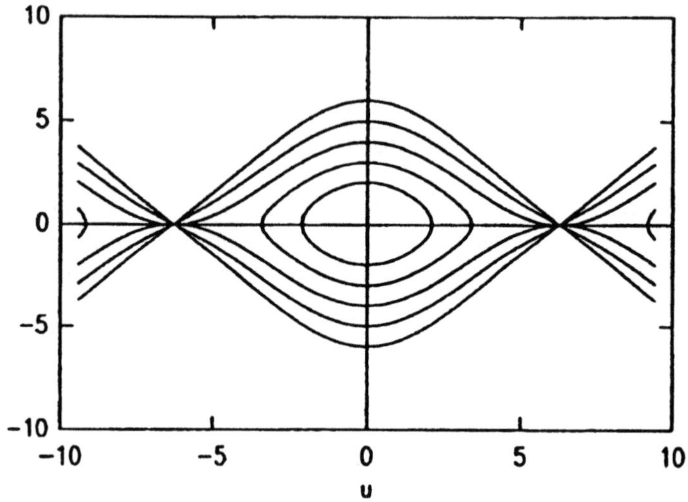

Fig. 8. Phase map of the parameters $(u, \tilde{u})$ corresponding to a breather and a $k - \bar{k}$ pair state.

are the same for a breather and for a soliton-antisoliton pair, so it is easier the study of the transition from one to the other one. The energy in terms of $(u, \tilde{u})$ is:

$$E(u, \tilde{u}) \frac{8}{\cos(u/4)} \sqrt{\frac{\tilde{u}^2}{4} + \sin^2 \frac{u}{2}}$$

and from the analysis of the phase map $(u, \tilde{u})$ (see figure 8) we see a separatrix for $E = 16$, given by $\tilde{u} = \pm 4 \cos^2 u/4$, between the oscillatory states of the breather and that of a $k - \bar{k}$ pair. There are also two singular points, $(\pm 2\pi, 0)$, which are reached for $t \to \pm\infty$ by the $k - \bar{k}$ pairs.

Looking at the SG equation as describing a coupled chain of oscillators, what we are doing is to describe the behavior of the system as a whole by the behavior of the oscillator at the origin. The transition from a breather to a $k - \bar{k}$ pair is clear because the system goes from a closed cycle in the phase map $(u, \tilde{u})$ to one of the singular points $(\pm 2\pi, 0)$. The decay to radiation is not so clear, because as the breather decays (the cycle becomes a spiral), it can end up in a small radius cycle around $(0, 0)$, which should be studied in detail to determine if it corresponds to either a breather or a phonon (looking for instance at its frequency: if it is less than 1 is a breather; if greater, it is radiation). The advantage of using these coordinates is because graphically we can study the transitions we are interested in.

So this is the approach we have followed when studying the behavior of a breather under stochastic perturbations. The figure 9 compares the behavior of a breather of the pure SG equation, with another of the same characteristics $(\varphi = 0, \omega_B = 0.8, k = 0.6)$ under multiplicative stochastic perturbations of the same type of the soliton case.

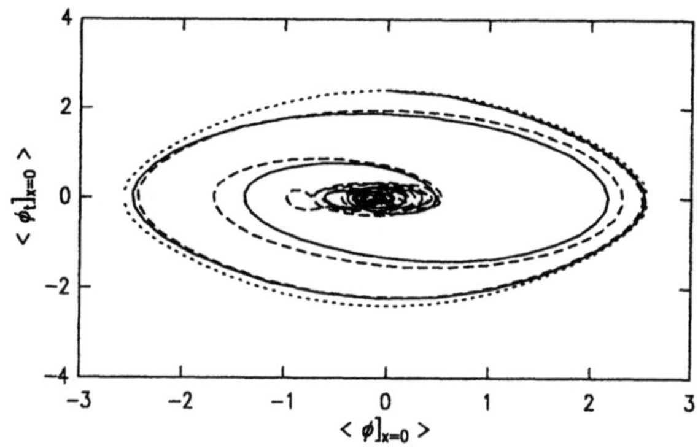

Fig. 9. Phase map of the parameters $(u, \tilde{u})$ of a breather without perturbation (dotted line) and under multiplicative noise (solid line for 100 realizations, dashed line for 50 realizations). Parameters of the breather: $\varphi = 0, \omega_B = 0.8(k = 0.6)$; noise: $\sigma = 1.0$.

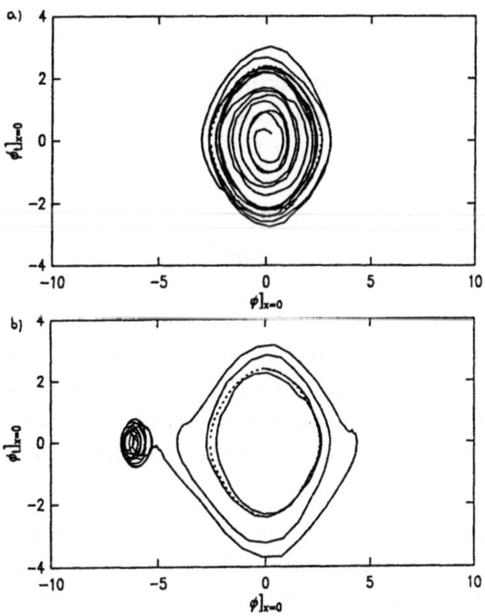

Fig. 10. Phase map of a breather. Phase map of the parameters $(u, \tilde{u})$ of a breather without perturbation (dotted line) and under multiplicative noise (solid line). Parameters of the breather: $\varphi = 0, \omega_B = 0.8(k = 0.6)$; noise: $\sigma = 1.0$. a) the breather decays into radiation; b) the breather decays to a $k - \bar{k}$ pair.

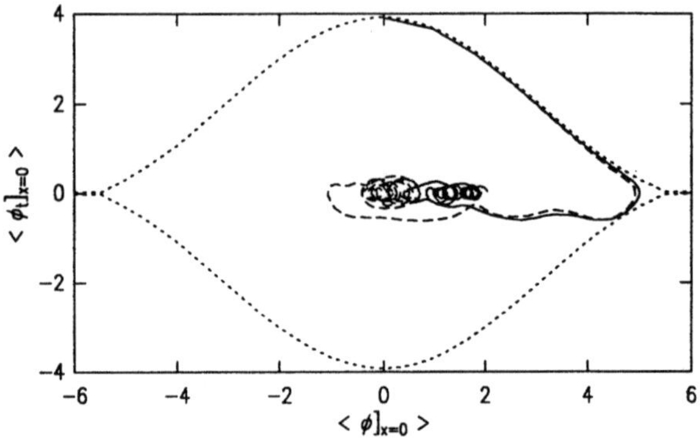

Fig. 11. Phase map of the parameters $(u, \tilde{u})$ of a breather without perturbation (dotted line) and under multiplicative noise (solid line for 100 realizations, dashed line for 50 realizations). Parameters of the breather: $\varphi = 0, \omega_B = 0.8 (k = 0.6)$; noise: $\sigma = 1.0$.

We observe how the mean values decay to the origin (cycle becomes a spiral, and amplitude and velocity decay), which suggests a decay to radiation. However we must keep in mind that we are taking mean values, and it could be the case that, though the breather does not loose its identity in every realizations, its phase changes randomly from one to the other, so that the mean value would tend to zero.

We tried to clarify this point by studying the evolution of a breather for different realizations of the noise. We found that it did not maintain its identity in any case; in most of the cases, it decayed to radiation, while in some others it became a $k - \bar{k}$ pair. Figure 10 shows examples of both cases: in a) the breather disappears in form of radiation and in b) we see a decay to a $k - \bar{k}$ pair, corresponding to the point $(-2\pi, 0)$ of the phase map. The decay to radiation was very different from one realization to another: in some cases it was very quick, while in some others it was very slow, as if the breather were standing up very well the noise.

Opposite to the previous case, where the breather had initially a small energy and was near initially to the radiation solutions ($\omega_B = 1.$), we studied also the case where the breather was initially near the $k - \bar{k}$ behavior (ω_B close to 0). Figure 11 shows the mean value of the trajectory for the same noise intensity and initial parameters of the breather. Now, we observe clearly how the mean values hide the individual behavior of each realization: the horizontal line to the origin is due to that the most frequent case now is the decay to $k - \bar{k}$ pairs, both at point $(2\pi, 0)$ and $(-2\pi, 0)$. In fact, it seems that the mean trajectory tends to a point to the right of the origin, perhaps because it is more likely the decay to $(2\pi, 0)$. Wwe must keep in mind that the breather always starts with positive velocity, so that this point is the $k - \bar{k}$ boundary closer in time. Besides these kinds of decay, there were some others in which the breather decayed to radiation as well.

5.3 Soliton under dissipation

When we have the linear dissipation term $-\alpha\phi_t$, the evolution equations of the soliton parameters become:

$$\begin{aligned}\dot{U} &= -\alpha U(1-U^2)\\ \dot{X} &= U - \frac{1}{2}U(1-U^2)\nu(t)\\ \dot{P} &= -\alpha P\end{aligned} \tag{5.22}$$

As for the case without dissipation, the evolution of momentum and velocity are deterministic:

$$\begin{aligned}P(t) &= P_0 e^{-\alpha t}\\ U(t) &= \frac{U_0 e^{-\alpha t}}{\sqrt{1-U_0^2(1-e^{-2\alpha t})}}\end{aligned}$$

The evolution of the velocity and the momentum corresponds to a relativistic dissipative system with a decay time $\tau = 1/\alpha$.

The equation for the evolution of the position remains unchanged when introducing a dissipative term, which only has an indirect effect, through $U(t)$, in the evolution of $X(t)$. The Fokker-Planck equation for X is similar to (5.10), where we must substitute U_0 by $U(t)$ because the velocity is no longer constant. The solution for $W(X;0) = \delta(X - X_0)$ is similar to (5.11)

$$W(X;t) = \frac{1}{\sqrt{4\pi \bar{D}_X(t)}} \exp\left\{-\frac{(X - x(t))^2}{4\bar{D}_X(t)}\right\}$$

where

$$\begin{aligned}x(t) &= X_0 + \int_0^t U(t)dt\\ &= X_0 + \frac{1}{\alpha}\left[\sinh^{-1}\left(\frac{U_0}{\sqrt{1-U_0^2}}\right) - \sinh^{-1}\left(\frac{U_0 e^{-\alpha t}}{\sqrt{1-U_0^2}}\right)\right].\end{aligned}$$

and

$$\begin{aligned}\bar{D}_X(t) &= \int D\left[\frac{1}{2}U(t)(1-U^2(t))\right]^2 dt\\ &= \frac{D}{16\alpha}(1-U_0^2)^2\left\{\frac{1}{[1-U_0^2(1-e^{-2\alpha t})]^2} - 1\right\}\end{aligned}$$

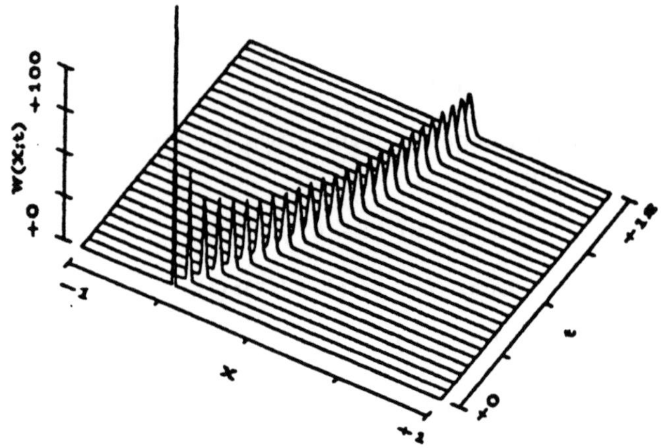

Fig. 12. Evolution of the probability density $W(X;t)$ of the position of a soliton under multiplicative noise and dissipation. $X_0 = -0.5, U_0 = 0.1, \alpha = 0.1, \sigma = 1.$.

Figure 12 shows the evolution of this distribution function. The soliton slows down in a time of order $\tau = 1/\alpha$, stopping at position $X_0 + \sinh^{-1}[U_0/\sqrt{1-U_0^2}]\alpha$.

So, while the noise only affects the evolution of the position, the effect of dissipation is on the velocity (and momentum), appearing only indirectly in the evolution of the position.

The evolution of the mean values is, assuming $W(X;0) = \delta(X)$:

$$\begin{aligned}
\langle U\rangle &= \frac{U_0 e^{-\alpha t}}{\sqrt{1-U_0^2(1-e^{-2\alpha t})}} \\
\langle U^2\rangle - \langle U\rangle^2 &= 0 \\
\langle X\rangle &= \frac{1}{\alpha}\left[\sinh^{-1}\left(\frac{U_0}{\sqrt{1-U_0^2}}\right) - \sinh^{-1}\left(\frac{U_0 e^{-\alpha t}}{\sqrt{1-U_0^2}}\right)\right] \\
\langle X^2\rangle - \langle X\rangle^2 &= 2\bar{D}_X(t) = \frac{D}{8\alpha}(1-U_0^2)^2\left\{\frac{1}{\left[1-U_0(1-e^{-2\alpha t})\right]^2} - 1\right\}
\end{aligned} \tag{5.23}$$

As for the case without dissipation, figures 13 and 14 show the comparison of the numerical simulation with equation (5.23), taking mean values over 100 realizations. We see a good qualitative fit and that the dissipation hides the discrepancies that showed before in the case without dissipation.

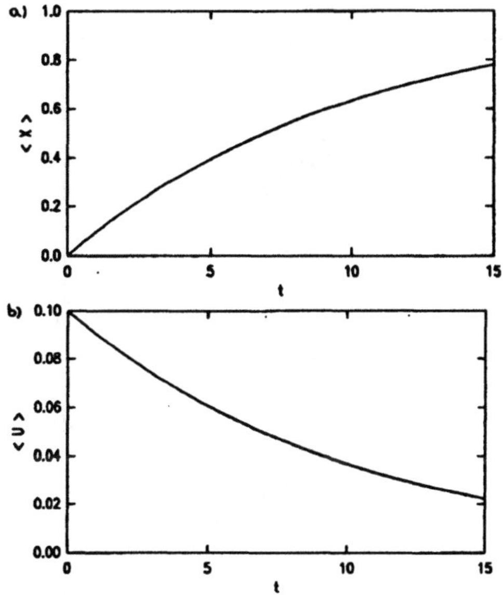

Fig. 13. Evolution of the mean values of the position and velocity of a soliton under multiplicative noise and dissipation. Comparison of a) $\langle X \rangle$, b) $\langle U \rangle$ obtained through numerical simulation (dotted line) with equation (5.23) (solid line) for a noise $\sigma = 0.5$; $X_0 = 0., U_0 = 0.1, \alpha = 0.1$. Mean values were taken over 100 realizations.

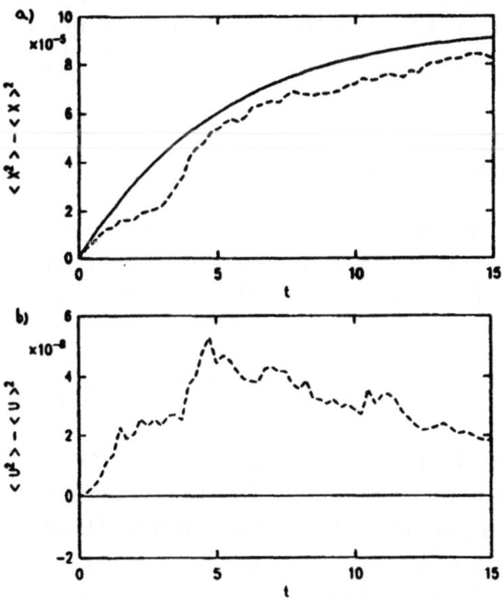

Fig. 14. Evolution of the mean deviations of the position and velocity of a soliton under multiplicative noise and dissipation. Comparison of a) $\langle X^2 \rangle - \langle X \rangle^2$, b) $\langle U^2 \rangle - \langle U \rangle^2$ obtained through numerical simulation (dotted line) with equation (5.23) (solid line) for a noise $\sigma = 0.5$; $X_0 = 0., U_0 = 0.1, \alpha = 0.1$. Mean values were taken over 100 realizations.

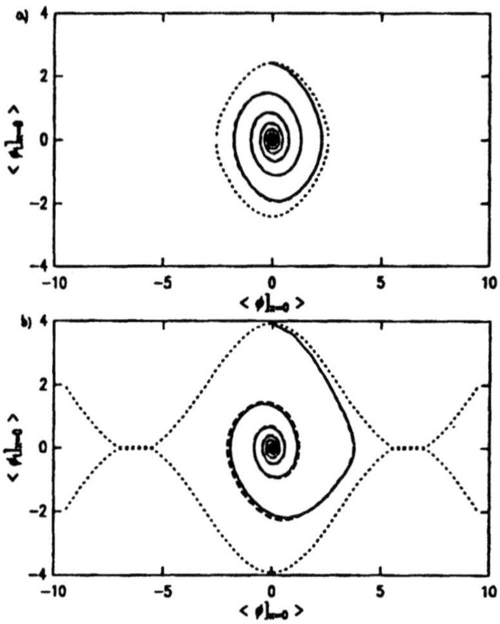

Fig. 15. Phase map of a breather. Phase map of the parameters $(u, \tilde{u})$ of a breather without perturbation (dotted line) and under multiplicative noise and dissipation (solid line for 100 realizations, dashed line for 50 realizations). Parameters of the breather: $\varphi = 0$, dissipation: $\alpha = 0.1$; noise: $\sigma = 1.0$. a) $\omega_B = 0.8(k = 0.6)$ b) $\omega_B = 0.2(k = 0.98)$.

5.4 Breather under dissipation

Regarding the case of a breather under dissipation and multiplicative noise, we would expect that the decay into radiation should be very strong now. However, in (Grønbech-Jensen et al., 1991) the perturbations of the form:

$$\phi_{tt} - \phi_{xx} + \sin\phi = -\alpha\phi_t - \epsilon_n \sin\omega t \sin\frac{\phi}{n}$$

are studied (ϵ_n measures the intensity of the perturbation, of different kind for each n), and it is observed the breather stabilization when it has a frequency half of the parametric force frequency. That work contains theoretical explanations of this behavior as well as numerical simulations to check the estimation of the threshold field ϵ_n, from which the breather does not decay and maintains itself.

The most similar case to our perturbation corresponds to $n = 1$; as white noise may be regarded as containing every frequency, we could imagine the breather absorbing energy corresponding to the mode $\omega = 2\omega_{\text{br}}$, so that it does not decay, as long as the perturbation is strong enough.

Figure 15 shows the mean trajectory for 100 (50) realizations of the phase map $(u, \tilde{u})$. It is still valid that the approaching of the trajectory to the origin is not necessarily evidence of a decay to radiation instead of a sign of the randomness in the phase. However, we see now less difference between the mean value of 50 trajectories and 100, which indicates that the behavior now is not so determined by the stochastic perturbation but by the dissipation. We checked this by studying the phase map of different realizations: the decay to radiation was very similar from realization to realization, and we did not find any creation of a $k-\bar{k}$ pair. In this study, we varied the intensity of the noise and the parameters of the breather so that they were similar to the values in which the breather was stable under a parametric force.

6. Sine-Gordon Equation Under Additive Noise and Periodic Perturbations

In this section, we will study the behavior of the solutions of the SG equation in the presence of stochastic and time periodic perturbations, with and without dissipation in the system:

$$\phi_{tt} - \phi_{xx} + \sin\phi + \alpha\phi_t = \nu(t) + F\cos(\omega t)$$

We will follow the scheme developed in the previous section, and in particular, we will use the equations (5.8) to find the analytical approximations to the evolution of the mean values of the parameters associated with the soliton.

The results of this section will be useful when facing the CDW problem, where more complex couplings show up. Another case of interest is also when the system is periodically perturbed by an *ac* electric field.

This study expands the results obtained for the case of additive noise with dissipation ($F = 0$) in (Pascual, 1984; Pascual and Vázquez, 1985).

6.1 The soliton

In this case, the perturbed SG equation (5.2) is:

$$\begin{aligned} \phi_{tt} - \phi_{xx} + \sin\phi &= f(\phi, \phi_t; t) \\ f(\phi, \phi_t; t) &= -\alpha\phi_t + \epsilon R(\phi; t) \end{aligned}$$

with

$$\epsilon R(\phi; t) = \nu(t) + F\cos(\omega t)$$

When there is no dissipation, ($\alpha = 0$), and with the ansatz (5.6), we get the evolution equations:

$$\begin{aligned} \frac{dU}{dt} &= -\frac{\pi}{4}(1-U^2)^{3/2}\left(\nu(t) + F\cos(\omega t)\right) \\ \frac{dX}{dt} &= U \\ \frac{dP}{dt} &= -2\pi(\nu(t) + F\cos(\omega t)) \end{aligned} \tag{6.1}$$

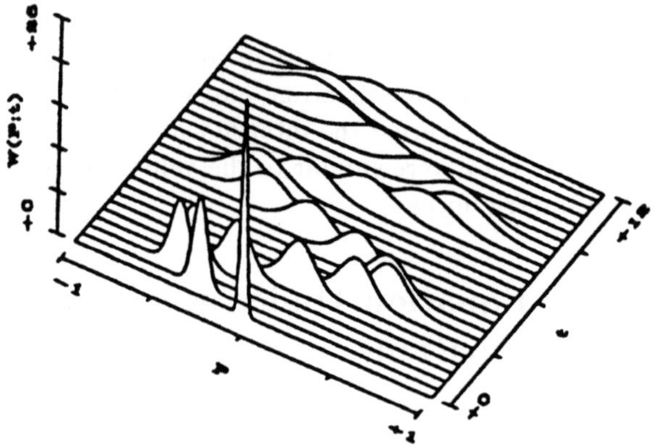

Fig. 16. Evolution of the probability distribution function $W(P,t)$ of the momentum corresponding to a soliton under additive noise and periodic perturbations; $P_0 = 0$, $F = 0.1$, $\omega = 0.4\pi$, $\sigma = 0.05$.

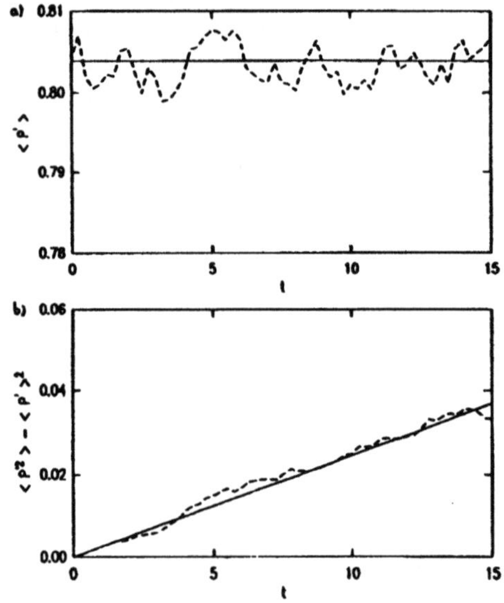

Fig. 17. Evolution of the mean values of the momentum corresponding to a soliton under additive noise and periodic perturbations. Comparison of a) $\langle P\rangle$, b)$\langle P^2\rangle - \langle P\rangle^2$ obtained through numerical simulation (dotted line) with equation (6.3) (solid line); $\sigma = 0.05$, $F = 0.05$, $\omega = 1.2$, $X_0 = 0$, $U_0 = 0.1$. We show the mean values over 100 realizations.

In the case of the momentum, we can consider the following change of variable:

$$P' = P + \frac{2\pi F}{\omega} \sin(\omega t)$$

so that the evolution equation become:

$$\dot{P}' = -2\pi\nu(t) \ , \ (P_0' = P_0)$$

The Fokker-Planck equation for the distribution function $W(P';t)$ associated with this problem is (Risken, 1984):

$$\frac{\partial W}{\partial t} = D' \frac{\partial^2 W}{\partial P'^2} \ , \ D' = 4\pi^2 D$$

where D is defined through the equation:

$$\langle \nu(t)\nu(t+\tau) \rangle = 2D\delta(\tau) \tag{6.2}$$

and whose solution for $W(P';0) = \delta(P' - P_0')$ (i.e., the initial state is the same for every realization, P_0') is a gaussian distribution function:

$$W(P';t) = \frac{1}{\sqrt{4\pi D' t}} \exp\left[-\frac{(P' - P_0')^2}{4D't}\right]$$

Coming back to the original momentum:

$$W(P;t) = W(P';t)\frac{dP'}{dP} = \frac{1}{\sqrt{4\pi D' t}} \exp\left[-\left(P - P_0 + \frac{2\pi F}{\omega}\sin(\omega t)\right)^2 / (4D't)\right]$$

Figure 16 shows an example of the evolution of this probability function, and we can see how it spreads out while it oscillates.

From these results we may obtain the evolution of the mean values:

$$\begin{aligned} \langle P' \rangle &= P_0' \\ \langle P'^2 \rangle - \langle P' \rangle^2 &= 2D't \\ \langle P \rangle &= P_0 - \frac{2\pi F}{\omega} \sin(\omega t) \\ \langle P^2 \rangle - \langle P \rangle^2 &= 8\pi^2 Dt \end{aligned} \tag{6.3}$$

Figure 17 shows the comparison of these equations with the results of the numerical simulation. The mean value of the momentum P' remains constant, while the deviation $\langle P^2 \rangle - \langle P \rangle^2$ has a linear behavior in t.

Regarding velocity, there is also a change of variables which allows to move to the random walk problem. If we consider the change

$$U' = \frac{U}{\sqrt{1 - U^2}} + \frac{\pi F}{4\omega} \sin(\omega t)$$

we have:

$$\dot{U}' = -\frac{\pi}{4}\nu(t)$$

which is again the random walk problem, with a distribution function:

$$W(U';t) = \frac{1}{\sqrt{4\pi D_U t}} \exp\left[-\frac{(U'-U'_0)^2}{4D_U t}\right] \ , \ D_U = \frac{\pi^2}{16}D$$

Coming back to our original velocity:

$$W(U;t) = W(U';t)\frac{dU'}{dU} =$$
$$= \frac{\gamma^3}{\sqrt{4\pi D_U t}} \exp\left[-\left(\gamma U - \gamma_0 U_0 + \frac{\pi F}{4\omega}\sin(\omega t)\right)^2 / 4D_U t\right]$$
$$\gamma_0 = \frac{1}{\sqrt{1-U_0^2}}$$

This expression can be obtained also directly from the distribution function of the momentum $W(P;t)$, by substituting the expression of the momentum (5.7), and keeping in mind that $W(U;t) = W(P;t)dP/dU$.

In the newtonian limit for low velocities, $|u| \ll 1$, we may approximate the evolution equations with:

$$\frac{dU}{dt} = -\frac{\pi}{4}(\nu(t) + F\cos(\omega t))$$
$$\frac{dX}{dt} = U \tag{6.4}$$

and there is also a change of variables to obtain the random walk problem:

$$X' = X - \frac{\pi F}{4\omega^2}\cos(\omega t)$$
$$U' = U + \frac{\pi F}{4\omega}\sin(\omega t)$$
$$\dot{U}' = -\frac{\pi}{4}\nu(t) \tag{6.5}$$
$$\dot{X}' = U'$$

Then, we may use the expression of the Fokker-Planck equation for the joint probability distribution of X', and U', $W(X',U';t)$,:

$$\frac{\partial W}{\partial t} = -U'\frac{\partial W}{\partial X'} + \frac{1}{2}S'_a\frac{\partial^2 W}{\partial U'^2} \ ; \ S'_a = \frac{\pi^2}{8}D$$

to obtain the evolution of the mean values of X' and U' through the equation:

$$\langle \frac{d}{dt} f(X',U') \rangle = \int\int f(X',U') \frac{\partial W(X',U';t)}{\partial t} dX'dU'$$

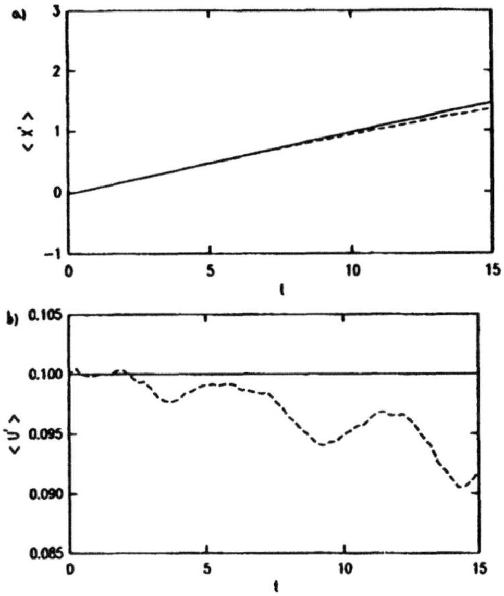

Fig. 18. Evolution of the mean values of the velocity and position of a soliton under additive noise and periodic perturbations. Comparison of a) $\langle X' \rangle$, b)$\langle U' \rangle$ obtained through numerical simulation (dotted line) with equation (6.6) (solid line); $\sigma = 0.05$, $F = 0.05$, $\omega = 1.2$, $X_0 = 0$, $U_0 = 0.1$. We show the mean values over 100 realizations.

We obtain then:

$$\begin{aligned} \langle U' \rangle &= \langle U' \rangle_0 \\ \langle U'^2 \rangle - \langle U' \rangle^2 &= S'_a t \\ \langle X' \rangle &= \langle X' \rangle_0 + \langle U' \rangle_0 t \\ \langle X'^2 \rangle - \langle X' \rangle^2 &= \frac{1}{3} S'_a t^3 \end{aligned} \tag{6.6}$$

These expressions may be translated to the original position and velocity.

In figures 18 and 19 we compare these expressions with the results obtained through numerical simulation. We observe a good qualitative agreement and the decrease of the velocity, we have already discussed for the multiplicative noise. We also see that $\langle X'^2 \rangle - \langle X' \rangle^2$ has a t^3 behavior as we mentioned in the multiplicative case, while $\langle U'^2 \rangle - \langle U' \rangle^2$ is linear in t, as it was the behavior of X for that case. We also observe how the correction for the diffusion constant D, which is now $\pi^2/16D$ for the position X', is independent of the velocity. Finally, we must remark that the intensity of the noise for this case was an order of magnitude smaller than in the multiplicative case, but the effects were of the same order.

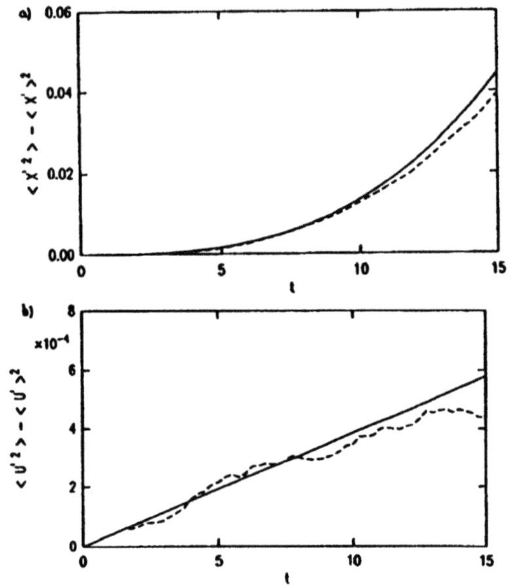

Fig. 19. Evolution of the mean deviations of the velocity and position of a soliton under additive noise and periodic perturbations. Comparison of a) $\langle X'^2 \rangle - \langle X' \rangle^2$, b)$\langle U'^2 \rangle - \langle U' \rangle^2$ obtained through numerical simulation (dotted line) with equation (6.6) (solid line); $\sigma = 0.05$, $F = 0.05$, $\omega = 1.2$, $X_0 = 0$, $U_0 = 0.1$. We show the mean values over 100 realizations.

6.2 The breather

The study of the behavior of a breather in the presence of an external oscillating field has a special interest because of the possibility of resonances between the internal frequency and the external one, and also because the real applications of such perturbations in the real world: *ac* fields applied to physical systems like Josephson junctions and charge density waves, as we will see later on.

We will study the joint effect of temporal noise with an *ac* field in the presence of dissipation in the more general case. While the noise has not been studied to date, for the periodic perturbation with and without dissipation there are some known results that we will outline here:

1. When there is only dissipation, the breather decays to extinction, because there is no topological obstacle, as for the case of the soliton, and the amplitude decreases to zero.
2. When there is only a periodic perturbation, the system may enter into a chaotic regime (Kivshar et al., 1989; Nozaki, 1982), because of a stochastic instability that leads to the creation of soliton-antisoliton pairs in a chaotic manner.
3. When both are present, dissipation and periodic perturbation, the energy supplied through the external force may balance the losses through dissipa-

tion, and a self-maintained breather may occur (Lomdahl and Samuelsen, 1986), and whose frequency may have a modulation in the presence of small perturbations, around the stationary state of the self-maintained breather (Lomdahl and Samuelsen, 1988). We have already seen another case of breather stabilization in a dissipative system, when there exists a parametric external force (Grønbech-Jensen et al., 1991), $\epsilon_n \sin\omega t \sin\phi/n$. In this case, the frequency of the self-maintained breather is half of the external field frequency.

We will describe briefly the stochastic instability that appears when there is only an oscillating external force (Nozaki, 1982). The parametrization chosen in this work is:

$$\phi(x,t) = 4\tan^{-1}\left(\frac{T(t)}{\cosh 2k(x-x_0)}\right)$$

which resembles the equation (5.19), if $Y(t) = T(t)$ and k becomes $2k$. We already explained the problems of this kind of parametrization to describe the transition between the breather and a soliton-antisoliton pair. As a matter of fact, it has the same characteristic as (5.19) that may cause instability, because:

$$\frac{\partial\phi}{\partial k} = -\frac{8T(t)\sinh[2k(x-x_0)]}{\cosh^2[2k(x-x_0)] + T^2(t)}$$

and again, $\partial\phi/\partial k$ mode becomes zero when $T(t) = 0$, i.e., when the breather becomes flat, or, from another point of view, when the soliton-antisoliton pair collides. As we discussed in the previous section, this may cause problems to determine the evolution of k because the Lagrange equations are being projected over a zero mode which gives no information, because the evolution equation of k holds independently of the evolution of $(T(t), k)$ in time. However, the instability is obtained from the hamiltonian, while the evolution equation are used only in the numerical simulation.

$T(t)$ and k may be related to the characteristic parameters of either the breather (frequency ω_B) or the soliton-antisoliton pair (velocity u). We can use this ansatz to obtain the perturbed hamiltonian, and using the appropriate canonical variables, we find the evolution equations of $(T(t), k)$.

Putting $T(t)$ as a function of ψ, the phase of the breather, $T(t) = \tan\theta \sin\psi$ (where $k = \frac{1}{2}\sin\theta$), and by obtaining the evolution equation of this phase, we can find that the breather has a resonance with frequencies equal to its internal frequency, $\omega_B = \cos\theta$ for low energy breathers, and with this frequency and its odd higher harmonics, $(2n+1)\cos\theta$, for high energy breathers ($H_B = 16\sin\theta$, with $\theta \simeq \pi/2$).

The chaotic behavior or stochastic instability of high energy breathers has its origin in the overlapping of these resonances. In order to find the condition under which this overlapping occurs, we expand the hamiltonian as a Fourier series in ψ, estimating the width of each resonance region as $\delta\theta_n \approx (\pi F/n)^{1/2}$, and the separation between resonance islands as $\Delta\theta_n \approx \omega/2n^2$. If we impose the overlapping condition, $\delta\theta_n > \Delta\theta_n$, we obtain a stochastic region of width

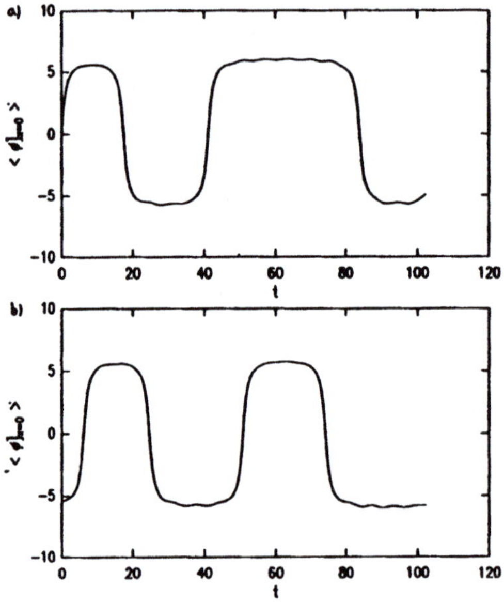

Fig. 20. Evolution of $\phi(x=0,t)$ for a breather of frequency $\omega_B = 0.2$, zero velocity, under a periodic perturbation $F \sin \omega t$, with $F = 0.01$ and $\omega = 0.9$. Breather initial phase a) $\psi_0 = 0$, b) initial phase $\psi_0 = 5.03$.

$\delta\theta \approx (\pi F \omega/2)^{1/2}$ around $\theta = \pi/2$ ($\omega_B = \cos\theta = 0$, the boundary between the breather and soliton-antisoliton pair behavior).

Holes are found, through numerical simulation of the evolution equations, at $\cos\theta = 0, \psi = 0, \pi, 2\pi$; through these holes, the breather becomes a soliton-antisoliton pair. We must keep in mind that for the breather $T(t) = \tan\theta \sin\psi$, and these are the only points of transition to the other regime, where $T(t)$ becomes undetermined. That is why $k - \bar{k}$ are randomly created, and that is what is called stochastic instability. In some sense, this is related to the remarks we made to the parametrization (5.19), because as the breather develops into a $k - \bar{k}$ pair, $T(t)$ becomes undetermined. In the above case, the system becomes unstable when the pair collapsed, and the separation parameter Y becomes zero; in this case this does not happen, though we have no details on the numerical simulation. We must also note that this stochastic behavior only happens to high energy breathers, where the approximation $\mid T(t) \mid = \mid \tan\theta \sin\psi \mid \gg 1$, as for low energy breathers there is only one resonance and there is no possibility of overlapping. The soliton-antisoliton pair, once created, behaves as a relativistic particles, interchanging energy and momentum periodically as a result of the external field.

We tried to verify these results through direct numerical simulation of the SG equation, and then study the behavior of the system under noise. We took values similar to those in (Nozaki, 1982): $\omega_B = 0.2$ for the breather frequency, $\omega = 0.9$ for the external oscillating field frequency and $F = 0.01$, though our perturbation was $F \sin \omega t$, instead $F \cos \omega t$ (a $\pi/2$ phase difference).

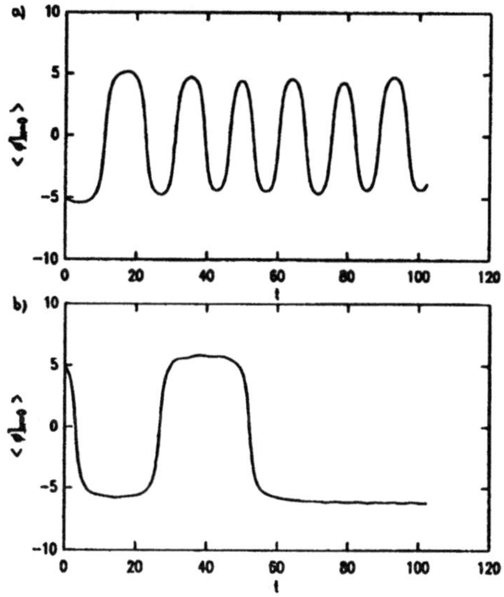

Fig. 21. Evolution of $\phi(x=0,t)$ for a breather of frequency $\omega_B = 0.2$, zero velocity, under a periodic perturbation and additive noise. a) Possible appearance of a breather with higher frequency and smaller amplitude for additive noise $\sigma = 0.1$ without periodic perturbation ($F = 0$). b) Soliton-antisoliton pair creation for additive noise $\sigma = 0.1$ and periodic perturbation $F = 0.01$, $\omega = 0.9$.

We made simulations without noise with different breather initial phases (from 0 to 2π in $2\pi/10$ steps) looking for the creation of soliton-antisoliton pairs. We did not observe any clear transition. In the figure 20 where $\phi(x=0,t)$ is represented for two different initial phases; it seems that the system trics to stabilize at $\phi \pm 2\pi$, what would suggest a soliton-antisoliton pair transition.

Then, we studied the effect of noise, without the periodic perturbation, of similar or higher intensity in term of its deviation σ. We studied the evolution $\phi(x=0,t)$ for different realizations of the noise and the initial phases used before, and we compared to the non perturbed breather and to the case with periodic perturbation. We found that for $\sigma = 0.01$ the breather almost did not change its frequency and amplitude with respect to the non perturbed case, and we only found a noticeable frequency decay in one case, but without a soliton-antisoliton pair transition. We did observe $k - \bar{k}$ pair transitions for higher intensity noise, $\sigma = 0.1$, and also the creation of breathers of higher frequency and smaller amplitude as it is illustrated in figure 21. However, for both intensities of the noise, we did not find that the results changed noticeably for different breather initial phases.

Finally, we studied the joint action of additive noise and periodic perturbation. For small noise, $\sigma = 00.01$, we found clear transitions to soliton-antisoliton pairs (see figure 21) which were more likely (there were more realizations with a transition) for the initial phases that suggested a transition without noise. For higher noise, $\sigma = 0.1$, we found a behavior similar to the case without periodic perturbation, with more transitions and again the appearance of breathers of higher frequency and smaller amplitude, though this only happened clearly for particular initial phases.

6.3 Soliton with dissipation

When there is a linear dissipation, the perturbed SG equation becomes:

$$\phi_{tt} - \phi_{xx} + \sin\phi + \alpha\phi_t = \nu(t) + F\cos(\omega t)$$

and then, the perturbative term (5.2) is $f(\phi, \phi_t; t) = -\alpha\phi_t + \nu(t) + F\cos(\omega t)$, and substituting into the soliton position and velocity evolution equations (5.8), we obtain:

$$\begin{aligned}\frac{dU}{dt} &= -\frac{\pi}{4}(1-U^2)^{3/2}(\nu(t) + F\cos(\omega t)) - \alpha U(1-U^2)\\ \frac{dX}{dt} &= U\\ \frac{dP}{dt} &= -\alpha P - 2\pi(\nu(t) + F\cos(\omega t))\end{aligned} \tag{6.7}$$

Through the change of variable:

$$P' = P + A\sin(\omega t) + B\cos(\omega t)\ ;\ A = \frac{2\pi\omega F}{\omega^2+\alpha^2}\ ,\ B = \frac{2\pi\alpha F}{\omega^2+\alpha^2}$$

we arrive to a Langevin equation for P':

$$\dot{P'} = -\alpha P' - 2\pi\nu(t)$$

with the associated Fokker-Planck equation:

$$\frac{\partial W}{\partial t} = \alpha\frac{\partial(P'W)}{\partial P'} + 4\pi^2 D\frac{\partial^2 W}{\partial P'^2}$$

which, under our normal initial conditions, it has the solution:

$$W(P';t) = \frac{1}{\sqrt{4\pi\bar{D}_P(t)}}\exp\left\{-\frac{(P' - P'_0\exp -\alpha t)^2}{4\bar{D}_P(t)}\right\}$$

where

$$\bar{D}_P = \frac{2\pi^2 D}{\alpha}(1 - e^{-2\alpha t})$$

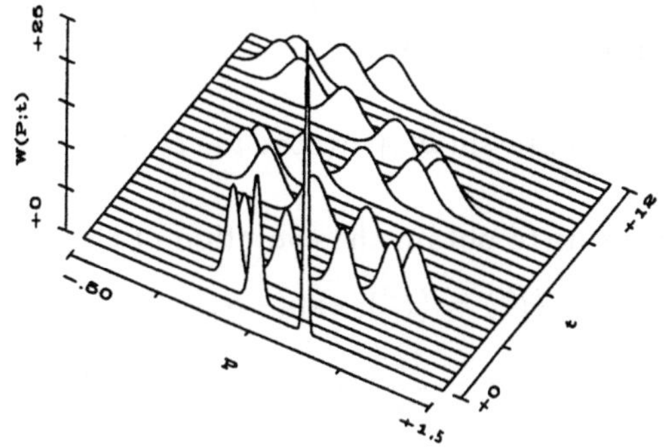

Fig. 22. Evolution of the probability density $W(P;t)$ of the momentum of a soliton under additive noise and periodic perturbation with dissipation; $P_0 = 8.04$, $F = 0.1$, $\omega = 0.4\pi$, $\sigma = 0.05$, $\alpha = 0.1$.

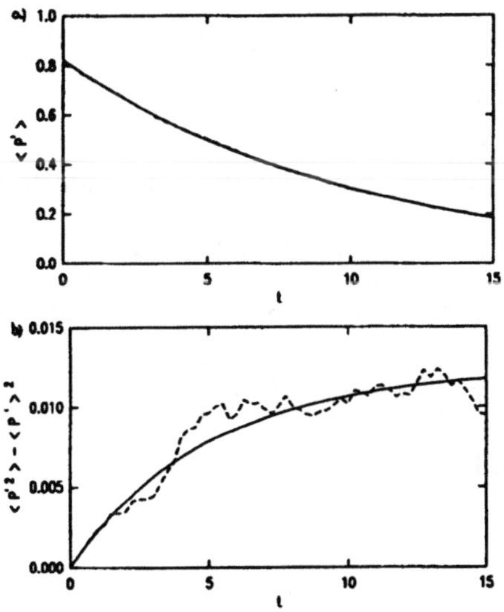

Fig. 23. Evolution of the mean values of the momentum of a soliton under additive noise and periodic perturbation with dissipation. Comparison of a) $\langle P' \rangle$, b) $\langle P'^2 \rangle - \langle P' \rangle^2$ obtained through numerical simulation (dotted line) with equation (6.8) (solid line) for a noise $\sigma = 0.05$ and $F = 0.05$, $\omega = 1.2$, $\alpha = 0.1$; $X_0 = 0.$, $U_0 = 0.1$. We show the mean values over 100 realizations.

Figure 22 shows the evolution of this probability density for the original momentum:

$$W(P;t) = \frac{1}{\sqrt{4\pi \bar{D}_P(t)}} \exp\left\{-\left[P - P_0 e^{-\alpha t} + \frac{2\pi F}{\omega^2+\alpha^2}(\omega \sin\omega t + \alpha(\cos\omega t - e^{-\alpha t}))\right]^2 / 4\bar{D}_P(t)\right\}$$

This Fokker-Planck allows us to obtain the evolution of the mean values:

$$\begin{aligned} \langle P' \rangle &= \langle P' \rangle_0 \exp(-\alpha t) \\ \langle P'^2 \rangle - \langle P' \rangle^2 &= \frac{4\pi^2}{\alpha} D(1 - \exp(-2\alpha t)) \end{aligned} \tag{6.8}$$

assuming that $\langle P \rangle_0^2 = \langle P^2 \rangle_0$.

Figure 23 shows the comparison of this equation with the results obtained through numerical simulation. The mean value of the momentum goes to zero in a decay time $1/\alpha$, and its deviation $\langle P'^2 \rangle - \langle P' \rangle^2$ tends to a constant inversely proportional to α $(4\pi^2 D/\alpha)$, in a time $1/2\alpha$.

Regarding the position and velocity of the soliton, if we consider only the newtonian case where $|U| \ll 1$, the evolution equations are:

$$\begin{aligned} \frac{dU}{dt} &= -\alpha U - \frac{\pi}{4}(\nu(t) + F\cos(\omega t)) \\ \frac{dX}{dt} &= U \end{aligned} \tag{6.9}$$

and we can convert them to Langevin equations through the change of variables:

$$\begin{aligned} U' &= U + A\sin(\omega t) = B\cos(\omega t) \\ X' &= X + \frac{1}{\omega}(-A\cos(\omega t) + B\sin(\omega t)) \\ A &= \frac{\pi}{4}\frac{\omega F}{\omega^2+\alpha^2} \; ; \; B = \frac{\pi}{4}\frac{\alpha F}{\omega^2+\alpha^2} \end{aligned} \tag{6.10}$$

The evolution equations become:

$$\begin{aligned} \dot{X}' &= U' \\ \dot{U}' &= -\alpha U' - \frac{\pi}{4}\nu(t) \end{aligned} \tag{6.11}$$

The Fokker-Planck equation for the joint probability distribution $W(X', U'; t)$ is:

$$\frac{\partial W}{\partial t} = -U' \frac{\partial W}{\partial X'} + \alpha \frac{\partial (WU')}{\partial U'} + \frac{1}{2} S_a' \frac{\partial^2 W}{\partial u'^2} \; ; \; S_a' = \frac{\pi^2}{8} D$$

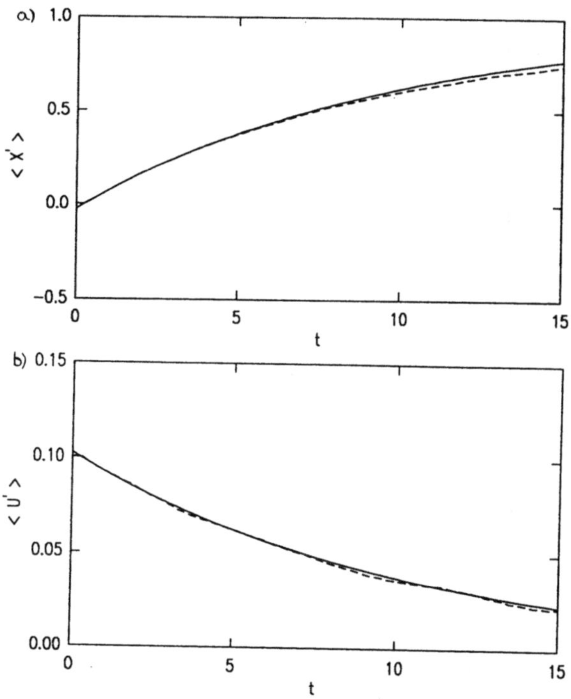

Fig. 24. Evolution of the mean values of the position and velocity of a soliton under additive noise and periodic perturbation with dissipation; Comparison of a) $\langle X' \rangle$, b) $\langle U' \rangle$ obtained through numerical simulation (dotted line) with equation (6.12) (solid line) for a noise $\sigma = 0.05$ and $F = 0.05, \omega = 1.2, \alpha = 0.1; X_0 = 0., U_0 = 0.1$. We show the mean values over 100 realizations.

and allows us to obtain the evolution of the mean values:

$$
\begin{aligned}
\langle U' \rangle &= \langle U' \rangle_0 \exp(-\alpha t) \\
\langle U'^2 \rangle - \langle U' \rangle^2 &= \frac{S'_a}{2\alpha}(1 - \exp(-2\alpha t)) \\
\langle X' \rangle &= \langle X' \rangle_0 + \frac{\langle U' \rangle_0}{\alpha}(1 - \exp(-\alpha t)) \\
\langle X'^2 \rangle - \langle X' \rangle^2 &= \frac{S'_a}{2\alpha^2}\left(2t + \frac{4}{\alpha}\exp(-\alpha t) - \frac{1}{\alpha}\exp(-2\alpha t) - \frac{3}{\alpha}\right)
\end{aligned}
\tag{6.12}
$$

Figures 24 and 25 shows the comparison with the numerical simulation. The velocity behaves like the momentum (we must keep in mind that we are in the newtonian approximation), while the position tends to a constant mean value, and its deviation $\langle X'^2 \rangle - \langle X' \rangle^2$ is asymptotically linear in time (it had a t^3 behavior when there was no dissipation).

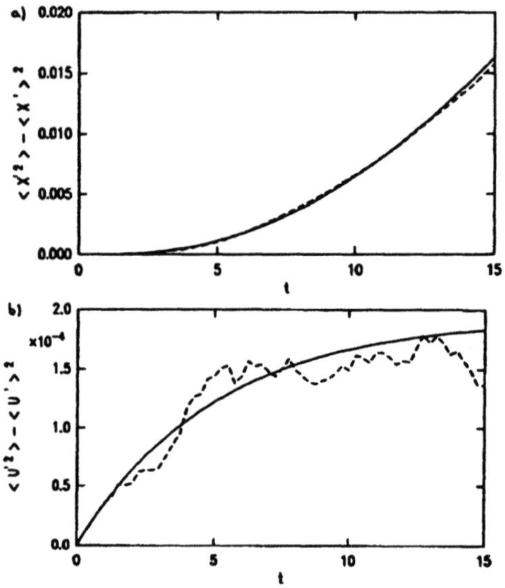

Fig. 25. Evolution of the mean deviations of the position and velocity of a soliton under additive noise and periodic perturbation with dissipation; Comparison of a) $\langle X'^2\rangle - \langle X'\rangle^2$, b) $\langle U'^2\rangle - \langle U'\rangle^2$ obtained through numerical simulation (dotted line) with equation (6.12) (solid line) for a noise $\sigma = 0.05$ and $F = 0.05, \omega = 1.2, \alpha = 0.1; X_0 = 0., U_0 = 0.1$. We show the mean values over 100 realizations.

6.4 Breather with dissipation

We have already mentioned that a breather may self maintain through time (Lomdahl and Samuelsen, 1986) due to the energy balance between the dissipation and the external periodic force, with frequency equal to the breather internal frequency. In such a context, and depending on the intensity of the external force and the dissipation, the breather may disappear, through radiation, maintain its identity or decay to a soliton-antisoliton pair. There are also some other possible, and more complex, transitions (intermittency, chaos, ...) that we will not consider here. Thus, we have a threshold field F_{thr} above which the breather maintains itself.

Reference (Lomdahl and Samuelsen, 1986) gives an explanation, with a perturbative analysis, of the mechanism through which the breather does not decay, and the results are tested by numerical simulation of the SG equation. The analysis is based on the fact that the wave function $\phi(x,t)$ consists of two parts: one corresponding to the breather, and the other to a homogeneous vacuum field ($\phi^{\text{vac}} = 0$) which is stimulated by the external perturbation.

By studying the variation of the energy of the system (the expression of the energy corresponding to the SG equation without any perturbation), it turns out that it comes from different terms: the breather takes energy from the external force and from its interaction with the vacuum field and losses energy through dissipation. On the other hand, the vacuum field takes energy from the external force and gives part of this energy to the breather and losses another part through dissipation.

If we calculate the energy supplied to the breather and the one it delivers, and use the analytical expression of the energy of the unperturbed SG equation with a breather frequency equal to the external one and an initial phase as discussed before, we can find an analytical expression for $(F_{\text{thr}}/\alpha)(\omega)$, which fits very well the data obtained numerically (see figure 26).

The perturbed SG equation under study is:

$$\phi_{tt} - \phi_{xx} + \sin\phi = F\sin\omega t + \alpha\phi_t$$

(we have changed the initial phase of the external field to match reference (Lomdahl and Samuelsen, 1986)), and the initial data:

$$\begin{aligned}\phi(x,t) &= \phi^{\text{br}} + \phi^{\text{vac}} \\ &= 4\tan^{-1}\left[\frac{\sqrt{1-\omega^2}}{\omega}\frac{\cos(\omega t + \psi_0^{\text{br}})}{\cosh(\sqrt{1-\omega^2}x)}\right] + \frac{F}{[(1-\omega^2)^2 + x^2\omega^2]^{1/2}}\sin(\omega t - \theta)\end{aligned}$$

being $\psi_0^{\text{br}} = \pi - 2\theta$ the initial phase of the breather and $\tan\theta = \alpha\omega/(1-\omega^2)$. The breather maintains its identity above F_{thr}, and assuming that its analytical expression remains valid, we can obtain the function $(F_{\text{thr}}/\alpha)(\omega)$.

Figure 27 shows one of these self-maintained breathers in the presence of an external oscillation and dissipation.

We must stress the importance of considering the apparition of a vacuum field and its interaction with the breather, because this interaction is the responsible of obtaining an initial phase for the breather so that it effectively absorbs energy from the external field. This is important because the energy flow, in both senses and in a period, is high enough to perturb significatively its parameters, which may produce its decay to radiation.

Starting from

$$\frac{dE}{dt} = -\alpha\int\phi_t^2 dx + F\sin\omega t\int\phi_t dx$$

where E is the energy corresponding to the unperturbed SG equation, we find that the injection of energy to the breather coming from the external force and the vacuum field is:

$$\begin{aligned}\Delta E_+^{\text{br}} &= F\int dt\sin(\omega t - 2\theta)\int dx\phi_t^{\text{br}} \\ &= \int dt F\sin\omega t\int\phi_t^{\text{br}}dx - \alpha\int dt\phi_t^{\text{vac}}\int\phi_t^{\text{br}}dx\end{aligned}$$

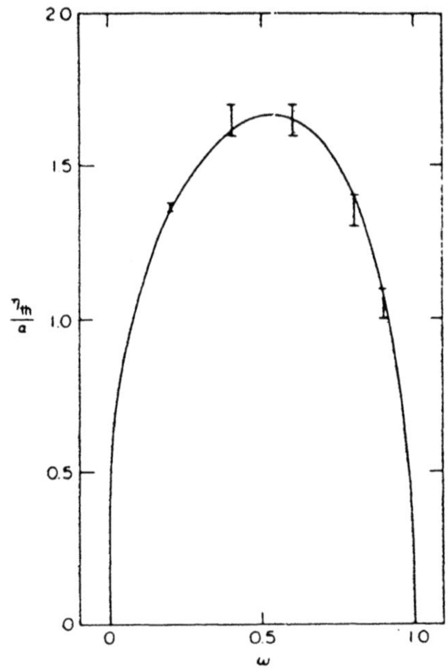

Fig. 26. Variation of F_{thr}/α with ω from (Lomdahl and Samuelsen, 1986) for $\alpha = 0.1$. We observe a good match between the analytical expression (solid line) and the data obtained numerically (bars).

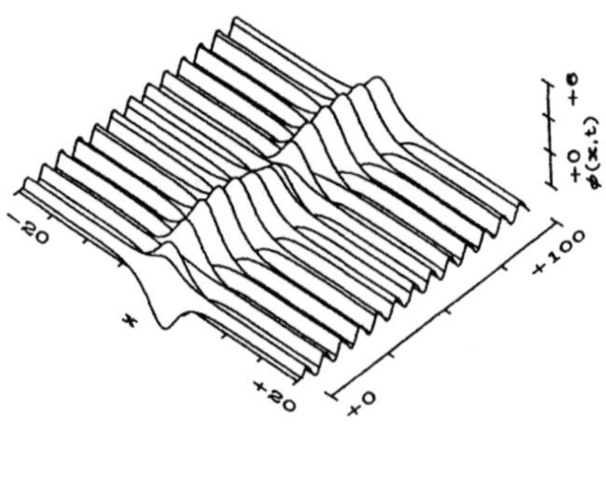

Fig. 27. Self-maintained breather under dissipation and an *ac* field. This figure shows how the breather maintains its identity for a periodic perturbation with $F = 0.15$, $\omega = \omega_B = 0.8$ in the presence of dissipation ($\alpha = 0.1$).

If we keep in mind that $\phi^{\mathrm{br}} \sim \cos(\omega t + \psi_0^{\mathrm{br}})$, $\phi_t^{\mathrm{br}} \sim \sin(\omega t - \pi + \psi_0^{\mathrm{br}})$, the energy supplied to the breather during one period depends on its initial phase (assuming that it does not change) through the product $\sin(\omega t - 2\theta)\sin(\omega t + \psi^{br_0} - \pi)$. For the threshold field, this phase must be optimal so that the breather obtains the mean value over a period, and from that $\psi_0^{\mathrm{br}} = \pi - 2\theta$.

For fields smaller than the threshold value, though this optimal phase, the energy supplied to the breather can overcome the losses through dissipation, and the breather extinguishes itself. What it is not so clear is what happens when the field is higher than the threshold value. If the breather has an optimal phase to absorb energy, and it does not change, it will continue getting energy until it will decay, for example, to a soliton-antisoliton pair. This invalidates the hypothesis that the parameters of the breather remain constant, because this absorption means a reduction of the frequency value, so that the breather will not absorb energy resonantly from the external field. The difference between the optimal phase for absorption, $\pi - 2\theta$, and that for which the energy absorbed balances the loss through dissipation for a field higher than the threshold may be estimated as $\Delta\psi_0^{\mathrm{br}} = F_{\mathrm{thr}}/F$. From the simulations carried out, starting from an optimal phase $\psi_0^{\mathrm{br}} = \pi - 2\theta$, it seems that when F is close to the threshold, the breather adjusts its phase to maintain itself. However, when F increases significatively, it seems that it can not adjust itself, though theoretically and under our assumptions it is possible, and it decays not necessarily to a soliton-antisoliton pair. The breather may adjust its phase either to absorb or to maintain its energy, but it is also possible that the change in the phase could lead to energy losses, high enough to extinguish, besides the fact already mentioned that a net increase in its energy would change its frequency and make the resonance with the external field to disappear.

Once studied and tested the behavior of the system without noise, we introduce it under the same conditions where the breather maintains itself, with an external field F high enough; thus, we had as initial data the breather plus the vacuum field.

We found that the breather did not disappear for a relatively high noise. For example, for the case shown in figure 27, where $F = 0.15$, $\alpha = 0.1$ and the noise $\sigma = 0.1$ ($D = 1.2510^{-4}$) there is no effect on the evolution of the breather. We must keep in mind that, though the parameter with a physical meaning is D, when we introduce the noise numerically, the values given by the random number generator are of order σ in absolute value, and σ is of the same order as F and α in this case. We should mention also that for the same noise $\sigma = 0.1$ and without dissipation and external field, the breather almost had no change.

When we increased the intensity of the noise, $\sigma = 1$ ($D = 1.2510^{-2}$), we found that the breather vanished emitting radiation, without generation of soliton-antisoliton pairs, as figure 28 shows. On the other hand, for noise stronger than the other perturbations, we observed that, though it was not able to supply energy when the field was below the threshold so that the breather would maintain itself, the decay took more time and the breather lasted longer. Figure 29 shows that; there wee represent $\phi(x = 0)$ (the parameter $u(t)$ used before) as a function

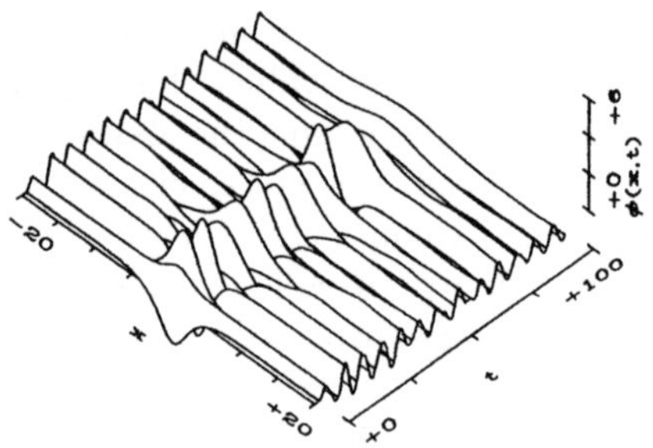

Fig. 28. Extinction of a breather under strong noise, $\sigma = 1$, an external *ac* field $F = 0.15, \omega = \omega_B = 0.8$ and dissipation, $\alpha = 0.1$.

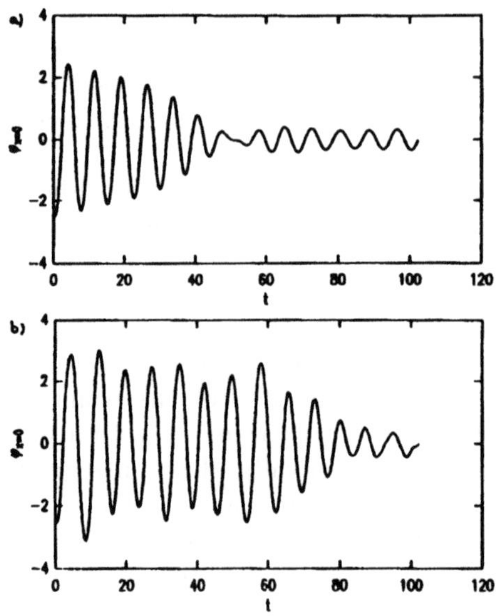

Fig. 29. Extinction of a breather under strong noise. Extinction of a breather without, (a), and with, (b), strong noise, $\sigma = 1$, under an external *ac* field $\omega = \omega_B = 0.8$ a) $F = 0.12$ b) $F = 0.10$, under the threshold field and dissipation, $\alpha = 0.1$.

of t for a perturbation below the threshold value without noise, and for another one even smaller but with strong noise. It is observed how the breather stood longer in the second case.

We also tested if only the noise was able to maintain the breather in the presence of dissipation, without external field, with an intensity comparable, in terms of σ, to the external field. The results were negative as it was expected from the previous results with external field.

We must also mention that when using as initial data a pure breather, without a vacuum field, with initial phase $\phi^{\mathrm{br}} \sim \sin\omega t$ and perturbation $F\cos\omega t$, we observed a decay to soliton-antisoliton pairs clearly when the field were high. This was not the case when the initial phase was optimal and the initial data had a vacuum field that should be generated by the external field. When we increased the external field above the threshold value, the breather always vanished without a clear formation of soliton-antisoliton pairs.

7. Charge Density Waves Modelling

In this section we will study a physical phenomenon related to the SG equation: Charge Density Waves (CDWs). We will not focus on the influence of the nonlinearity in the system and the soliton appearance, but on a more basic task: to find a model which reproduces what is found experimentally. In this sense, we have chosen a modification of the one proposed in (Sherwin and Zettl, 1986) with the introduction of stochastic perturbations to try to reproduce the characteristic phenomena, experimentally known, as the mode-locking.

7.1 General aspects

The CDWs are a collective transport phenomenon, whose origin lies in the interaction between electrons and phonons in a solid (see for example (Gruner and Zettl, 1985)). Depending on the occupation of the electronic bands of an one dimensional metal, the more stable configuration may have a periodic distortion of the lattice, so that there is a modulation of the electronic charge density and this is the origin of the CDW. The wave vector turns out to be $Q = 2k_F$, where $2k_F$ is the Fermi wave vector. The electronic density becomes:

$$\delta\rho = \rho_0 \cos(2k_F x + \phi)$$

In this context, a gap at the Fermi level shows up, and the conduction electrons lower its kinetic energy. Fröhlich suggested that if Q was not commensurable with the lattice constant, the CDW energy would be independent of the phase ϕ, and thus, an electrical current would appear under any electric field, independently of its intensity.

If the translational invariance of ϕ is broken, then there is a phase for which the CDW energy is the lowest, and there is also a minimum threshold field to overcome this energy reduction and to start the conduction. A possible cause

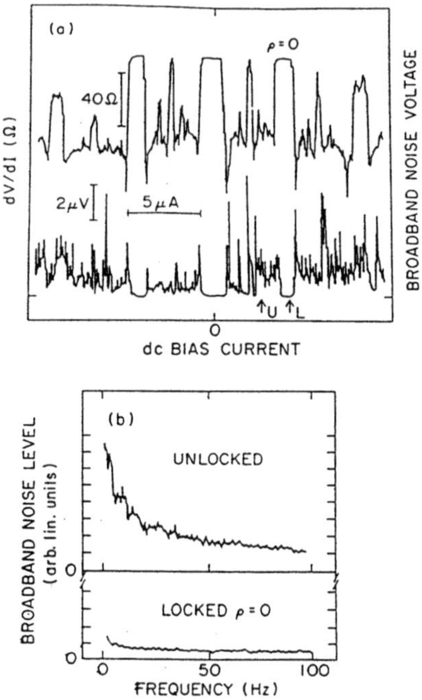

Fig. 30. Differential resistance and broad band noise in $NbSe_3$ as obtained in (Sherwin and Zettl, 1985). (a) Differential resistance dV/dI (in ohms) and rms broad band noise level (in mV) of a CDW as a function of a external control parameter, showing the step structure characteristic of mode locking dynamics. (b) Low frequency power spectra showing suppression of noise within a locked interval.

of the invariance break could be that the CDW were commensurate with the lattice, though this is very unusual; in this case, and for a CDW with a period quasi-multiple of the lattice constant, there may be solitons in the form of constant phase zones, separated by abrupt change areas. However, there is strong foundation to think that the origin of the pinning of the CDW to the lattice, and the appearance of a threshold field, lies in the impurities which are present in the materials which show this behavior (dependence of the threshold field on the impurity density, etc.)

There is a number of materials which show CDW behavior (transition metal trichalcogenides, halogen transition metal tetrachalcogenides, etc.). Evidence of this kind of transition is detected through magnitudes affected by the gap at the Fermi level: magnetic susceptibility, resistivity, thermoelectric power, etc. and scattering experiments where the CDW wave vector manifests itself.

The conductivity is one of these parameters that is of interest in this study. The Hall and thermoelectric effects in these materials suggest that their conductivity consists of an ohmic linear term and a CDW nonlinear one.

We have already mentioned the existence of a threshold field. The response of the CDW to a field higher than this threshold value is twofold: first, there appears a high frequency coherent current, or narrow band noise, which seems to be due to the displacement of the CDW over the pinning potential. Secondly, a low frequency broad band noise, incoherent response, is also detected.

It is also found that the conductivity saturates for high values of the external field; it seems that this is due to electrons leaving the CDW region or the elimination of $2k_F$ phonons.

When an *ac* field is present, the CDW exhibits a strong dependence on the field frequency, and its conductivity, σ_{ac}, also saturates for high frequencies. There also appear an induced conductivity, σ_{dc}, which increases when V_{ac} increases, and some interference phenomena between the narrow band noise and the *ac* current. The external field $E_{ac} \cos \omega_e t$ causes oscillations of the current at frequency ω_e; if there is also a *dc* field, E_{dc}, there are also oscillations at frequency ω_i corresponding to the narrow band noise. These two frequencies may interact and produce mode locking phenomena when they are commensurate: $n\omega_i = m\omega_e$. In the CDW systems, this locking shows up in the step structure of the differential resistance as a function of the *dc* field, as figure 30 shows. As the external *dc* current changes, the non linearity of the system keeps constant the relation between both frequencies, ω_i/ω_e, over a finite interval of the external parameter, corresponding to the intervals where dV/dI is constant in figure (a). When the external parameter gets far from the locking region, the system carries a transition to an unlocked state, which is quasi periodic, with two incommensurate frequencies. The interference between the internal frequency, ω_i and the external one ω_e is the origin of the coherent and incoherent responses of the system. Usually, the low frequency region of the power spectra consists of a broad band noise, while the narrow components show up at high frequencies as narrow band noise. Sherwin and Zettl found in their experiments on CDWs (Sherwin and Zettl, 1985) a systematic elimination of the broad band noise when the CDW entered mode locking, and a reinforcement of this noise in the unlocked regime, as it is shown in figure 30.

7.2 Models

There are several models trying to explain the CDW behavior, but none of them is completely satisfactory. The classical model takes the CDW as a rigid carrier, without any internal degree of freedom, using the forced oscillator equations with some analogy to the Josephson junctions. The tunneling model focus on the gap in the excitation spectrum of the CDW, explaining the non linear conductivity and the scale relationship between $\sigma_{ac}(\omega)$ and $\sigma_{dc}(E)$. However, none of these models explains the interference phenomena between the narrow band noise and the external field frequency ω.

There are some other models which consider the internal degrees of freedom of the CDW, segmenting the CDW, either through a hydrodynamical description or the Kelmm-Schrieffer model, but none of them completely explains the phenomenology observed in a CDW.

On the other hand, the reference (Wiesenfeld and Satija, 1987) faces the phenomena associated to mode locking between two frequencies, one internal ω_i and another external ω_e, from a general point of view in dynamical systems. They study the Poincaré section (circle map) associated with the torus which appears in the phase map of the dynamics under these two frequencies. For a mode locked state, when these frequencies are commensurable, the Poincaré section consist of a finite number of points in the circumference (if $\omega_i = \omega_e$ there will be only a point, etc.), while for the unlocked state, the section is a closed curve, that may be translated to a circumference, and the system will pass through every point of the curve. The behavior of the systems exhibiting mode locking, though they may have a large number of dimensions, is usually described through the motion in this torus; this is even more valid for systems with dissipation, as it is the case for the CDWs, where the orbit quickly establishes in this torus.

The work in (Wiesenfeld and Satija, 1987) studies the mapping from a phase point in the circumference corresponding to the Poincaré section, ϕ_n, to the next one, ϕ_{n+1}. They simulate perturbations, like the actual multidimensional character of the system, by introducing noise in the mapping:

$$\phi_{n+1} = f(\phi_n) + N_n$$

where N_n is a stochastic term. With approximations valid for strong either locked or unlocked regimes, they predict the lowering of the low frequency broad band noise observed in CDW systems (Sherwin and Zettl, 1985). Numerical simulations of the nonlinear mapping, with a term proportional to $\sin 2\pi\phi_n$, also show this lowering.

However, this study is generally applicable to any system exhibiting mode locking between two frequencies, without any reference to any explicit model of the system, but considering only the simulation of the locking and unlocking process through the study of the evolution in the Poincaré section, perturbed with a noise term.

We will use explicit models of the CDW trying to reproduce the mode locking, and with noise terms, as suggested in the study of general dynamical systems. We must keep in mind that in the above study, ϕ_n corresponds to the phase in the circle map, while in our study, ϕ is the phase of the modulation of the electronic density of the CDW, though the final equations are very similar when there is a non linear perturbation $\sin 2\pi\phi_n$ to the circle map.

There are more recent and complex models, but our aim is not to try to explain every aspect of the CDW, but to get closer to a real problem from the numerical simulation perspective.

Next, we will simulate the CDW with a classical model described through a SG equation. First, we will only focus on the electronic properties of the CDW, and afterwards we will take into account internal degrees of freedom through a model which considers the interaction of the CDW with the underlying lattice, so that it reflects elastic properties of the system (Bourne et al. 1986; Sherwin and Zettl, 1986). In both cases we will study the presence of noise, as well as constant and periodic terms, which corresponds to the electric fields applied in the experimental setup to these materials, and we will investigate the mode locking shown by these materials.

7.3 CDW overdamped model

First, we will study a classical simplified model of the CDW, so that we only take into account the electronic properties of the system, and then it may be described through the phase $\theta(x,t)$ of the CDW.

The suggested evolution equation is (Zettl, 1994):

$$\frac{\partial^2\theta}{\partial t}^2 - u_p^2\frac{\partial^2\theta}{\partial x}^2 + \eta\frac{\partial\theta}{\partial t} + \omega_0^2\frac{\partial V}{\partial\theta} = \frac{2k_F eE}{m^*} + N(x,t) \tag{7.1}$$

where $u_p = (m/m^*)^{1/2}\, v_F$, $E = E_{dc} + E_{ac}\cos(\omega t)$is the applied electric field, $\eta = 1/\tau$ is the dissipative term and $N(x,t)$ is the noise term.

Usually the potential takes the form $V = (1 - \cos Q\theta)$, so that the equation becomes a perturbed SG equation with dissipation.

From the experimental field (Zettl, 1994), it is suggested that the dissipative term is the most important one, and that the internal degrees of freedom (x dependence) and the second derivative in t are not so relevant. So (7.1) becomes an overdamped equation that may be rewritten:

$$\frac{1}{\omega_0\tau}\frac{d\theta}{dt'} + \sin\theta = e_{dc} + e_{ac}\cos(\omega' t') + N'(t') \tag{7.2}$$

where $t' = \omega_0 t$, $\omega' = \omega/\omega_0$, $e_{dc} = E_{dc}/E_T$, $E_T = m^*\omega_0^2/e2k_F$ being the expression of the threshold field in this model, and analogously for e_{ac}.[12]

Let us study this equation in the absence of noise. When there is not *ac* field, the solutions of the equation are obtained integrating:

$$\int_{\theta_0}^{\theta}\frac{d\theta'}{e_{dc} - \sin\theta'} = \omega_0\tau t$$

and it turns out to be:

$$\begin{aligned}
&(e_{dc} = 0) \quad \tan\frac{\theta}{2} = \tan\frac{\theta_0}{2\exp(-\omega_0\tau t)} \\
&(0 < e_{dc} < 1) \quad f(\theta/2) = \frac{e_{dc}\tan\frac{\theta}{2} + 1 - \sqrt{1-e_{dc}^2}}{e_{dc}\tan\frac{\theta}{2} + 1 + \sqrt{1-e_{dc}^2}} \\
&\qquad\qquad = f(\theta_0/2)\exp(-\omega_0\tau t) \\
&(e_{dc} = 1) \quad \cot(\frac{\pi}{4} - \frac{\theta}{2}) = \cot(\frac{\pi}{4} - \frac{\theta_0}{2}) + \omega_0\tau t \\
&(1 < e_{dc}) \quad g(\theta/2) = \tan^{-1}\left(\frac{e_{dc}\tan\frac{\theta}{2} - 1}{\sqrt{e_{dc}^2 - 1}}\right) \\
&\qquad\qquad = g(\theta_0/2) + \frac{\sqrt{e_{dc}^2 - 1\omega_0\tau t}}{2}
\end{aligned} \tag{7.3}$$

[12] For the sake of simplicity, we will refer to this equation without the apostrophes appearing due to the change of variables.

From these formulae we see that for $0 \leq e_{dc} \leq 1$, θ tends, as $t \to \infty$, to the constant value $\sin^{-1} e_{dc}$ exponentially, except for the case $e_{dc} = 1$. In this case, $\dot{\theta}$ decays, exponentially except for $e_{dc} = 1$, to $\dot{\theta} = 0$, so that (7.2) holds finally for the trivial solution, independently of the initial state.

When $e_{dc} > 1$, i.e. for fields above the threshold value, there is not the trivial solution $\sin^{-1} e_{dc}$. If we study $\dot{\theta}$, we see that, from (7.2), $\dot{\theta} = (e_{dc} - \sin\theta)\omega_0\tau$ and from the formula for θ:

$$\sin\theta = \frac{2}{h(t) + \dfrac{1}{h(t)}}$$

$$h(t) = \frac{1}{e_{dc}}\left\{1 + \sqrt{e_{dc}^2 - 1} \quad \tan\left(\frac{1}{2}\omega_0 t\sqrt{e_{dc}^2 - 1} + g(\theta_0/2)\right)\right\}$$

$h(t)$ is, except for a constant term, a tangent (π period) function, and its period is $(2\pi)/\omega_0\tau\sqrt{e_{dc}^2 - 1}$, and so it is for $1/h(t)$; so, $\sin\theta$ has a fundamental frequency $\omega_i = \omega_0\tau\sqrt{e_{dc}^2 - 1}$, which is also the frequency of $\dot{\theta}$. We see how this model reproduces the experimental result when we apply an electric *dc* field above a threshold value, an *ac* current (narrow band noise) shows up in the CDW system; we must keep in mind that $I_{CDW} \propto v_{CDW} \propto \dot{\theta}$. Besides this frequency, $\dot{\theta}$ has a constant component (zero frequency) equal to $e_{dc}\omega_0\tau$; that is why θ grows continuously while it oscillates.

When $e_{ac} \neq 0$, if $e_{dc} = 0$ and $\theta \ll 1$, after some transient state, we have:

$$\theta \simeq \frac{e_{ac}\omega_0\tau}{\sqrt{(\omega_0\tau)^2 + \omega^2}} \sin(\omega t - \psi) \; ; \; \tan\psi = -\frac{\omega_0\tau}{\omega}$$

This approximation is then valid when $[(e_{ac}\omega_0\tau)/\sqrt{(\omega_0\tau)^2 + \omega^2}] \ll 1$; after a transient state which is solution of the homogeneous equation $(1/\omega_0\tau)\dot{\theta} + \theta = 0$, $\dot{\theta}$ oscillates as the *ac* external field with a phase difference ψ. If $e_{dc} \neq 0$, but $\theta \ll 1$, $\dot{\theta}$ does not change, while θ has an additional term due to e_{dc}. When $\theta \ll 1$ is no longer valid, it is known that the original system, not overdamped and with the $\ddot{\theta}$ term, exhibits a chaotic behavior.

As for the case when we discretize the SG equation with all its terms, the numerical scheme used is:

$$\frac{1}{\omega_0\tau}\frac{\theta^{n+1} - \theta^n}{\Delta t} + \frac{\cos\theta^{n+1} - \cos\theta^n}{\theta^{n+1} - \theta^n} = e_{dc} + e_{ac}\cos(\omega t^n) + N^n \tag{7.4}$$

From the experimental data, we estimate the values of the constants appearing in the model, as we will see later in the next section. It turns out that $1/\omega_0\tau \sim 20$, e_{dc} is slightly above the threshold value $e_T = 1$, and e_{ac} is of the same order of e_{dc}. Figure 31 shows the evolution of $\dot{\theta}$ as a function of t for different values of e_{dc}, keeping constant the *ac* field and the dissipation $\omega_0\tau$. We observe how $\dot{\theta}$ seems to try to enter into mode locking, with a slight variation over e_{dc}, around $e_{dc} = \sqrt{5}$, which corresponds to $\omega_i = \omega$; however, this locking is not clear enough, as we observed when studying the evolution of $\dot{\theta}$ for longer times.

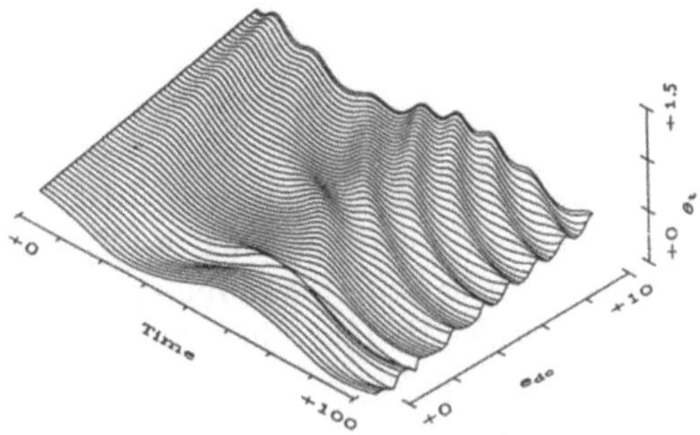

Fig. 31. $d\theta/dt$ evolution in the absence of noise. $d\theta/dt$ as a function of time for different values of e_{dc} and $1/\omega_0\tau = 20$, $e_{ac} = 3.$ and $\omega = 1.$.

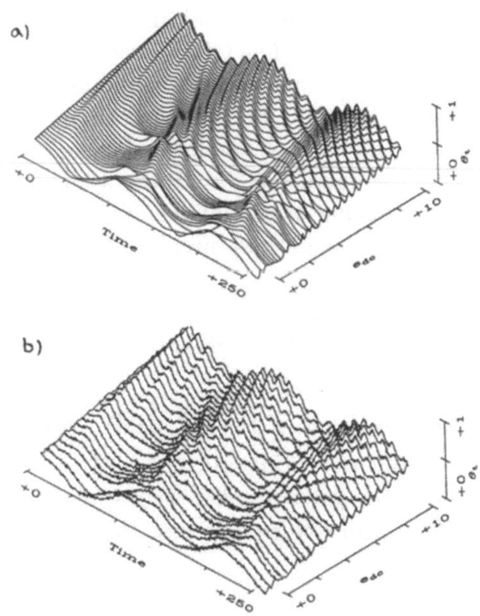

Fig. 32. $d\theta/dt$ evolution in the presence of noise. $d\theta/dt$ as a function of time for different values of e_{dc} and $1/\omega_0\tau = 20$, $e_{ac} = 3.$ and $\omega = 0.06$. (a) without noise, (b) with noise $\sigma = 4.$.

Regarding the effect of the noise, from (Wiesenfeld and Satija, 1987) one would expect that it should lower the low frequency component when entering locking, and increases it again when returning to the unlocked state. In figure 32 we observe that a noise not very large comparing to the dissipative term, but relatively strong with respect to the *dc* and *ac* fields, seems to destroy or hide the possible locking at $\omega_i = \omega$ and $\omega_i = 2\omega$.

7.4 Underlying elastic lattice model

Here, we will continue the study of the locking of CDW systems, but using a different model which takes into account the coupling with the underlying lattice. This model was suggested in (Zettl, 1994; Sherwin and Zettl, 1986) to explain the effects that the locking had on the elastic properties of the system. We will also study the effect of the noise.

This model is an extension of the Frenkel-Kontorova (Frenkel and Kontorova, 1938, 1939) one, and treats the CDW and the lattice as coupled elastic media. In the Frenkel-Kontorova model the internal degrees of freedom of the CDW are considered by treating it as consisting of harmonically coupled particles which lie in a rigid and periodic pinning potential. Then, the motion of these particles is described by the discrete SG equation.

In the present model we incorporate the lattice elasticity by discretizing it, and the associated pinning potential, in rigid units of mass M, harmonically coupled through springs of constant K. The CDW is represented by discrete particles of mass m elastically coupled to their closest neighbors through a spring of constant κ. This model can describe the commensurate and incommensurate states, and may be extended to the case of random pinning. However, for simplicity reasons, we will focus on the equations describing a CDW commensurate with a wave length equal to the spacing of the lattice: $\lambda/a = 1$.

Assuming only the closest neighbors interactions, the energy potential function is:

$$U = \sum_j \left(\frac{\kappa}{2}(r_j - r_{j-1})^2 + \frac{K}{2}(x_j - x_{j-1})^2 + V(1 - \cos[Q(r_j - x_j)]) \right)$$

where r_j and x_j are the positions of the jth CDW mass and lattice unit, V is the intensity of the pinning potential and $Q = \frac{2\pi}{\lambda} = 2k_F$, with λ being the CDW wave length and k_F the wave number corresponding to the Fermi surface. If we add internal friction and external forces, the evolution equations are the following:

$$\begin{aligned} m\frac{d^2 r_j}{dt^2} &+ \gamma\frac{d(r_j - x_j)}{dt} + \kappa(2r_j - r_{j+1} - r_{j-1}) \\ &+ QV\sin[Q(r_j - x_j)] = f_j(t) \\ M\frac{d^2 x_j}{dt^2} &+ \Gamma\frac{d(2x_j - x_{j+1} - x_{j-1})}{dt} + \gamma\frac{d(x_j - r_j)}{dt} \\ &+ K(2x_j - x_{j+1} - x_{j-1}) + QV\sin[Q(x_j - r_j)] = F_j(t) \end{aligned} \tag{7.5}$$

where Γ is the internal friction of the lattice and γ represents the friction between the CDW and the lattice; $f_j(t)$ is the external force on the jth CDW particle through the action of external electric fields, and $F_j(t)$ is the external mechanical force acting over the jth lattice unit. In the limit $\kappa, K \to \infty$, equations (7.5) become SG ones, continuous and with dissipation in the form of terms ϕ_t and ϕ_{xxt}.

In the usual experimental context, the screening of the normal (not condensed) electrons in the CDW crystal makes the applied field E be relatively uniform all along the crystal, so that $f_j(t) \simeq f(t) = E_{dc} + E_{ac}\cos(\omega t)$. On the other hand, $F_j(t)$ is zero except for $j = 1$, where $F_1(t) = F\cos(\Omega t)$, in case we were interested in studying the elastic properties of the system. The boundary conditions are free for $j = 1$ and fixed in the other extreme.

To simplify the simulations, it is easier if the equations have an adimensional form as follows:

$$
\begin{aligned}
&m'\frac{d^2\theta_j}{d\tau^2} + \frac{d(\theta_j - \varphi_j)}{d\tau} + \kappa'(2\theta_j - \theta_{j+1} - \theta_{j-1}) + \sin(\theta_j - \varphi_j) = f'(\tau) \\
&M'\frac{d^2\varphi_j}{d\tau^2} + \Gamma'\frac{d(2\varphi_j - \varphi_{j+1} - \varphi_{j-1})}{d\tau} + \frac{d(\varphi_j - \theta_j)}{d\tau} \\
&\qquad + K'(2\varphi_j - \varphi_{j+1} - \varphi_{j-1}) + \sin(\varphi_j - \theta_j) = F_j'(\tau)
\end{aligned} \tag{7.6}
$$

where

$$
\begin{aligned}
\tau &= \frac{Q^2V}{\gamma}t \quad \Gamma' = \frac{\Gamma Q^2 V}{\gamma^2} \\
\theta_j &= Qr_j \quad M' = \frac{MQ^2V}{\gamma^2} \\
\varphi_j &= Qx_j \quad m' = \frac{mQ^2V}{\gamma^2} \\
\kappa' &= \frac{\kappa}{Q^2V} \quad f'(\tau) = \frac{f}{QV} = e_{dc} + e_{ac}\cos(\omega'\tau) \\
K' &= \frac{K}{Q^2V} \quad F_j'(\tau) = \frac{F_j}{QV} = F\cos(\Omega\tau)
\end{aligned} \tag{7.7}
$$

To obtain the overdamped equations, (7.1) and (7.2), it is enough to consider a rigid underlying lattice (the x_j equation is removed) and keep in mind that

$$
\begin{aligned}
\omega_0^2 &= \frac{Q^2V}{2} \quad Q = 2k_f \\
\eta &= \frac{\gamma}{m} \quad eE_T = QV
\end{aligned}
$$

Time τ in equation (7.6) is $\tau = \omega_0^2/\eta t = \omega_0/\eta t'$, compared to time t' in (7.2) (we use η instead of $\tau = 1/\eta$ to avoid confusion with the different meaning of τ in equations (7.2) and (7.6)); if we put equation (7.2) as a function of τ, we would have:

$$
\frac{d\theta}{d\tau} + \sin\theta = e_{dc} + e_{ac}\cos(\omega'\tau) + N'
$$

As we have already said, we obtain the interesting values of the parameters appearing in the equations from the experimental field; we had $\eta/\omega_0 \sim 20$, and for the rest of parameters:

- $\omega_0 \sim 10^{-6}\, seg^{-1}$
- $m' \sim 0$, i.e., overdamped CDW
- $\kappa' \sim 10^{-3} K' \sim 10^{-3}$
- $M' \sim 10^3$. The internal frequency of the lattice, $\sim 1/M'$, must be much smaller than the CDW one, $\sim \sqrt{e_{dc}^2 - 1}$
- $K' \sim 1$. For only one particle, $K' \sim 10^{-6}$; when scaling to n particles, K' grows as n^2, and in our case, $n \sim 10^3$.
- $\Gamma' \sim 10^{-6}$. Very small compared to the rest.

and, as before, e_{dc}, e_{ac} in the range 1 to 10, one order of magnitude over the threshold value.

Regarding the mechanical external force, $F' \cos \Omega t$, if we are interested in measuring the elastic properties of the system, as the Young modulus Y, F' must be small, ~ 0.01, so that the system remains in the linear regime of the lattice, and we must vary Ω until we find a resonance Ω_r, so that $x_j \sim A_r \cos(\Omega_r t - \psi)$; we have then that $Y \propto \Omega_r$. This was done in (Sherwin and Zettl, 1986), and found experimentally that $\Omega_r \sim 10^{-3}\omega$. In that work, they used a three particles model, with fixed boundary conditions, so that the system could be studied focusing on only one particle through simulation on an analog computer. They studied Y and obtained the same anomalies found experimentally when there was locking and for the depinning when $e_{dc} > e_T =$ and $e_{ac} = 0$; they also looked at the behavior of the system as a function of e_{ac} when $e_{dc} = 0$. They observed that Y followed the same behavior as the differential resistance dV/dI as a function of e_{dc}, and how mode locking, even subharmonic ($\omega_i = \omega/2$), appeared for joint *ac* and *dc* fields.

We used in our simulations 800 particles for the CDW (and another 800 for the lattice) and observed that, after a transient period, they oscillated homogeneously, so that we could focus on the evolution of one of them. Though this model was suggested to study the elastic behavior of a CDW system, which is experimentally measurable, we focused, as in the overdamped case, in the study of the locking to an external frequency, which is the origin of one of the anomalies detected experimentally in the behavior of the elastic properties of a CDW. As we have mentioned, we also studied the influence of noise in the system.

As in the section before, the interesting magnitude is $d\theta/dt$, as the CDW current depends on it. Figure 33 shows the dependence of the modulus of the Fourier transform of $d\theta/dt$, $F(d\theta/dt)$, on the dc field. In a), $e_{ac} = 0$, and it is shown the peaks corresponding to the internal frequency of the system, $\omega_i = \sqrt{e_{dc}^2 - 1}$, and its harmonics. In b), $e_{ac} = 2$, and the external frequency is equal to ω_i when $e_{dc} = 3$; this figure shows the locking between the two frequencies when they get close, while varying e_{dc}, so that around $e_{dc} = 3$, ($\omega_i = \omega$, there is only one peak in the frequency. Figure 34 shows the dependence of the range of e_{dc} where locking exists on e_{ac}. In a), $e_{ac} = 1$, it is difficult to see the locking in the fundamental frequency, though it is clear in the first harmonic, because

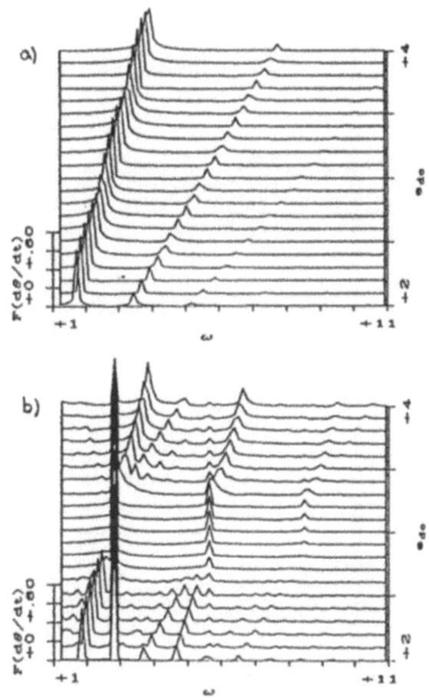

Fig. 33. $F(d\theta/dt)$ as a function of e_{dc} in the absence of noise. The modulus of the Fourier transform of $d\theta/dt$, $F(d\theta/dt)$, as a function of e_{dc}. a) $e_{ac} = 0$, b) $e_{ac} = 2$, with an external frequency ω equal to the internal one ω_i for $e_{dc} = 3$; this figure shows the locking of the internal frequency to the external one when we vary e_{dc} and thus ω_i.

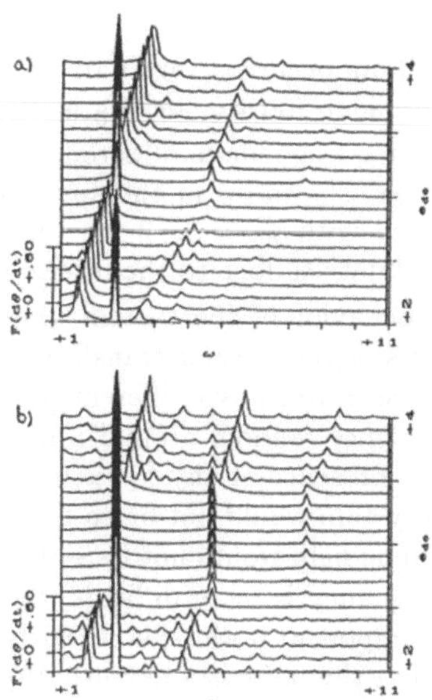

Fig. 34. Dependence of the locking on e_{ac}. The modulus of the Fourier transform of $d\theta/dt$, $F(d\theta/dt)$, as a function of e_{dc}. a) $e_{ac} = 1$, b) $e_{ac} = 2$, with an external frequency ω equal to the internal one ω_i for $e_{dc} = 3$.

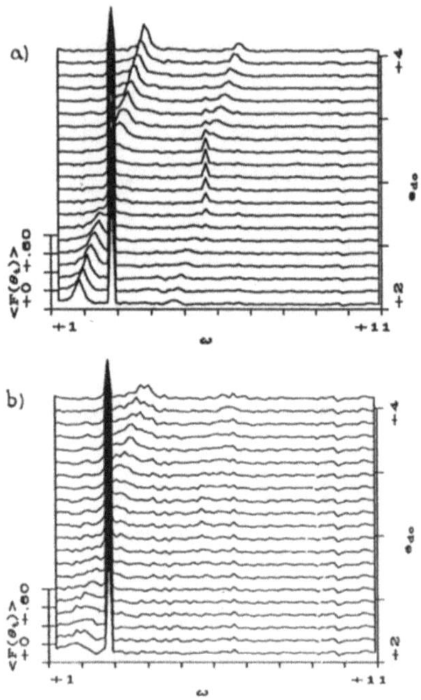

Fig. 35. Dependence of the locking on e_{ac} and noise. The mean value of the modulus of the Fourier transform of $d\theta/dt$, $F(d\theta/dt)$, as a function of e_{dc} for $e_{ac} = 2$ with an external frequency ω equal to the internal one ω_i for $e_{dc} = 3$. a) weak noise, $\sigma = 2$, b) strong noise $\sigma = 4$..

the range in e_{dc} where locking occurs is very narrow. In b), $e_{ac} = 3$, we observe that this range is much broader and the locking is very clear.

It is interesting that some frequencies, different from the internal and external ones or their harmonics, shows up in the spectrum. We also tried to find subharmonic mode locking, with $\omega_i = \omega/2$ for this range of values, but the frequency component candidate to lock to the internal frequency was too small.

Figure 35 shows the behavior of the system under noise, representing the mean value of the modulus of the Fourier transform over different realizations. The effect of noise can be seen from two different points of view: it may add to the *dc* field, so that it becomes random around a mean value, or it may colour the *ac* field, increasing the presence of any frequency in it. In the first case, the internal frequency of the system would not be so defined, while from the second point of view, the external *ac* field would not be so monochromatic. In any case, locking to the internal frequency is more difficult, though it could be easier for frequencies with a small component in the system without noise. From the figure, it seems better to think that the noise is a perturbation of the *dc* field, while the component of the external frequency ω remains clear, the one corresponding to ω_i gets broader when the intensity of the noise grows, and it hides in the white noise background; so the locking disappears for strong noise ($\sigma = 4$ compared to $e_{dc} \sim 3$ and $e_{ac} = 2$), and it is difficult to identify it for weaker noises ($\sigma = 2$); it is again much clear to see locking in the first harmonic. We also looked for subharmonic locking, but we did not find it, though the white noise reinforces every frequency, including the one candidate for this kind of locking, but the destructive overall effect seems predominant.

Acknowledgments: L. Vázquez wishes to thank Prof. J. Marro for the invitation to participate in the stimulating and interesting Meeting in the charming city of Granada. This work has been partially supported by the Dirección General de Investigación Científica y Técnica of Spain (Grant PB92-0226) and the Human Capital and Mobility Program (network CT93-0417).

References

Ablowitz, M. Kruskal, M., and Ladik, J.: SIAM Appl. Math. **36**, 428.

Bass, F.G., Kivshar, Yu.S., Konotop, V.V. and Sinitsyn, Yu.A. (1988): Phys. Rep. **157**, 63.

Berestycki, H. and Lions, P.L. (1983): Arch. Rat. Mech. Anal. **82**, 313.

Bishop, A., McLaughlin, D. and Salerno, M. (1989), Phys. Rev. A **40**, 6463.

Bourne, L., Sherwin, M. and Zettl, A. (1986): Phys. Rev. Lett. **56**, 1952.

Brezis, H. and Lieb, E.H. (1984): Comm. Math. Phys. **96**, 97.

Cazenave, T. and Vázquez, L. (1986): Comm. Math. Phys. **105**, 35.

Caputo, J. and Flytzanis, N. (1990): *Kink-Antikink collisions in Sine-Gordon and ϕ^4 problems in the variational approach* preprint.

Dodd, R.K., Eilbeck, J.C., Gibbon, J.D. and Morris, H.C. (1982): *Solitons and Nonlinear Wave Equations* (Academic Press, London).

Frenkel, J. and Kontorova, T. (1938): Zh. Eksp. Theor. Fiz. **8**, 1340.

Frenkel, J. and Kontorova, T. (1939): J. Phys.(USSR) **1**, 137.

Grønbech-Jensen, N., Kivshar, Yu.S. and Samuelsen, S. (1991): Phys. Rev. B **43**, 5689.

Grüner, G. and Zettl, A. (1985): Phys. Rep. **119**, 117.

Guo, Ben-Yu, Pascual, P., Rodríguez, M.J. and Vázquez, L. (1986): Appl. Math. Comput. **18**, 1-14.

Haraux, A. (1981): *Nonlinear Evolution Equations. Global Behavior of Solutions*, Lecture Notes in Mathematics 841 (Springer-Verlag, Berlin).

Helfand, E. (1979): Bell Syst. Tech. J. **58**, 2289.

Jiménez, S. and Vázquez, L. (1989): Appl. Math and Comp. **25**, 207.

Jiménez, S. and Vázquez, L. (1990): Appl. Math and Comp. **35**, 61.

Jiménez, S. and Vázquez, L. (1991): in *Nonlinear Problems in Future Particle Accelerators*, pp. 151-162. Edited by W. Scandale and G. Turchetti (World Scientific, Singapore).

Jiménez, S. (1994): Appl. Math and Comp. **64**, 13.

Jiménez, S. and Malavasi, M. (1994): *Conservative schemes for Lagrangian systems with finite degrees of freedom*, preprint.

Kivshar, Yu.S. and Malomed, B.A. (1989): Rev. Mod. Phys. **61**, 763.

Konotop, V.V. and Vázquez, L. (1994): *Nonlinear Random Waves* (World Scientific, Singapore).

Legrand, O. and Reinisch, G. (1987): Phys. Rev. A **35**, 3522.

Legrand, O. (1987): Phys. Rev. A **36**, 5068.

Lomdahl, P. and Samuelsen, M.R. (1986): Phys. Rev. A **34**, 664.

Lomdahl, P. and Samuelsen, M.R. (1988): Phys. Lett. A **128**, 427.

Nozaki, K. (1982): Phys. Rev. Lett. **49**, 1883.

Parrondo, J.M., Mañas, M. and Rubia, F.J. (1990): J. Phys. A **23**, 2363.

Pascual, P. (1984): *Comportamiento de un soliton de Sine-Gordon en presencia de perturbaciones estocásticas pequeñas*, Universidad Complutense de Madrid, Madrid.

Pascual, P. and Vázquez, L. (1985): Phys. Rev. B **32**, 8305

Pascual, P. and Vázquez, L. (1986): Hadronic Journal **9**, 307.

Pascual, P., 1992, *Ondas no lineales bajo perturbaciones estocásticas*, Ph.D. Thesis, Universidad Complutense, Madrid.

Perring, J. and Skyrme, T. (1962): Nucl. Phys. **31**, 550.

Potter, D. (1977): *Computational Physics* (Wiley-Interscience, New York).

Risken, H. (1984): *The Fokker-Planck Equation* (Springer-Verlag, Berlin).

Ritchmyer, R. and Morton, K. (1967):*Difference Methods for Initial-Value Problems* (Interscience Publishers, New York).

Rodríguez, M.J. (1988): *Estabilidad de ciertas ondas solitarias sometidas a perturbaciones estocásticas*, Ph.D. Thesis, Universidad Complutense, Madrid.

Rodríguez, M.J. and Vázquez, L. (1990): Phys. Rev. B **41**, 11437.

Sánchez, A. and Vázquez, L. (1989): in *Nonlinear Coherent Structures*. Lecture Notes in Physics **353**, pp. 251-260. Edited by M. Barthes and J. Léon (Springer-Verlag, Berlin).

Sánchez, A. and Vázquez, L. (1991): Int. J. Mod. Phys. B **5**, 2825.

Sherwin, M. and Zettl, A. (1985): Phys. Rev. B **32**, 5536.

Sherwin, M. and Zettl, A. (1986): Physica D **23**, 62.

Strauss W.A. and Vázquez, L. (1978): J. Comp. Phys. **28**, 271.

Stubbe, J. and Vázquez, L. (1988), in *Mathematics + Physics. Lectures on Recent Results*, pp. 81-123, Vol. 3, edited by L. Streit (World Scientific, Singapore).

Van Kampen, G. (1981): J. Stat. Phys. **24**, 175.

Vázquez, L., (1982): Il Nuovo Cimento B **69**, 63.

Vázquez, L. (1984): Anales de Física **83A**, 254.

Wiesenfeld, K. and I. Satija (1987): Phys. Rev. B **36**, 2483.

Zettl, A. (1994): private communication.

Topics on Chaotic Dynamics

G. Gallavotti

Dipartimento di Fisica
I^a Università di Roma
P.le Moro 2, 00185, Roma, Italia.

Abstract: Various kinematical quantities associated with the statistical properties of dynamical systems are examined: statistics of the motion, dynamical bases and Lyapunov exponents. Markov partitions for chaotic systems, without any attempt at describing "optimal results". The Ruelle principle is illustrated via its relation with the theory of gases. An example of an application predicts the results of an experiment along the lines of Evans, Cohen, Morriss' work on viscosity fluctuations. A sequence of mathematically oriented problems discusses the details of the main abstract ergodic theorems guiding to a proof of Oseledec's theorem for the Lyapunov exponents and products of random matrices.

1. Dynamical systems and their statistics

A dynamical system $(\mathcal{C}, S)$ will consist of a piecewise smooth compact manifold and of a piecewise diffeomorphic map S of $\mathcal{C}$ into itself with a piecewise diffeomorphic inverse S^{-1}.[1] The set $\mathcal{N}$ of the singularities of either S or of the manifold $\mathcal{C}$ will be called the set of the singularities of $(\mathcal{C}, S)$.[2]

---

[1] By piecewise smooth manifold I mean a manifold that can be regarded as the union of a finite number of C^∞ compact manifolds, with any two of them having in common only boundary points: the union of the boundaries of such manifolds will be called the set of singularities of $\mathcal{C}$. Similarly a map of $\mathcal{C}$ into itself is said to be piecewise smooth if $\mathcal{C}$ can be regarded as the union of finitely many compact C^∞ manifolds and S is a diffeomorphism as a map of the interior of each such manifold and its image. A compact C^∞ manifold, of dimension d, is a C^∞ manifold with boundary consisting of finitely many compact C^∞ manifolds of dimension $< d$ with only boundary points in common; the inductive definition is started by declaring that a point is a C^∞ manifold of dimension 0.

[2] Sometimes one may wish to consider as singular points some points where S is in fact regular: in this case such points will also be supposed to lye on piecewise smooth submanifolds of $\mathcal{C}$ (of lower dimension) and will be included in the set of singularities. Furthermore, since the action of S on the singular points will not be studied, one may wish to require that S^{-1} be undefined or arbitrarily defined on the points $x \ni S(\mathcal{C}/\mathcal{N})$ so that in such points the property $S(S^{-1}x) = x$ may even fail. And likewise one may avoid defining S on $\mathcal{N}$. I will not try to be so general: see (Pesin, 1992).

The reader disturbed by the above generality can simply think that all I am discussing is the case of a C^∞ diffeomorphism of a C^∞ compact manifold without boundary. The concession to the generality is due to the fact that some of the most important dynamical systems really show physically significant singularities, like the billiards or the hard sphere gases.

Example 1: A Hamiltonian system with l degrees of freedom is observed at the instants in which a certain event, *the timing event*, happens. This is the event in which the point representing the system passes through a predefined surface in phase space (*e.g.* one particle of the system passes through an ideal wall in physical space or the distance between two particles takes some prefixed value r). If $\mathcal{C}$ is the family of timing events with a prefixed energy H_0, and S is the transformation mapping one event in $\mathcal{C}$ into the following, then $(\mathcal{C}, S)$ is a dynamical system. The probability distribution μ_L defined by assigning to a (measurable) set of events $E \subset \mathcal{C}$ the Liouville measure of the set of phase space points on the energy surface experiencing the first timing event in E, is an invariant measure (*i.e.* $\mu_L(E) = \mu_L(S^{-1}E)$). The μ_L will still be called the Liouville measure.

Example 2: consider a box, $[-\frac{1}{2}L, \frac{1}{2}L]^2$, with side L and periodic boundary conditions (*i.e.* opposite sides identified) and with a few circular regions C_i, "obstacle of radius" r_i, in it. A particle moves freely among them and collides elastically, at velocity v. The space $\mathcal{C}$ consists of the *collisions*, parameterized by the point α on the obstacle where the collision takes place and by the angle φ formed between the incoming velocity and the outer normal to the collision point: hence $\varphi \in [-\frac{\pi}{2}, \frac{\pi}{2}]$. We see that $\mathcal{C}$ consists in the union of $\partial C_i \times [-\frac{\pi}{2}, \frac{\pi}{2}]$. A point in $\mathcal{C}$ is determined by 2 coordinates (α, φ), with $\alpha \in [0, 2\pi]$ being the angular position of the collision point and a label i denoting the obstacle on which the collision takes place. The obstacles will be supposed so distributed so that no collisionless trajectory is possible. The map S is smooth everywhere except on the boundary of $\mathcal{C}$ (*i.e.* $\varphi = \pm\pi$) and on the collisions $x = (i, \alpha, \varphi)$ such that Sx is a tangent collision. This is the *billiard system.* A natural metric on $\mathcal{C}$ is $d\alpha^2 + d\varphi^2$. The system being hamiltonian it is easy to find the Liouville distribution: in the coordinates (α, φ) it is given by $\Gamma r_i d\alpha \cos\varphi \, d\varphi$ where (α, φ) represent collisions with the obstacle i and Γ is defined so that $\mu_L(\mathcal{C}) = 1$.

Example 3: many particles, say N, in a periodic box as in the example 2. The particles interact with a radial pair potential v (with range smaller than L) and with an external potential represented by the hard core due to the obstacles. The phase space is $\mathcal{C}^N$ and the equations of motion are:

$$\dot{\underline{q}}_j = \frac{1}{m}\underline{p}_j, \quad \dot{\underline{p}}_j = \underline{F}_j + E\underline{i} - \alpha\underline{p}_j \tag{1.1}$$

where $\underline{i}$ is the 1–axis unit vector, E is an external constant field, $\underline{F}_j$ is the force generated by the total potential Φ on the j–th particle; the obstacles are taken into account by the elastic reflection rule and α is so defined that the energy

$H = \sum_j \frac{1}{2m}\underline{p}_j^2 + \Phi$ is a constant of motion (*gaussian thermostat*): *i.e.* $\alpha = E\underline{i} \cdot \sum_j \underline{p}_j / \sum_j \underline{p}_j^2/2m$, as a simple calculation proves; m is the particles mass. The phase space $\mathcal{C}$ consists of the energy H_0 configurations in which one particle (any one) is colliding with some obstacle. The time evolution S maps one such configuration into the following configuration of the same type ("next collision configuration"). The phase space thus defined is a $4N-2$ dimensional subspace of the total $4N$ dimensional phase space; and a simple calculation of the divergence of the r.h.s. of (1.1) shows us that the phase space volume changes at a rate $(2N-1)\alpha$ in the total $4N$ dimensional phase space $\mathcal{F}$. This example has a rather involved set of singularities: nevertheless it is piecewise smooth and it defines a dynamical system in the above sense. Physically it is a model for electric conductivity (*Lorentz gas conductor*).

Example 4: The map of the torus $\mathcal{C} = [0, 2\pi]^2$ defined by: $S(\varphi_1, \varphi_2) = (\varphi_1 + \varphi_2, \varphi_1 + 2\varphi_2) \bmod 2\pi$. One also writes:

$$S\begin{pmatrix}\varphi_1\\ \varphi_2\end{pmatrix} = \begin{pmatrix}1 & 1\\ 1 & 2\end{pmatrix}\begin{pmatrix}\varphi_1\\ \varphi_2\end{pmatrix} \bmod 2\pi \tag{1.2}$$

Then S preserves the distribution $\mu_0(d\underline{\varphi}) = d\underline{\varphi}/(2\pi)^2$. This system has *no singularities.*

Given a dynamical system $(\mathcal{C}, S)$ we shall consider initial data chosen randomly with respect to a given probability distribution μ_0. It is customary to consider the "Liouville distribution": the latter means a probability distribution with a positive density with respect to the volume on the phase space, as one often calls the space $\mathcal{C}$. Sometimes it could literally coincide with the Liouville measure of analytical mechanics; *e.g.* when the system $(\mathcal{C}, S)$ is obtained by timing the evolution of a hamiltonian system: this is the case in examples 1,2.

The *qualitative theory* of $(\mathcal{C}, S)$ studies the asymptotic properties of the motions following initial data chosen randomly with distribution μ_0, which we call "μ_0-random data".

More generally one could investigate the properties enjoyed by motions with initial data chosen randomly with respect to other probability distributions μ not absolutely continuous with respect to the volume measure. Sometimes this is regarded as a less interesting question on the grounds that the Liouville distribution is the *natural one* to use in selecting initial data. This is a preconception; in fact it is well known that it is equally easy (or, rather, difficult) to produce random initial data with a distribution singular with respect to the uniform distribution: therefore such data are equally interesting; or at least one has to find better reasons to regard them less interesting. This being not the place to undertake a learned philosophical discussion on the *prominence* of the Liouville measure μ_0, I shall concentrate on the most studied question of which is the asymptotic behavior of motions with initial data chosen randomly with distribution μ_0. Many of the properties that will emerge will be relevant also for the other random choices with respect to less appreciated distributions.

The key notion is that of *statistics*:

1 Definition: *Given a dynamical system $(\mathcal{C}, S)$ and a probability distribution μ attributing probability 1 to the set of points x which never in their evolution fall on the singularity set $\mathcal{N}$ (i.e. $\mu(\cap_{k=0}^{\infty} S^{-k}(\mathcal{C}/\mathcal{N})) = 1$), we say that μ_0 has a statistics with respect to the evolution S if for any continuous function F on $\mathcal{C}$, called observable, the time average of F on the motion generated by μ-almost all points $x \in \mathcal{C}$ exists, and has the form:*

$$\lim_{T\to\infty} \frac{1}{T} \sum_{k=0}^{T-1} F(S^k x) = \int_{\mathcal{C}} F(y)\overline{\mu}(dy) \tag{1.3}$$

where $\overline{\mu}$ is a suitable probability distribution on $\mathcal{C}$, called the statistics of μ in the dynamical system $(\mathcal{C}, S)$. If $\mu = \mu_0$ is the Liouville distribution then $\overline{\mu}$ will be called simply the statistics of $(\mathcal{C}, S)$ and we shall say that "$(\mathcal{C}, S)$ has a statistics" without reference to μ_0.[3]

Remarks:

(i) Usually the map S is such that $S^{-k}\mathcal{N}$ is contained in a lower dimensional compact submanifold of $\mathcal{C}$: therefore almost all points (randomly chosen with respect to the Liouville measure) will never fall on $\mathcal{N}$ and it will make sense to ask if the Liouville measure has a statistics.

(ii) Note that $\overline{\mu}$ *does depend on* μ_0. It makes no sense to talk about the statistics of a dynamical system without specifying the distribution with which the initial data are randomly selected.

(iii) If $(\mathcal{C}, S)$ has a statistics (*i.e.* if the Liouville distribution has a statistics) then one says also that $(\mathcal{C}, S)$ has a *unique attractor*. It might happen, and in some interesting cases it does happen, that the space $\mathcal{C}$ can be represented as a union of several open sets U_i and of some 0 μ_0-measure sets N_k: for almost all points in U_i the average of F is given by a formula like (1.3) but with some $\overline{\mu}_i$ replacing $\overline{\mu}$. In this case one would say that $(\mathcal{C}, S)$ has several attractors each with its own statistics. In the present analysis I shall not consider such cases: often they can be reduced to cases with unique statistics simply by redefining the phase space to be $\overline{U}_i$ rather than $\mathcal{C}$; although this is not the most general case.

A further general notion is that of *attractor*:

2 Definition: *The set $A \subset \mathcal{C}$ is an "attractor" for μ-random data if μ has a statistics in the dynamical system $(\mathcal{C}, S)$ and (1): $SA = A$ ("A is invariant"), (2): $\overline{\mu}(A) = 1$ (A has "probability 1 with respect to the statistics), (3): the fractal (Hausdorff) dimension of A is minimal among the A's with the properties (1),(2).*

Remarks:

[3] Adhering, only for lack of time, to the prejudice that the volume measure is somewhat privileged.

It is not convenient to require that the attractor be a closed set (as sometimes done). In many cases the attractors A in the above sense are dense in $\mathcal{C}$ but have a fractal dimension strictly less than that of $\mathcal{C}$. So the above notion is subtler than the ones that imply or require that an attractor be closed. It also stresses that an attractor will usually have a non trivial fractal dimension. On the other hand the attractor in this sense is not unique (*e.g.* one can usually remove from it countably many points, except in special cases like when A is a fixed point or a periodic orbit, or even larger sets with zero Hausdorff dimension and zero $\overline{\mu}$ measure).

With the above notions in mind we can proceed to define some of the main qualitative features of the motions following data chosen randomly with respect to a distribution μ_0.

2. Dynamical Bases and Lyapunov Exponents

A given motion analysis can starts with the attempt at understanding the behavior of nearby trajectories: this means understanding the linearization of the motions taking place near it.

For this purpose the notion of regular point is necessary:

3 Definition: *A point x is regular if the map S is regular in the vicinity of $S^k x$ for all $k \geq 0$. Recalling that S is always supposed to be piecewise regular this simply means that $S^k x$ does not fall into the set $\mathcal{N}$ of singularities of S: i.e. $x \notin \cup_{k=0}^{\infty} S^{-k}\mathcal{N}$ or $x \in \cap_{k=0}^{\infty} S^{-k}(\mathcal{C}/\mathcal{N})$.*

Let $S^k x$ be a trajectory starting at a regular point x, $k = 0, 1, \ldots$: the trajectory $S^k(x+dx)$ with dx infinitesimal will stay close forever to that of x departing from it by an infinitesimal amount simply given by $\partial S^k(x)dx$, where $\partial S^k(x)$ is the jacobian matrix of S^k evaluated at the point x. It is a $d \times d$ matrix if d is the dimension of $\mathcal{C}$; by the chain rule $\partial S^k(x)$ is :

$$T_k(x) \equiv \partial S^k(x) = \partial S(S^{k-1}x) \cdot \ldots \cdot \partial S(x) \equiv \prod_{j=1}^{k} \partial S(S^{k-j}x) \tag{2.1}$$

where the factors appear ordered from left to right as j increases.

I shall assume that for some $\bar{C}, \bar{\varepsilon} > 0$ it is $|\det \partial S(x)| \geq \bar{\varepsilon} > 0$ *and* $|\partial S(x)| < \bar{C}$, for all regular points x. This assumption could be greatly weakened, see the appended problems and (Pesin, 1992), but it is convenient continuing the discussion without having to worry about such matter.

Therefore the square of the *stretching* of dx will be described by the ratio:

$$\frac{(\partial S^k dx, \partial S^k dx)}{(dx, dx)} \tag{2.2}$$

where the scalar products are evaluated in the metric defined on $\mathcal{C}$. If $G(x)$ is the metric tensor in x and if the *dilatation* matrix $(M_k)_{ij}$ is defined by setting: $M_k =$

$(\partial S^k(x))^* \partial S^k(x)$, we see that in the local coordinates the matrix is expressed by: $(M_k)_{ij} = (\partial S^k(x))^T G(S^k x)(\partial S^k(x))_{ij}$. So that (2.2) can be explicitly written as:

$$\frac{\sum_{ij}(M_k)_{ij}\, dx_i\, dx_j}{\sum_{ij} G(x)_{ij} dx_i dx_j} \tag{2.3}$$

Suppose that the symmetric matrix $M_k^{1/2k}$ has a limit $D_+(x)$, as $k \to \infty$. Then there is $\tilde{c}$ such that $\tilde{c} > |\det D_+(x)| \geq \bar{\varepsilon}$ by the assumption, and we can define $l_1(x) \geq l_2(x) \geq \ldots l_d(x) > 0$ to be its ordered eigenvalues, counted according to multiplicity. Let $\underline{v}_1(x), \ldots, \underline{v}_d(x)$ be the corresponding eigenvectors, which can be taken orthonormal with respect to the "naive" scalar product $(a, b) = \sum_i a_i b_i$ (which is the scalar product in which M_k and D_+ are symmetric). Then $M_k \underline{v}_j(x)$ grows as $l_j(x)^k$, in the sense that $\frac{1}{k} \log |M_k \underline{v}_j(x)| \xrightarrow[k \to \infty]{} l_j(x)$, as an elementary estimate shows.

We also call $\bar{l}_1(x) > \bar{l}_2(x) > \ldots > \bar{l}_{s(x)}(x)$, respectively $\overline{\lambda}_1(x) > \overline{\lambda}_2(x) > \ldots > \overline{\lambda}_{s(x)}(x)$, the *distinct eigenvalues* (respectively their logarithms) of $D_+(x)$ and $n_1(x), \ldots, n_{s(x)}(x)$ their multiplicities, *i.e.* the dimensions of the eigenspaces $U_1(x), \ldots, U_{s(x)}(x)$ spanned by the eigenvectors with the same eigenvalue. The $l_j(x), \lambda_j(x)$ can be called the spectral or scaling coefficients at x and, respectively, the spectral or scaling exponents at x.

Then it is clear that the vector space R^d can be regarded as containing the planes $V_1(x) \equiv R^d \supset V_2(x) \supset V_3(x) \supset V_{s(x)}(x)$ where $V_j(x)$ is the plane spanned by the eigenvectors of $D_+(x)$ with eigenvalue $\leq \bar{l}_j(x)$: the dimensions of $V_j(x)$ are $d_j = n_j(x) + \ldots + n_{s(x)}(x)$ and the planes $V_j(x)$ are $V_j(x) = U_j(x) \oplus \ldots \oplus U_{s(x)}(x)$.

It is very important to realize that the vectors $\underline{v}_j$ *do depend on the choice of coordinates (and of the metric* G): by changing the metric or the coordinates the vectors $\underline{v}_j$ may, in general, change. However the eigenvalues $\bar{l}_j(x)$, and their multiplicities $n_j(x)$ cannot change, and *also the sequence of decreasing subspaces* $V_j(x)$ *cannot change*, as one can see by noting that the scaling properties of the vectors in such spaces are not metric properties, but have an intrinsic geometric meaning. They can be *characterized* by the property that $\frac{1}{k} \log |\partial S^k(x) u| \xrightarrow[k \to +\infty]{} \overline{\lambda}_j(x)$ if $u \in V_j(x)/V_{j-1}(x)$.

One could try to define the planes with given contraction rate $\overline{\lambda}_j$ to be just $U_j(x)$. But such spaces would not have an intrinsic dynamical meaning: because one could "tilt" $U_j(x)$ sligthly and still keep the property that all the vectors contract at the rate $\overline{\lambda}_j$ (provided the tilting does not generate components along the $U_{j+p}(x)$ with $p > 0$, of course).

Likewise one cannot assign a special meaning to the planes $U_1(x) \oplus \ldots \oplus U_j(x)$: this would in fact generate a system of planes such that if u is a vector in the j-th plane and not in the $(j-1)$-th then the exponent of dilatation will be $\overline{\lambda}_j$; *but such system is not uniquely determined*, for the same reasons discussed in the previous paragraph. On the other hand the system $V_j(x)$ has the property of being uniquely determined by the action of S, and it defines a system of planes along which the scaling size becomes weaker.

4 Definition: *Given the dynamical system $(\mathcal{C}, S)$ a point $x \in \mathcal{C}$ admits a "system of scaling (or contracting) planes" for the forward motion, or for S, if:*
(1) x is a regular point.
(2) there exist numbers $\overline{\lambda}_j(x)$ and positive integers $s(x)$ and $n_j(x)$, with $j = 1, \ldots, s(x)$, with the properties:
(3) the space R^d of the infinitesimal vectors out of x can be regarded as containing a sequence of $s(x)$ subspaces $V_1(x) \equiv R^d \supset V_2(x) \supset \ldots \supset V_{s(x)}(x) \supset V_{s(x)+1} \equiv \underline{0}$ with $V_j(x)$ having dimension $n_{s(x)} + \ldots + n_j(x)$ and
(4) the following limits hold:

$$\lim_{k \to +\infty} \frac{1}{k} \log |\partial S^k(x) v| = \overline{\lambda}_j(x) \qquad \text{if } v \in V_j(x)/V_{j+1}(x) \tag{2.4}$$

Then one says that the numbers $\overline{\lambda}_j(x)$ are the scaling exponents *of S at x in the forward direction, their exponentials $l_j(x) = e^{\overline{\lambda}_j(x)}$ are the* scaling coefficients, *the number $n_j(x)$ is the multiplicity of the j-th coefficient. Sometimes the coefficients are repeated according to the multiplicity: in this case their number is, of course, exactly the phase space dimension. The scaling exponents (coefficients) for S are often called the* forward Lyapunov *exponents (coefficients).*
Remarks:
(i) There is no reason why a point should admit a system of scaling planes.
(ii) A sufficient condition for the existence of contracting planes is that the limit of the sequence of matrices $(\partial S^k(x)^* \partial S^k(x))^{1/2k}$ exists and is a positive matrix D, see the discussion preceding the definition 4,
(iii) If $\overline{\lambda}_1(x) > 0$, $\overline{\lambda}_{s(x)} < 0$ and $\overline{\lambda}_j(x) \neq 0$ for all j's one says that x is a *hyperbolic point*. In this case if r_- is such that $\overline{\lambda}_j > 0$ for $j < r_-$ and $\overline{\lambda}_j < 0$ for $j \geq r_-$ the plane $V^s = V_{r_-}$ will be called the *contraction*, or *stable, plane*. The motion of a hyperbolic point is very unstable in, essentially, all possible senses.
(iv) If a point x admits a system of scaling planes so do all the points on the trajectory generated by x. Such points do have the same coefficients (multiplicities included) and the spaces $V_j(x)$ and $V_j(Sx)$ are related by: $V_j(Sx) = \partial S(x) V_j(x)$, for all j's. The spaces $U_j(x)$ instead *do not have, in general, any covariance property* because they are not intrinsically defined by the dynamics (see above).
(v) If x admits a system of scaling planes then it may not admit such a system in the dynamical system $(\mathcal{C}, S^{-1})$; and even if it does there is no reason why there should be any relation between the two systems or the relative exponents, c.f.r. problem (32).
(vi) Note, once more, that the subspaces in which the contraction exponent is $< \lambda$ for some real λ, intermediate between the exponents values, are well defined (and coincide with $V_j(x)$ if $\overline{\lambda}_{j-1} > \lambda > \overline{\lambda}_j$). But the ones in which the rate is $> \lambda$ are *not* well defined and this explains the impossibility of associating expanding planes to the forward motion.

(vii) Note that the scaling planes for S at a point, when existent, concern the *forward* motion: *i.e.* they are properties of the trajectory $S^k x$, $k \geq 0$. Likewise the scaling planes for S^{-1} concern only the *backward* motion.

In view of the above remarks it is important to try to establish some general results about the existence of systems of scaling planes.

A simple case concerns the regular hyperbolic fixed points x: the latter always admit a system of scaling planes for the forward and backward motion, *i.e.* for S and for S^{-1}. The linearization of the map S around x, *i.e.* the matrix $T = \partial S(x)$ defines the V_x^s plane spanned by the spectral planes of the eigenvalues of T with modulus < 1 and the subspace V_x^u spanned by the planes relative to the eigenvalues of T with modulus > 1. The two subspaces V_x^s, V_x^u can be continued into two small regular connected manifolds tangent to them in x, Δ_x^s, Δ_x^u, such that if $y \in \Delta_x^s$ then $|S^k y - S^k x| \xrightarrow[k\to\infty]{} 0$ bounded proportionally to l_-^k, if l_- denote the absolute values of the eigenvalue of T closest to the unit circle and < 1. Or, similarly, if $y \in \Delta_x^u$ or > 1 then $|S^{-k}y - S^{-k}x| \xrightarrow[k\to\infty]{} 0$ bounded proportionally to l_+^{-k}, if l_+ denotes the absolute value of the eigenvalue of T closest to the unit circle and > 1.

In this case the forward scaling coefficients coincide with the absolute values of the eigenvalues of D, called in stability theory the *Lyapunov coefficients* of the fixed point. They can also be defined as the eigenvalues of the matrix D (whose existence is not completely trivial, see problems): $\lim_{n\to\infty}((T^*)^n T^n)^{1/2n} = D$. The backward scaling coefficients can likewise be identified with the eigenvalues of the matrix $\lim_{n\to\infty}((T^{-k})^* T^{-k})^{1/2k} = D_-$ and, therefore, they are the reciprocals of the forward coefficients. The contracting plane V_x^u for S^{-1} is, also, spanned by the eigenplanes of D_- corresponding to the eigenvalues < 1.

It is convenient to set up the following definition for future use:

5 Definition: *If x is a fixed point the set $\cup_{k=0}^{\infty} S^k \Delta_x^u \equiv W_x^u$ will be called the "global unstable manifold", while the set $\cup_{k=0}^{\infty} S^{-k} \Delta_x^s \equiv W_x^s$ will be called the global stable manifolds of x.*

The two manifolds are locally regular around all their points that do not fall on the singularities of S^k or S^{-k}, respectively, for some $k \geq 0$. But they may be even disconnected, in general.

The subspace tangent to Δ_x^s at x coincides with the plane generated by the vectors associated with the Lyapunov numbers less than 1 which, as said above, is intrinsically associated with S. It consists of the infinitesimal vectors that contract at exponential rate under the action of S.

One cannot identify "similarly" the subspace tangent to Δ_x^u, at x, as the plane generated by the infinitesimal vectors expanding at exponential rate under S because, as already pointed out, such vectors are not unambiguously defined (given a vector expanding at exponential rate one can add to it a vector that expands at a lower rate, getting a vector expanding at the same rate). However one can identify Δ_x^u in the same way as Δ_x^s by replacing S with S^{-1}.

The extension of the analysis to the case of a periodic point is very easy and well known and it will be skipped. In the case x is not fixed, nor periodic, the discussion above leads to a natural extension expressed by the definition:

6 Definition: *A point is said to admit a dynamical base* $(Z_1, \ldots, Z_s)$, *consisting of s mutually transversal planes, if:*
(1) it is regular for S *and* S^{-1} *and it admits, both in the forward and in the backward motions, scaling systems of planes with opposite Lyapunov exponents.*
(2) no Lyapunov exponent vanishes.
(3) the following limit relation holds as $k \to +\infty$ *as well as* $k \to -\infty$*:*

$$\lim_{k\to\pm\infty} \frac{1}{k} \log |\partial S^k u| = \overline{\lambda}_j(x) \qquad \text{for } u \in Z_j \tag{2.5}$$

In this case the planes $V_j = Z_s \oplus \ldots \oplus Z_j$ *will be called the system of contraction planes for* S *and the* $\tilde{V}_j = Z_1 \oplus \ldots \oplus Z_j$ *will be the system of expanding planes for* S. *If* s_- *is such that* $\overline{\lambda}_j < 0$ *for* $j \geq s_-$ *and* $\overline{\lambda}_j > 0$ *for* $j < s_-$ *the planes* $V^s = V_{s_-}$ *and* $V^u = \tilde{V}_{s_- - 1}$ *will be called the stable and unstable planes of* S *at* x.

Remarks: (a) The corresponding notions for the backward motion (*i.e.* for S^{-1}) are trivially related to those for S.
(b) If $u \in V_j(x)/V_{j+1}(x)$ then (2.4) holds.
(c) The points $S^k x$ also admit a dynamical base and $Z_j(S^k x) = \partial S^k(x) Z_j(x)$ for all (signed) integers k.

Finally one more definition is useful to simplify the language:

7 Definition: *A point* x *is called normal if it admits a dynamical base.*

Hence a normal point is a generalization of a periodic point. And it is remarkable that such points do exist and in fact abund, in some sense. The first classical result concerns invariant ergodic distributions μ on $\mathcal{C}$: here invariance means that $\mu(\cap_{-\infty}^{\infty} S^k \mathcal{N})) = 0$ and $\mu(S^{-1}E) = \mu(E)$ for all (Borel)[4] sets E and ergodicity means that there are no non trivial constants of motion which are not μ–almost surely constant.

I Theorem: *Let* μ *be an invariant ergodic distribution for* $(\mathcal{C}, S)$*: then* μ *almost all points are normal for* S. *Furthermore the Lyapunov exponents and their multiplicities are almost everywhere constant.*
If no contraction or expansion coefficient has value 1 *and if* V_x^s *and* V_x^u *are the contracting and expanding planes, then under further "weak" regularity assumptions, there exist (integrability property) manifolds* W_x^u *and* W_x^s *tangent in* x *to* V_x^u *and* V_x^s, *smooth near* x, *such that:*

---

[4] This is in parenthesis because all the sets I will mention are Borel sets or differ from a Borel set by a set of measure 0 with respect to the measure that is being considered and which I also call for brevity Borel sets if no ambiguity arises. In fact, I cannot even conceive the other sets.

$$|S^k x - S^k y| \xrightarrow[k\to\infty]{} 0, \quad for\ y \in W_x^s$$
$$|S^{-k} x - S^{-k} y| \xrightarrow[k\to\infty]{} 0, \quad for\ y \in W_x^u \tag{2.6}$$

and the approach to 0 *takes place exponentially fast bounded above proportionally to* $e^{-\lambda k}$, *where* $\lambda > 0$ *is a suitable constant (such that no Lyapunov exponent is in the interval* $[-\lambda, \lambda]$*).*

A discussion of the "further regularity assumptions" would lead us away from the themes chosen here for discussion: I just mention that they are assumptions on the speed at which a regular point x can approach, in its evolution, the singularities, (Pesin, 1992). The idea is that such speed should be slower than any exponential (or at least slower than the speed corresponding to a bound λ on the contraction and expansion rate).

The above first statement is the *Oseledec theorem*: a "guided" proof to it is described in the problems, the statements concerning the integrability property are part of Pesin's theory, (Pesin,1992). The theorem shows that if the data are picked up randomly with respect to an ergodic distribution (any one!) then they provide (non constructive) examples of points with dynamical bases. It also shows that the scaling properties of S and those of S^{-1} are intimately related and the two maps show the "same" properties (or, rather, properties that are trivially related), provided the data are randomly distributed with an *invariant* ergodic distribution giving zero measure to data visiting, in their evolutions, the singularities. "Nothing about data randomly chosen with an ergodic invariant distribution can be learnt by running the motion backwards, that cannot be learnt by running it forward". In other words one can say that the motion is "reversible" *on such data.*[5]

The case in which there are 0 contraction or expansion exponents is more involved and it will not be discussed. It certainly arises when the system has non trivial smooth constants of motion, each of which generates a vanishing Lyapunov exponent. But such cases can be easily eliminated because one usually can restrict the phase space to the data for which the constants of motion have a fixed value. When zero Lyapunov exponents are not related to smooth constants of motion, however, a genuinely more complicated situation arises: I shall not deal with it here, besides saying that some of the above kinematical properties *do not depend on the assumption of the absence of* 0 *Lyapunov exponents.* In particular the first sentence in the theorem I above *does not require* that the exponents be non zero.

Also the ergodicity, in some sense, is not really necessary and it has been introduced only to simplify the exposition. The case in which ergodicity is missing can be often reduced to the ergodic case by using the "ergodic decomposition theorem": a theorem and a discussion that I avoid here.

[5] Remember, however, that such data are very special.

A more interesting question is what can be said when the data are picked up with a distribution that is *not* invariant. The case of non invariant distribution of the initial data will be analyzed, under suitable further assumptions, in the next section. Here I add only a few remarks on a frequently misrepresented procedure.

Note that the Lyapunov exponents for S and those of S^{-1} are opposite *if the data are chosen randomly with respect to an invariant distribution.* However usually one selects the data with respect to a *non invariant distribution*: in this case the above theorem does not say much. *In particular one should expect that the asymptotic properties of the motions in the future and in the past are different and described by different statistics.* For instance in cases in which there is a *time reversal symmetry* one will find that the Lyapunov exponents for the motions towards the past and the future are identical (rather than opposite). This will not contradict the fact that if the data are chosen randomly with respect to the *future* statistics then the Lyapunov exponents for the motion towards the past would be the *opposite* of the ones for the motion towards the future!

How comes that often a method to measure the minimum Lyapunov exponent is *said* to be: "just measure the maximum Lyapunov exponent for the backward motion and change sign"? In most cases this would seem wrong: for instance when the system is time reversible *because* one would, instead, get again the maximum Lyapunov exponent.

The "paradox" is understood if one examines what is really meant by the above statement, *i.e.* what is the actual measurement performed. One finds that the measurement consists (schematically) in picking at random with a distribution μ_0 (non invariant) a point x, following its trajectory for a long time T, together with the trajectory of a nearby point y until it reaches a point $S^T x$. The rate of separation of the two points gives[6] a measurement of the maximum Lyapunov coefficient. The trajectory of x is memorized and that of y "thrown away". One then starts from $S^T x$ and from a nearby point y', and runs the two motion backwards: of course the motion of $S^T x$ is already known and one computes that of y'. The motions of both points y, y' is computed as a perturbation of the motion of x or of $S^T x$ (forward and backwards respectively) thus it is very convenient to use twice the same trajectory as this saves considerable time.

But this is not merely a matter of convenience: we see that $S^T x$ is *no longer a random point with the initial distribution.* It is, rather, approximately a random point with respect to the statistics $\overline{\mu}$ of the distribution μ_0. Therefore it will show in the backward motion, Lyapunov exponents opposite to the ones exhibited in the forward motion (which have the *same* value if the initial data are chosen with respect to μ_0 or $\overline{\mu}$), by the above theorem. Hence one gets the wanted result: in the *improper sense just discussed* the maximum exponent in the backward motion is the opposite of the minimum in the forward motion. This property is

[6] Unless the two points are very special, *e.g.* if y lies on the lower dimensional contracting manifold W^s_x.

very reminiscent of the build up of correlations that appears in the theory of the molecular chaos in the Boltzmann equation, see (Cohen and Berlin, 1960).

3. Chaotic Motions

The definition that I shall use for a chaotic system is:

8 Definition: *A dynamical system* $(\mathcal{C}, S)$ *will be called "chaotic" if:*
(1) There exists a periodic hyperbolic point O *such that the global stable and unstable manifolds of* O *consist of regular points for* $S^{\pm 1}$*, and are smooth, connected and dense on* $\mathcal{C}$*, (*instability axiom*).*
(2) The restrictions of S^k *to* W_O^u*, and of* S^{-k} *to* W_O^s*, are uniformly expansive for* $k \geq 0$ *and uniformly contractive for* $k \leq 0$*, (*expansivity axiom*).*
(3) If x_n, x'_n *are points on* W_O^u *and* $|x_n - x'_n| \xrightarrow[n\to\infty]{} 0$ *then the tangent planes to* W_O^u *at such points become parallel uniformly at speed* $|x_n - x'_n|^\beta$ *for some* $\beta > 0$*, together with the matrix that linearizes the evolution* S *on* W_O^u*: a similar property holds for* W_O^s*; (*continuity axiom*).*
(4) The manifolds W_O^s, W_O^u *form, everywhere they cross, an angle uniformly bounded away from* 0*, (*transversality axiom*).*
(5) Given any open sets $G, \Gamma_1, \ldots, \Gamma_n$ *there is a* k *such that* $S^h G$ *has intersection with all the* Γ_j *for* $h \geq l$ *(*topological mixing axiom*).*

Remarks:
(i) To understand the role of the various conditions in the above definition note that from the hyperbolicity of O, and from the definitions of the previous section, it follows that the tangent planes to the manifolds W_O^u and W_O^s are the expansive and contractive scaling manifolds of O. However it does not follow from this only fact that the action of S on W_O^u, nor that the action of S^{-1} on W_O^s, are expansive. The latter property, as well as any statement on the Lyapunov exponents on W_O^u, concern properties of the *future* evolution on W_O^u, while the knowledge that W_O^u is the unstable manifold of the fixed point gives us only information about the motion towards the past. Hence it cannot follow that $(\partial S^k(x)_u^* \partial S^k(x)_u)^{1/2k}$, where $\partial S(x)_u$ denotes the jacobian of the transformation S regarded as a map of W_O^u, has a limit as $k \to \infty$. The role of (2) is to require such properties, or at least the part of them that we need, explicitly (as well as the corresponding ones for W_O^c). Its quantitative meaning is expressed as: there exist $D_0, \lambda > 0$ such that for all $x, y \in W_O^u$ and $k \geq 0$ the distance $d_u(S^{-k}x, S^{-k}y)$ between $S^{-k}x, S^{-k}y$, *measured along* W_O^u, is:

$$d_u(S^{-k}x, S^{-k}y) \leq D_0 e^{-\lambda k} d_u(x, y) \tag{3.1}$$

and the corresponding properties are required to hold for W_O^s for $k \leq 0$.
(ii) Property (3) on the jacobians means that there exists $D_1, \beta > 0$ such that the "unit vector orthogonal to W_O^u at x", which is in fact a tensor that we can denote $\partial W_O^u(x)$ verifies:

$$|\partial W_O^u(x) - \partial W_O^u(y)| \le D_1 d(x,y)^\beta \tag{3.2}$$

for all $x, y \in W_O^u$ and that there exists $D_2, \beta > 0$ such that the jacobian $\partial S(x)_u$ of S as a map of W_O^u into itself verify, for all $x, y \in W_0^u$:

$$|\partial S(x)_u\,(\partial S(y)_u)^{-1} - 1| \le D_2 d(x,y)^\beta \tag{3.3}$$

The analogous conditions are also imposed on W_O^s and on S^{-1}.
(iii) the property (4) can be put in a quantitative form as follows: if dw^u and dw^s are two surface elements tangent to W_O^u and W_O^s, respectively, at a common point x then the Liouville volume of the parallelogram generated by them is $\mu_0(dw^u dw^s) = b(x) dw^u dw^s$ and $e^{-B_0} < b(x) < e^{B_0}$ holds for some constant B_0.
(iv) The continuity above is a "Hölder continuity"; it is convenient here being generous on the weakness of this assumption and *not* requiring Lipshitz continuity or higher smoothness. Even very smooth dynamical systems may have just Hölder regularity of the foliation of phase space into stable or unstable manifolds.
(v) The definition essentially yields what is usually called an Anosov system. A more general notion (of axiom A-system) could be envisaged. However in the cases considered here the attractors will always be dense in the full phase space and the above generality will be sufficient. All the items in the definition are essential; but in a sense the axiom (1) is the major one among them.

In this section I suppose the system to be chaotic in the above sense. The following analysis (essentially due to Sinai, Bowen and Ruelle, (Sinai, 1968a, 1968b,1994; Bowen, 1975,1978; Ruelle, 1976), will show that *there is a well defined statistics for the Liouville measure* and the attractor will be dense on C. The statistics on it can be determined quite explicitly and it can be shown to have strong ergodic properties. The existence of a well defined statistics plays the role of the ergodic hypothesis and it will therefore be called *ergodicity property*: it will appear that it has been introduced here under the disguise of the chaoticity assumptions (essentially the smoothness and density of the stable and unstable manifolds of O).

We first describe how the measures $\overline{\mu}^{\pm}$, the past and future statistics of the Liouville measure, can be characterized. An apparently involved geometric construction is necessary: it is very simple in the case in which W_O^s and W_O^u are 1*-dimensional*, hence the phase space C has dimension $d = 2$. I shall, however, provide a sketch of the construction in the general case: but one should first understand it (via drawings) in the trivial $d = 2$ case, to realize later that the general case is its natural generalization. The geometrical construction leads to the definition 9 below of *Markov pavement*: the construction that follows is somewhat different from the classical constructions of (Sinai, 1968a, 1968b,1994; Bowen, 1975) and, admittedly sketchy (except in the $d = 2$ case, in which it becomes a well known construction whose generalization is attempted here): its details will not be really used in the following and one could invoke here the theorem of Sinai stating the existence of a Markov pavement, *i.e.* proceed to the paragraph preceding definition 9 below.

Given a small ball of radius δ (small compared to the curvature radii of the manifolds W^u_O, W^s_O) centered at $x \in W^u_O$ the manifold W^u_O will intersect the ball on a (dense) family of connected surfaces: only one of them will contain x and it will be called "the disk on W^u with center x and radius δ". The disks with radius δ on W^s are likewise defined.

The assumed density and continuity of W^s_O and W^u_O allows us to define "disks on W^u or W^s" centered at any point x: they are the surfaces obtained as limits of a sequence of disks of radius δ on W^s_O (or W^u_O) centered at the points $x_n \in W^s_O$ (or $x_n \in W^u_O$) of a sequence of converging to x as $n \to \infty$. And the union of the disks that match smoothly with a disk on W^s (or W^u) centered at x will define the manifold W^s_x (or W^u_x).

Let Δ, Δ' be two small disks on W^u_O and, respectively, W^s_O centered at O. Let ϑ_0 be so large that the web generated by $S^{\vartheta_0}\Delta$ and $S^{-\vartheta_0}\Delta'$, fills the phase space $\mathcal{C}$ so densely that there is no point further away from the web than a prefixed $\delta > 0$, and so that the disk of radius δ on W^s centered at any $x \in S^{\vartheta_0}\Delta$ has at least another intersection with $S^{\vartheta_0}\Delta$ (and "viceversa", exchanging Δ, Δ' and S, S^{-1}).

On $S^{\vartheta_0}\Delta$ we consider all the points $x_1, x_2, \ldots$ common to $S^{-\vartheta_0}\Delta'$: they form a rather dense set of points on $S^{\vartheta_0}\Delta$, and by taking ϑ_0 large we can suppose that there is a triangulation $\mathcal{T}$ of $S^{\vartheta_0}\Delta$ with base on such points[7] consisting of

[7] A smooth triangle (one should call it pyramid but I use the more reassuring word triangle to make the object look simpler) T is d–dimensional smooth manifold W consisting in a compact set with connected interior containing $d+1$ points, the *vertices*, and a boundary formed by joining them with d smooth surfaces of dimension $d-1$ each, *faces*, containing a different d–ple of the $d+1$ vertices and with the further property that the set is *conical* (see below) around each vertex and with the intersections of any pair of them forming a smooth $d-1$–dimensional triangle, *face*. The definition is recursive if one declares that a 0 dimensional triangle is a point and a 1 dimensional triangle is a smooth arc. Here conical means that at every vertex one can draw an open cone: this also requires a definition, as the surface may be not flat; for instance one can consider a family of smooth curves emerging from the vertex and tangent at the vertex to a true open right cone with apex at the vertex and opening angle $\varepsilon > 0$, and with all the points close enough to the vertex entirely contained inside the interior of the triangle. If the boundaries of the various triangles are required to be only Hölder continuous rather than smooth one obtains the more general notion of triangle: the key transversality property remains, however, unchanged for such more general triangles and we call $\varepsilon(T), \eta(T)$ the minimum of the opening angles at the vertices of the various faces of the triangle and, respectively, the minimum Hölder continuity exponents of the various faces. Note that all the d dimensional triangles have boundaries with 0 d–dimensional volume. A triangulation of a compact manifold with base on a given family of points is a covering by a family of triangles with the vertices on the base, no interior points in common and no base point in their interior. All this is quite trivial in the case the triangle is 1

smooth triangles T_j^u which have vertices at distances $\le \delta$. By slightly deforming Δ we can also suppose that the boundary $\partial(S^{\vartheta_0}\Delta)$ consists of sides of some of the triangles of $\mathcal{T}$.[8] Let a be the *smallest distance between distinct triangles vertices* and let ε be the smallest opening angle at the vertices.

Let ∂_δ^s be the surface formed by drawing through *each* x on the *boundary* of *each* triangle a disk on W^s centered at x and of radius δ. This is a codimension 1 surface. It is not difficult to see that if the expansion rate λ of S is large enough (so that $\delta D_0 e^{-\lambda} \ll a$) then we can slightly modify the sides of the triangles (hence Δ as well) so that the image of ∂_δ^s is contained in itself. The expansion rate will, however, be in general much weaker than needed. In that case we imagine, for the time being, replacing S with a high iterate of S so that the new transformation meets the expansivity requirement.

The construction can be done by successive approximations, recursively, as every change of a given triangle boundary performed to impose the wanted condition, will ruin the validity of the condition for some other triangle (but by a smaller amount). One will only be able to infer that the final triangles have Hölder continuous boundaries, even when the W_O^u and W_O^s are (locally) flat (it is not difficult to see that the exponent of Hölder continuity should be bounded below by $\eta = e^{-(\lambda_+ - \lambda_-)}$ if λ_+ is the maximal expansion rate and $\lambda_- \ge \lambda$ is the minimal expansion rate on W^u, which will be in general different: just think at what happens to an angle when the abscissa and the ordinate are contracted by a different scaling factor). In the case of one dimensional Δ, however, the construction is clearly completed "in one stroke" and the triangles are just smooth segments on W_O^u (and their boundaries are isolated points, hence necessarily smooth (0-dimensional manifolds). The reason why in general the construction leads to triangles with mildly regular boundaries can be understood through the celebrated elementary Bowen's example, (Bowen, 1978), ultimately relating it to the direction dependence of the expansion rate on the unstable manifold, when it has dimension > 1 and at least two distinct positive Lyapunov exponents (so that $\lambda_+ > \lambda_-$ as expected from the above bound on the Hölder continuity exponent). The deformation of the triangles, during the construction, is small (at most of the order of $\delta e^{-\lambda\vartheta_0}$) but the regularity of the deformation is quite out of control (because of the mechanism in (Bowen, 1978)).

An identical construction can be performed on the stable manifold leading to a triangulation $\mathcal{T}^s$ of $S^{-\vartheta_0}\Delta'$, with Hölder continuous triangles.

If $\mathcal{T}^u$ and $\mathcal{T}^s$ are the above triangulations of W_O^u and W_O^s we can consider the pairs of triangles T_h^u, T_k^s with one point in common and consider the sets

---

dimensional. Consider a family of points, distributed on a compact manifold with smooth boundary, dense enough so that every point has another one closer than δ to it. Then it is possible to build a triangulation based on the given points by trimming the manifold boundary by an amount not exceeding, say, twice δ if δ is small enough. I take this as obvious, although a proof is likely to be somewhat verbose.

[8] The deformation is at most of the order of $\delta e^{-\lambda\vartheta_0}$.

$E_j = T^u_h \times T^s_k$ built with the points obtained by fixing $\xi \in T^u_h$ and $\eta \in T^s_k$ and by drawing the disk with radii 2δ on W^u_η centered at η and on W^u_ξ centered at ξ and by considering their intersection $x = x(\xi,\eta)$ as ξ, η vary in T^u_h and T^s_k, (here δ may have to be required to be small enough, always with respect to the curvature radii of the manifolds, so that the intersection defining x is certainly non empty).

Such sets will be called *prisms*: the name "parallelogram" or "rectangle" is used in the literature: but the construction here is somewhat different from the original one of (Sinai, 1968a, 1968b, 1994; Bowen, 1975, 1978) (which is not based on triangulations). Furthermore even in the original construction the sets called parallelograms look to me, no matter how I attempt to draw them, prisms except of course in the two dimensional case, when they really look like parallelograms. The sets T^u_h and T^s_k will be called, respectively, the *horizontal* and the *vertical axis* of the prism $E_j = T^u_h \times T^s_k$. Therefore a natural name for the sets $\partial^u E_j \equiv T^u_h \times \partial T^s_k$ and $\partial^s E_j \equiv \partial T^u_h \times T^s_k$ will be *horizontal* boundary of the prism E_j and, respectively, *vertical boundary* of E_j (or unstable and stable boundary of E_j). The set $\partial^s = \cup_j \partial^s E_j$ $\partial^u = \cup_j \partial^u E_j$ define the *total vertical* or the *total horizontal boundary* of the family of prisms E_j which themselves constitute a *pavement* of the phase space with small *prisms*.

The above notion of prism as a set having the form $E = T^u \times T^s$, see the above two last paragraphs, where T^u and T^s are small connected surfaces, with Hölder continuous boundaries, on the stable and unstable manifolds of a common point x is more general; so that one can consider a pavement $\mathcal{E}$ of the phase space $\mathcal{C}$ with prisms, $\mathcal{E} = (E_1, \ldots)$ and define, likewise, the stable and unstable boundaries $\partial^u_j, \partial^s_j$ of the prisms E_j and the total horizontal and the total vertical boundaries ∂^u and ∂^s: note that ∂^s, ∂^u have zero volume (by the Hölder continuity of the boundaries). Given such concepts the following definition is useful:

9 Definition: *A pavement $\mathcal{E} = (E_1, E_2, \ldots)$ by prisms is called a Markov pavement (or a Markov partition) if the total stable boundary is mapped into itself under the action of S and the total unstable boundary is mapped into itself under the action of S^{-1} and they have both 0 volume.*

The key property of the special pavement $\mathcal{E}$ constructed above, and motivating the last definition, is that by construction the total vertical boundary of the prisms is mapped into (a small portion of) itself under the action of S and likewise the total horizontal boundary ∂^u is mapped into (a small portion of) itself under the action of S^{-1}. Hence we have just seen that special Markov partitions, associated with the stable and unstable manifolds of a hyperbolic fixed point, exist for a chaotic system at least if S is replaced by a high enough power S^K of S. Such special Markov partitions are particularly suited for numerical applications, see (Franceshini and Zironi, 1985).

But it follows immediately from the definition 9 just given (and the covariance of W^s_x, W^u_x: $SW^s_x = W^s_{Sx}$, $SW^u_x = W^u_{Sx}$) that if $\mathcal{E}$ is a Markov partition for S^K then the partition $S^p\mathcal{E}$ (whose prisms are obtained by transforming with the map

S^q those of $\mathcal{E}$) is *also* a Markov partition for the iterate S^K of S. Furthermore the partition obtained by intersecting the prisms of the partitions $S^p\mathcal{E}$ with p taking K consecutive values is a Markov partition for S. Hence any dynamical system which is chaotic in the sense of definition 8 admits a Markov partition $\mathcal{E}$, (Sinai, 1968a, 1968b, 1994).

We can form, given any integer ϑ, and a Markov partition $\mathcal{E}$ a more general and finer Markov partition $\mathcal{E}_\vartheta$ simply by "intersecting" the partitions $S^{-\vartheta}\mathcal{E}, S^{-\vartheta+1}\mathcal{E}, \ldots, S^{\vartheta}\mathcal{E}$. We shall only consider sequences of Markov partitions of the above form $\mathcal{E}_\vartheta$.

Given $\mathcal{E}$ we consider the map $\Sigma:\ x \to \underline{\sigma}(x)$ mapping x to the sequence σ_j defined by $S^j x \in E_{\sigma_j}$, the *history* of x on the partition $\mathcal{E}$.

This map is unambiguously defined for all x such that $S^k x$ does not fall on the boundary of any one of the parallelograms E_σ of $\mathcal{E}$. If we define the *compatibility matrix* element $M_{\sigma\sigma'}$ to be 1 if $SE^0_\sigma \cap E^0_{\sigma'} \neq \emptyset$, and 0 otherwise, where E^0_σ is the set of interior points of E_σ, then it is clear that the history $\underline{\sigma}(x)$ of a point x, for which it is unambiguously defined, is an *allowed sequence*: in the sense that $\prod_{j=-\infty}^{\infty} M_{\sigma_i,\sigma_{i+1}} = 1$. Viceversa, and this follows directly from the mapping properties of the boundaries of the prisms in definition 9 (attempt a proof only after finding via a drawing in the trivial $d = 2$ case), if $\underline{\sigma}$ is an allowed sequence there is always a point x producing $\underline{\sigma}$ as history, provided one chooses conveniently the definition of $\sigma_j(x)$ in the cases in which this quantity is ambiguously defined (because some iterate of x falls on the boundary of some E_σ). Item (5) of the chaoticity assumption, in definition 8, implies that the matrix M is "transitive", *i.e.* there is an integer p such that M^p has all entries non vanishing,(Sinai, 1968a, 1968b, 1994). This important property may be implied by other more topological properties, see the discussion of the axiom–A in (Smale, 1965), but here we do not discuss this point (in the same spirit animating what precedes and follows, of not trying to get "optimal theorems" operating with "minimal assumptions").

If $\underline{\sigma}$ is the history of x: $\underline{\sigma} = \Sigma x$, then the history of Sx is $\varphi\underline{\sigma}$ where φ is the *shift* to the left of $\underline{\sigma}$. This means: $\Sigma(Sx)_j = (\Sigma x)_{j+1}$.

Even when the history of a point is ambiguous it can be seen that the number of possible *allowed* sequences is never more than a number that can be explicitly bounded in terms of the maximum number of triangles with one vertex in common and the dimensions of W^s, W^u. But we do not have to worry about this, since the boundaries of the prisms of $\mathcal{E}_\vartheta$ have zero volume measure and we are interested only on properties valid for the data x which have probability 1 with respect to the Liouville measure.

Therefore, for our purpose, we can think that the points of C can be described by allowed (in the above sense) sequences of symbols $\underline{\sigma}$: *we shall denote K the space of such sequences*, defining on it a metric that sets to 2^{-q} the distance between two sequences which agree on the sites $-q, \ldots, q$ but not on any larger symmetric interval. Here 2 is arbitrarily chosen and any number > 1 could replace it, generating the same topology on the space of sequences. In other words we can use the sequences $\underline{\sigma}$ as a system of coordinates (a slight variation on the usual representation of the cartesian coordinates in some digital representation of the reals, *e.g.* base 10; but the present representation is intrinsically tied to the dynamics).

It is convenient to think of a history sequence $\underline{\sigma} = (\sigma_j)_{j=-\infty}^{\infty}$ as a *configuration of a one dimensional spin system*, to make more striking the analogy (that will emerge below) with the theory of one dimensional statistical mechanics : thus we call the labels j, labeling the times marking the history (in units of t_0), with the name of *sites*. Hence the value of $\underline{\sigma}$ at the site j will be σ_j.

The Liouville measure μ_0 becomes a probability distribution μ'_0 on the space of the allowed sequences. In fact consider the set of sequences denoted

$$G' = \begin{pmatrix} -q & . & . & . & q \\ \sigma^0_{-q} & . & . & . & \sigma^0_q \end{pmatrix},$$

consisting in the sequences $\underline{\sigma}$ whose values σ_j with $j = -q, \ldots, q$ coincide with the given σ^0_j. Such set G' will be naturally given a probability equal to the μ_0–measure of the set: $G = \Sigma^{-1}G'$, if Σ is the above history map.

Thus setting $\mu'_0(G') \equiv \mu_0(G)$, with $G = \Sigma^{-1}G'$, for any (Borel) set G' in the space K allows us to think of the dynamical system $(\mathcal{C}, S)$ as (K, φ) and the probability measure μ_0 becomes μ'_0.

The main point is that there is a simple formula expressing the probability $\mu'_0(G')$, (Sinai).

To find it just remark that the set G is (clearly) a prism for the Markov partition $\mathcal{E}_q$ (because $G \equiv \cap_{r=-q}^{q} S^{-r} E_{\sigma^0_r}$). If β^s_σ denotes the surface of the vertical axis T^s_σ of $E_\sigma = T^u_\sigma \times T^s_\sigma$ lying on W^s_O then the area of the vertical axis of G is essentially given by: $\beta^s_{\sigma_{-q}} \prod_{j=-q}^{-1} \Lambda_s(S^j x)$, where x is a point in G and $\Lambda_c(x)$ is the absolute value of the jacobian determinant of S, at x, as a map from W^s_x to W^s_{Sx}.

Likewise if β^u_σ is the surface of the horizontal axis T^u_σ of E_σ on W^u_O then the area of the horizontal axis of G is, essentially: $\beta^u_{\sigma_q} \prod_{j=0}^{q-1} \Lambda_u^{-1}(S^j x)$ where $\Lambda_u(x)$ is the absolute value of the jacobian determinant of S, at x, as a map from W^u_x to W^u_{Sx}.

Therefore the μ_0 measure of G is, essentially:

$$\beta^s_{\sigma_{-q}} \Big(\prod_{j=-q}^{-1} \Lambda_s(S^j x) \Big) \cdot \Big(\prod_{j=0}^{q-1} \Lambda_u^{-1}(S^j x) \Big) \beta^u_{\sigma_q} \cdot b(x) \tag{3.4}$$

where $b(x) dw dw'$ is the volume element (*i.e.* the $\mu_0(dw dw')$ corresponding to two surface elements dw, dw' tangent to, respectively, W^u_x, W^s_x (see remark (iv) above). The "essentially" means that there is an error in (3.4) due to the finite size of the sides of G: see below for its treatment.

To connect (3.4) with something more familiar we can go through the definitions and remarks that led Ruelle, (Ruelle, 1978, 1980a, 1980b,1989), to call what follows the *thermodynamic formalism* for strange attractors. Define:

$$\begin{aligned} h_+(\underline{\sigma}) &= \log \Lambda_u(x(\underline{\sigma})), \qquad h_-(\underline{\sigma}) = -\log \Lambda_s(x(\underline{\sigma})) \\ h_s(\underline{\sigma}) &= \log \beta^s(\sigma_0), \qquad h_u(\underline{\sigma}) = \log \beta_u(\underline{\sigma}_0), \qquad h_0(x) = -\log b(x(\underline{\sigma})) \end{aligned} \tag{3.5}$$

and remark that h_0, h_+, h_- as functions of σ_j depend very little on the σ_j's with j large in the sense that ther exist constants $H_\gamma, \kappa > 0$ such that for $\gamma = u, s, +, -, 0$:

$$|h_\gamma(\underline{\sigma}) - h_\gamma(\underline{\sigma}')| < H_\gamma e^{-\kappa q} \tag{3.6}$$

if $\sigma_i = \sigma_i'$ for $|i| < q$. This follows again from the the continuity and hyperbolicity assumption, because varying $\underline{\sigma}$ on the sites $j > k$-th means varying $x(\underline{\sigma})$ in

$$\Sigma^{-1}\begin{pmatrix} -k & . & . & . & k \\ \sigma_{-k} & . & . & . & \sigma_k \end{pmatrix}$$

which is a prism whose diameter has size of order $O(\delta e^{-\lambda k})$, see (3.1). Therefore, by the continuity assumption in definition 8, we see that the variation of h_j, $j = 0, \pm, u, s$, is bounded by $O(e^{-\lambda k \beta}\delta^\beta)$. This is usually quoted by saying that dependence of h_j on σ_k "vanishes exponentially" or h_j has "memory vanishing exponentially".

In terms of the functions $h_\gamma, \gamma = s, u, +, -, 0$ we can rewrite (3.4) as:

$$e^{-h_s(\varphi^{-q}\underline{\sigma}) - \sum_{j=-q}^{-1} h_-(\varphi^j\underline{\sigma}) - h_0(\underline{\sigma}) - \sum_{j=0}^{q-1} h_+(\varphi^j\underline{\sigma}) - h_u(\varphi^q\underline{\sigma})} \tag{3.7}$$

This shows, unless the approximation involved in the discussion of the word "essentially" above spoils everything, that the probability distribution μ_0' on K coincides with a Gibbs' state on K for the short range *non translation invariant* formal hamiltonian:

$$\sum_{j=-\infty}^{-1} h_-(\varphi^j\underline{\sigma}) + h_0(\underline{\sigma}) + \sum_{j=0}^{\infty} h_-(\varphi^j\underline{\sigma}) \tag{3.8}$$

Such a state, to the far right (*i.e.* in the far future) corresponds to the Gibbs state $\overline{\mu}_+'$ with hamiltonian $\sum_{-\infty}^{\infty} h_+(\varphi^j\underline{\sigma})$ and to the far left (*i.e.* in the far past) corresponds to the Gibbs state $\overline{\mu}_-'$ with hamiltonian $\sum_{-\infty}^{\infty} h_-(\varphi^j\underline{\sigma})$.[9]

The last statement defines unambiguously $\overline{\mu}_\pm'$ as probability measures on K, via the theory of one dimensional Gibbs states (Ruelle, 1968; Bowen, 1975, 1978; Gallavotti, 1981). This is so because the Gibbs states potentials have short range

[9] One should not forget that (3.8) is defined for the allowed sequences. Hence one should more precisely say that the hamiltonian of our spin system is (3.8) if the configuration is allowed and it is $+\infty$ otherwise: *i.e.* the spin system has a (short range) *hard core* corresponding to the compatibility condition.

(by the above remark on the shortness of the memory, the potential decreases exponentially, see the paragraph following (3.5)), and no phase transitions are possible, the system being one dimensional. By the map Σ the distributions $\overline{\mu}'_{\pm}$ are transformed into the natural candidates for the forward and backward statistics $\overline{\mu}_{\pm}$ for the distribution μ_0.

Hence the only problem is the check that μ'_0 is really the Gibbs state on the allowed sequences with formal energy (3.8): this means discussing quantitatively the error mentioned in connection with (3.4). All the properties of $\overline{\mu}^{\pm}$ would then follow from the well known theory of the one dimensional short range Gibbs states for spin systems, (Ruelle, 1968; Bowen, 1975), including the property that the $\overline{\mu}'_{\pm}$ are the forward and the backward statistics of μ'_0. The latter property holds simply because the Gibbs state with hamiltonian (3.8) really converges to $\overline{\mu}'_{+}$ or to $\overline{\mu}_{-}$ if it is observed to the far right or to the far left. Thus the characterization of $\overline{\mu}^{\pm}$ would have received a complete description.

To correct the error involved in the above use of the word "essentially" one can use the deeper properties of Gibbs states; namely their characterization in terms of the DLR equations. To use the DLR theory one needs an expression of the ratio of the probabilities that two sequences $\underline{\sigma}^1$ and $\underline{\sigma}^2$ have the value σ or σ' at site 0, conditioned to the two sequences having the same value in all the other sites. This is the ratio of the probabilities for the events in which the spin at site 0 is σ or σ', *having fixed* the same configuration for the spins at *all* the other sites.

This means considering the ratio of the probabilities:

$$\lim_{q\to\infty} \frac{\mu'_0\left(\begin{pmatrix} -q & . & . & -1 & 0 & 1 & . & . & q \\ \sigma_{-q} & . & . & \sigma_{-1} & \sigma & \sigma_1 & . & . & \sigma_q \end{pmatrix}\right)}{\mu'_0\left(\begin{pmatrix} -q & . & . & -1 & 0 & 1 & . & . & q \\ \sigma_{-q} & . & . & \sigma_{-1} & \sigma' & \sigma_1 & . & . & \sigma_q \end{pmatrix}\right)} \tag{3.9}$$

where $\underline{\sigma} = (\sigma_j)_{-\infty}^{\infty}$ is an arbitrary (allowed) sequence. Then clearly the error involved in (3.4) can be regarded as a multiplicative factor correction to the ratio in (3.9) approaching 1 as $q \to \infty$. If $\underline{\sigma}^1$ and $\underline{\sigma}^2$ denote two allowed sequences differing only in the entry with label 0, one gets that the value of the limit in (3.9) is, rigorously:

$$e^{-\sum_{-\infty}^{-1}(h_-(\varphi^j\underline{\sigma}^1)-h_-(\varphi^j\underline{\sigma}^2))-(h_0(\underline{\sigma}^1)-h_0(\underline{\sigma}^2))-\sum_0^{\infty}(h_+(\varphi^j\underline{\sigma}^1)-h_+(\varphi^j\underline{\sigma}^2))} \tag{3.10}$$

which is the DLR characterization of μ'_0 as the Gibbs state with formal hamiltonian (3.8): proving, by the uniqueness theory of the short range one dimensional Gibbs states and the DLR theory of the Gibbs states, that indeed μ'_0 *is the Gibbs state with energy (3.8)*, (Bowen, 1975; Gallavotti, 1981).

Hence the forward and backward statistics $\overline{\mu}^{\pm}$ are precisely identified and described, because the theory of one dimensional short range Gibbs states is very well developed. *It also follows that $\mu_\pm$ are ergodic (in fact isomorphic to a Bernoulli shift) measures*, (see (Gallavotti, 1973)).

We proceed to derive some properties of $\overline{\mu}^+$ by making full use of the short range (3.6) of the potential h_+ generating $\overline{\mu}^+$ as a Gibbs state: they will be used in the application in Section 4.

Let $\Gamma_T, T > 0$ denote the set of the allowed strings $\sigma_j^T, |j| < T$: these are the strings that can becontinued to infinite allowed strings $\hat{\underline{\sigma}}^T \in K$: we in fact imagine to continue each $\underline{\sigma}^T$ to some $\hat{\underline{\sigma}}^T$ (arbitrarily).

Then $\overline{\mu}^+$ can be constructed as a limit of the distributions μ_T defined by the average they assign to an arbitrary smooth function F:

$$\int \mu_T(dx)F(x) = \frac{\sum_{\underline{\sigma}^T \in \Gamma_T} e^{-\sum_{-T}^{T} h_+(\varphi^j \hat{\underline{\sigma}}^T)} F(x(\hat{\underline{\sigma}}))}{Z_{[-T,T]}} \tag{3.11}$$

where $Z_{[-T,T]}$ is the normalization factor, that should be called the "partition function" of the energy h_+ relative to the interval $[-T,T]$.

Since $\overline{\mu}^+$ is the Gibbs state with potential h_+ it follows that $\overline{\mu}^+$ is the limit of μ_T as $T \to \infty$, (Sinai, 1968a, 1968b, 1994). We now investigate the probability distribution of a random variable $w = \sum_{j=-t}^{t} f(S^j x)$ with f smooth and $t > 0$. This is done by considering the probability distribution $\overline{\mu}_{T,t}$ defined, on the smooth functions F, by:

$$\int \mu_{T,t}(dx)F(x) = \frac{\sum_{\underline{\sigma}^T \in \Gamma_T} e^{-\sum_{-t}^{t} h_+(\varphi^j \hat{\underline{\sigma}}^T)} F(x(\hat{\underline{\sigma}}))}{\text{normalization}} \tag{3.12}$$

We want to show:

Theorem II: *If f is smooth the probability distribution of $F = \sum_{-t}^{t} f(S^j x) = w$ computed by using the distribution $\overline{\mu}_{T,t}$, $T > t$ is different from that obtained by using $\overline{\mu}^+$ by a factor bounded, for each value of w, between e^{-B_f} and e^{B_f} for some B_f which is T,t independent.*

Remark: this is a simple consequence of the short range ((3.6) of h_+ and the general theory of Gibbs states on one dimensional lattices. Note that the set of strings in K (the space of the compatible strings) which agree with a given $\underline{\sigma}^T \in \Gamma_T$ is a prism of the Markov partition $\mathcal{E}_T = \cap_{k=-T}^{T} S^k \mathcal{E}$; so that the sum in (3.12) can be regarded as a sum on the prisms of $\mathcal{E}_T$.

Partial proof: since f is smooth the function $f'(\underline{\sigma}) = f(x(\underline{\sigma}))$ also has exponentially vanishing memory, as the h_γ in (3.6): hence there exist C, ν such that $|f'(\underline{\sigma}) - f'(\underline{\sigma}')| < Ce^{-\nu q}$ if $\underline{\sigma}, \underline{\sigma}'$ agree on the sites between $-q$ and q. Therefore we suppose first that f' depends *only* on the sites $j \in [-r, r]$, ("finite range assumption"). The the value of $w = \sum_{-t}^{t} f(S^j x(\underline{\sigma}))$ is entirely determined by $\underline{\sigma}_j^T$, $|j| < t + r$, so that the probability of w can be computed with the distribution:

$$e^{-\sum_{j=-t}^{t} h_+(\varphi^j \underline{\sigma}^T)} \cdot \frac{\sum_{\substack{\underline{\sigma}^T \in \Gamma_T \\ \underline{\sigma}^{t+r} \text{ fixed}}} e^{-\sum_{j=-T}^{-t-1} h_+(\varphi^j \underline{\sigma}^T)} e^{-\sum_{j=t+1}^{T} h_+(\varphi^j \underline{\sigma}^T)}}{Z_{[-T,T]}} \tag{3.13}$$

but since h_+ has short range the above numerator is essentially the product of the partition functions relative to the intervals $[-T, -t-1]$ and $[t+1, T]$, with some $\underline{\sigma}^t$ dependent "boundary condition" (in the sense of the Gibbs states terminology). Therefore the ratio is bounded *between*:

$$e^{-\sum_{j=-t}^{t} h_+(\varphi^j \underline{\sigma}^T)} \cdot \frac{Z_{[-T,-t-1]} Z_{[t+1,T]}}{Z_{[-T,T]}} . e^{\pm B_r} \tag{3.14}$$

where B_r is a T, f–independent constant: note that the ratio of the partition functions is large or small (exponentially in T, T) but *it is not random* (*i.e.* it is independent of $\underline{\sigma}^{t+r}$).

This proves the theorem if f has finite memory r: if f has exponentially vanishing memory the result has to be proven by approximating it with a finite memory function and passing to the limit; I do not discuss this problem further.

4. Time Reversible Systems. Ruelle's Principle

As an application I consider the system in example 3, Section 1, to show that the analysis of Section 3 can have some far reaching consequences and relevance for a fundamental theory of non equilibrium ensembles. Consider a 2 dimensional periodic box containing a few circular obstacles disposed so that there are no collisionless trajectories (this is the case called $0H$ in (Garrido and Gallavotti, 1994), "no horizon"). The box contains N particles and the equations of motion are the (1.1), *i.e.* with the usual symbols:

$$\dot{\underline{q}}_j = \frac{1}{m} \underline{p}_j, \quad \dot{\underline{p}}_j = -\partial_{\underline{q}_j} V_{ext}(\underline{q}_j) - \partial_{\underline{q}_j} V(\underline{q}_1, \ldots, \underline{q}_N) + E\underline{i} - \alpha(\underline{p})\underline{p}_j \tag{4.1}$$

where V_{ext} is the hard core potential (*i.e.* it is really a boundary condition imposing elastic reflections on the obstacles) and V is a short range pair potential energy. The α represents a friction mechanism to keep the total energy $H = \sum_i \underline{p}_i^2/2m + V(\underline{q}_1, \ldots, \underline{q}_N)$ bounded. We choose: $\alpha(\underline{p}) = E\underline{i} \cdot \sum_j \underline{p}_j / \sum_j \underline{p}_j^2$ which keeps H constant, exactly. This is called a "gaussian thermostat" from the meaning it has when $V = 0$. The case $V = 0, N = 1$ has been "completely" studied in (Sinai, 1970; Chernov et al., 1993).

The dynamical system generated by (4.1) can be regarded as defining a dynamics on the $4N$ dimensional full phase space $\mathcal{F}$, or on the $4N-1$ dimensional surface of constant energy $\mathcal{E}$, or on the $4N-2$ dimensional manifold $\mathcal{C}$ consisting in the phase space points of $\mathcal{E}$ in which one particle is exactly colliding with one of the hard obstacles. We denote S_t^0 the dynamics on $\mathcal{F}$, S_t that on $\mathcal{E}$ and S will denote the map defined, on $\mathcal{C}$, by mapping one collision to the next.

The system properties are:

(A) Dissipativity: the phase spaces $\mathcal{F}, \mathcal{E}, \mathcal{C}$ contract at an average rate $(2N-1)\langle\alpha\rangle$, where $\langle\alpha\rangle$ is the time average of α, at least if we suppose that the forward infinite time average $\langle\alpha\rangle$ is positive, as it seems intuitively correct and as some numerical experiments seem to support.

(B) Reversibility: the map $i : (\underline{q}, \underline{p}) \to (\underline{q}, -\underline{p})$ is such that, if $t \to x(t)$ is a solution of (4.1), then $t \to ix(-t) \equiv (ix)(t)$ is also a solution.

I will furthermore suppose that the system verifies the following property:

(C) Chaoticity: in the sense that "things go as if the system was chaotic".

The strict validity of this assumption in the sense of definition 8, Section 3, is *not true* for the model (4.1) already for $N = 1$, simply because the system is *not* smooth (Bunimovitch et al., 1990). For large N testing its *approximate* validity is very hard, but hopefully something will be done in the future. The best one can hope is that the failure to be verified is not "relevant" for the discussion of the thermodynamics of the system, in the same sense that the failure of ergodicity is believed to be, in many cases, irrelevant for the equilibrium thermodynamics (*i.e.* in the limit as $L \to \infty$ with fixed particle density and periodically repeated obstacles).

This is essentially Ruelle's principle: its operational meaning is to proceed as if the system was chaotic in the sense of the definition 8 of Section 3, or some similar definition, and then to suppose that the deductions are correct even though we have no way to check the chaoticity assumption or even though it does, strictly speaking, fail (as in the case at hand, (1.1)). One can hope that the assumptions should become more and more "true" as N grows. They fail, as just mentioned, for instance if $N = 1$, see (Chernov et al., 1993): but, in this particular case, some conclusions of Section 3 would nevertheless be correct *even if* $N = 1$, as it seems likely to be implied by combining the present analysis with that of (Chernov et al., 1993).

For instance an example of a property that is expected to hold, if the chaoticity assumption holds, is a strong, exponential, decay of the correlations between smooth observables, (Sinai, 1968a, 1968b, 1994; Ruelle, 1976; Bowen, 1975). The above assumption is therefore expected to correspond to a strong exponential decay of the correlations, at least over a time scale that is reachable by the experiments that can be conceived in order to check the picture that we are developing.

The latter decay has not been experimentally investigated (because it is really difficult). But after a decade of uncertainty it is becoming increasingly acceptable that indeed the correlations, in the $N = 1$, $E = 0$ case, may decay exponentially: in the paper (Garrido and Gallavotti, 1994) the "simple" cases $V = 0, N = 1$ and several choices for V_u are studied. No evidence for a non exponential decay could be detected, at least on time scales reliably attainable by numerical experiments (and, of course, for the simple observables considered). Essentially all the data confirm, instead, an exponential decay, both for the observables of the collision system $(\mathcal{C}, S)$ and for the continuous system $(\mathcal{E}, S_t)$. In this respect "things go as if the system was chaotic".[10] In spite of the evidence for the failure of the assumption, we have therefore an example, in the case least favorable for thermodynamic interpretations (as $N = 1$), strengthening the view that "things may go as if the system was hyperbolic" expressed here.

The existence of time reversal invariant periodic orbits can be deduced for many arrangements of the obstacles: hence we shall also suppose, for simplicity that the periodic point discussed in the definition of chaoticity is just a time reversal invariant periodic point. And to simplify further the analysis we assume that in fact O is a fixed point (it should be clear that the analysis that follows would not really change if O was just a periodic motion).

Other properties can be expected to hold for our model, on the basis of analogies with similar models in which they have been shown to hold. I make here a list, hoping that someone will find their experimental check worth of some effort. They *will not bee used in the following* but they make the general picture brighter.

(D) Density: following (Livi et al., 1986) we label the non negative Lyapunov exponents of S_t (on $\mathcal{E}$) as λ_j^+, $j = 2N-1, 2N-2, \ldots, n_+$, in decreasing order, and the negative ones as λ_j^-, $j = 2N-1, \ldots, n_-$ in increasing order, discarding from the enumeration the trivially zero exponent associated with the flow direction. Thus the maximum Lyapunov exponent is $\lambda_{\max} \equiv \lambda_{2N-1}^+$. Then in various models it has been experimentally[11] found, starting with the work (Livi et al., 1986) which has been followed by other interesting cases,(*e.g.* (Evans et al., 1990; Sarman et al., 1992)), that the graphs of $x = j/(2N-1) \to \lambda_j^\pm$ seem to have a smooth limit $f_\pm(x)$ as $N \to \infty$ and $x > x^\pm$. No experimental result seems to be available in the case (4.1): I shall assume this property to hold for it. The Lyapunov exponents of S (on $\mathcal{C}$) will be $t_0\lambda_j^\pm$, where t_0 is the average time between any two collisions, which is the time scale associated with the map S).

(E) Pairing: it has been noted that in various cases, starting with (Dressler, 1988; Evans et al. 1991), that:

$$\lambda_j^+ + \lambda_j^- = j - \text{independent} = -\langle\alpha\rangle \qquad j = 2N-1, \ldots \tag{4.2}$$

Where $\langle . \rangle$ denotes (forward) time average (over ∞ time).

The relation (4.2), that will be called *strong pairing property*, has been experimentally found to hold for many other models, as it emerges from the subsequent

---

[10] In our paper there is only one experimental result that can rise, in our opinion, doubts about the decay law. It is the result corresponding to the extremely long run leading to fig. 6 of (Garrido and Gallavotti, 1994): see the discussion there. More recent work, (Artuso et al, 1994), provides much stronger further evidence on the exponential decay.

[11] By *experimental measurement* I mean here also a computer experiment: as I think it does not conceptually differ from the older notions of experimental measurement.

analysis in (Sarman et al., 1992). It is not yet clear whether it holds strictly or up to corrections of $O(N^{-1})$.

Remarks:

(1) The pairing property appears to be related to several features of the equations of motion among which the reversibility. As essentially suggested in (Dressler, 1988) one may think that the property holds, for instance, for locally hamiltonian equations (as (4.1) with $\alpha = 0$) with a constraint imposed via a variational principle (like the Gauss' least constraint principle or some generalizations of it). But no proofs are available and, furthermore, the case (Evans et al., 1990) and others in (Sarman et al, 1992) do not fall into this category, although (4.2) holds for them.

(2) In other models considered in (Sarman et al., 1992) a weaker property seems to hold:

$$-C\langle\alpha\rangle \leq \lambda_j^+ + \lambda_j^- \leq 0 \tag{4.3}$$

for some $C > 0$, that I will call the *weak pairing property*, whose essential feature is the N–independence of C.[12] Neither (4.2) nor (4.3) have been tested numerically for the present model. I consider quite likely that the strong pairing rule holds in this case.

(3) The pairing property and the density property imply that the Kaplan Yorke fractal dimension of the attractor (also called the Lyapunov dimension), see (Eckmann and Ruelle, 1985), is *macroscopically smaller* than the full dimension $4N - 4$ of the phase space, by an amount $O(\lambda_{\max}^{-1}\langle\alpha\rangle N)$: the proportionality to N of the loss in dimension is sometimes called "dimensional reduction", (Posch and Hoover, 1992; Chernov et al., 1993).

Going back to the system (4.1), after the assumption (C) above, I want to discuss its consequences, with the main objective of finding some that could be tested and thus provide a test for Ruelle's principle for the statistics $\overline{\mu}$, and to see if any experimental consequences can be drawn from it (accessible by using the present day technical capabilities).

[12] In (Sarman et al., 1992) the Lyapunov exponents are considered in the full phase space $\mathcal{F}$. As discussed in (Sarman et al., 1992) the rules (4.2) or (4.3) may fail to be obeyed in a few cases: but this seems often related to the way the exponents are counted and computed. For instance in some systems there may exist observables called in (Sarman et al., 1992) "constants of motion" or "conservation laws" which, if initially having some special value, conserve it in the evolution, (note that this is a somewhat unusual notion of conservation law, as it holds only for special initial values), or have slow oscillations around it . For instance H or the flow direction in (4.1), or the total momentum $\underline{P}$ in the case $V_{ext} = 0$ (in the latter case this is conserved only if $\underline{P} = \underline{0}$). And one is naturally led to fix the values of such observables or to perform the measurements in some special way (*e.g.* by timing them appropriately when there is a periodic forcing). This is done to simplify the calculations, eliminating trivially behaving coordinates: but the end result might be in a odd number of exponents and in a consequent apparent "failure" of the pairing rule.

Under the assumptions (C) the "principle" is in fact a *theorem* (see theorem II, Section 3): the reason it is called a principle is that, as already mentioned, the hypotheses of definition 8 are not to be expected to hold rigorously for our model: they must be regarded to hold approximately, and this is the reason we say that to use them is a "principle", which we regard as correct for all the consequences it may have about thermodynamic quantities.

Note first that the time reversal symmetry implies that W^u_O and W^s_O have the same dimension, equal to half that of the phase space, *i.e.* $2N-1$.

A key idea behind all what is said in this paper, and on which intuition can be built, (Gallavotti, 1994), is that of regarding A, rather than as a fractal set of $\sim 4N$ dimensions, as a *smooth surface* of dimension $2N-1$ consisting in the unstable manifold W^u_O of some periodic point O. Note that, in this way, there is unification between the equilibrium ($E=0$) and the non equilibrium cases (in both cases W^u_O is smooth and it has the same dimension).

One should avoid thinking of A as a nasty fractal: following (Gallavotti, 1994) one must think of it as an infinite uncut *folio* (*i.e.* W^u_O) confined into a finite region (the phase space $\mathcal{C}$) and folded over and over again to fit into it, thus forming the (uncut) "book" A. Forcing simply introduces some (small if E is small) wrinkles on the folio accounting for the (mild) fractal nature of A. In particular the dimension of W^u_O stays $2N-1$ and does not change because of the introduction of the forcing (while the dimension of A changes by $O(N)$).

The unification between the equilibrium theory and the non equilibrium one is made possible by regarding the attractor in the phase space $\mathcal{C}$ as $2N-1$ dimensional (rather than $\sim 4N-2$ dimensional as one would be normally tempted to do).

The real question is whether, if $E\neq 0$, we can predict anything that could be experimentally checked as a test of Ruelle's principle.

The work (Evans et al., 1993) provides evidence in this direction. I apply here the ideas of (Evans et al., 1993; Cohen and Gallavotti, 1994), to show their relevance to the present case (as an example of the claim made in (Cohen and Gallavotti, 1994) about the generality of the ideas).

We can study the fluctuations, in the stationary state $\overline{\mu}$, of the average $(2N-1)\langle\alpha\rangle_t$ of the phase space contraction rate $(2N-1)\alpha$ over time stretches $t\,\tau_0$:

$$\langle\alpha\rangle_t(x)=\frac{1}{t\tau_0}\int_{-t\,\tau_0/2}^{t\tau_0/2}\alpha(S_\vartheta x)d\vartheta=\frac{1}{t\,\tau_0}\sum_{j=-t/2}^{t/2-1}\int_0^{\tau_0(S^jx)}\alpha(S_\vartheta S^j x)\,d\vartheta \tag{4.4}$$

where τ_0 is the average time elapsing between timing events, *i.e.* the average (over infinite time) of the time interval $\tau_0(x)$ between the event x and the successive Sx; the S_ϑ is the continuum time evolution on the surface of constant energy.

The quantity $(2N-1)\langle\alpha\rangle_t$ is also called the *entropy production rate* on the trajectory stretch between $S^{-t/2}x$ and $S^{t/2}x$: it is related to a transport coefficient (the conductivity, *i.e.* to the ratio between the particle current and the field E, see (Chernov et al., 1993)), at least in the present model (because the relation between entropy production and a suitable transport coefficient might only hold for a special class of examples, see (Chernov and Lebowitz, 1994)).

If the quantity t is small compared to the duration of the experiment the quantity $\langle\alpha\rangle_t$ *fluctuates* around the mean value $\langle\alpha\rangle$ (defined as the (forward) average of α over infinite time). If we write:

$$\langle\alpha\rangle_t(x) = \langle\alpha\rangle\, a_t(x) \tag{4.5}$$

we define a dimensionless random variable $a_t(x)$ with (forward) average 1. We can divide the axis a_t into intervals $I_{j\delta}$, $j = 0, \pm 1, \ldots$, with $I_{j\delta} = [j\delta, (j+1)\delta]$ for some small δ and measure the probability distribution of a_t by counting how many times a_t takes a value in each $I_{j\delta}$ when measured at the phase space points into which x evolves at times multiples of $t\,\tau_0$,

The measurement of $\langle\alpha\rangle_t(x)$ requires measuring the value of α points $S^{-t/2}x, \ldots, S^{t/2-1}x$.

The probability $\pi(p)dp$ that $a_t \in (p, p+dp)$ can be computed from (3.13) with $T \gg t$ by using the Markov partition $\mathcal{E}_T$ constructed in Section 3. The prisms of the partition $\mathcal{E}_T$ are naturally labeled by the allowed strings $\underline{\sigma} \in \Gamma_T$ (see the paragraph preceding (3.11)); for brevity we label them with a label denoted j. If E_j is a generic prism of this partition with axes intersecting at x_j, consider the measure $\mu_{T,t}$ defined by setting, for every smooth F:

$$\int_C F(x)\,\mu_{T,t}(dx) = \frac{\sum_j \overline{\Lambda}_{u,t}^{-1}(x_j)F(x_j)}{\sum_j \overline{\Lambda}_{u,t}^{-1}(x_j)} \tag{4.6}$$

where $\overline{\Lambda}_{u,t}(x) \equiv \prod_{j=-t/2}^{t/2-1} \Lambda_u(S^j x)$ is defined before (3.4) and x_j is a point in E_j. In the limit $T \to \infty$ the distribution μ gives the correct probability to the values of an observable F which is smooth and has a t–dependence like the one in (4.4), (*i.e.* like the one considered in theorem II, with $f(x) \equiv \tau_0^{-1}\int_0^{t_0(x)} \alpha(S_\vartheta x)d\vartheta$) up to an error which is a *factor bounded above and below uniformly in* t, by theorem II.

The partition $\mathcal{E}$ constructed in Section 3 turns out to be time reversal invariant, if the triangulation on W^s is taken to be the i image of the triangulation on W^u: this means that if $E_j \in \mathcal{E}$ then $iE_j = E_{j'} \in \mathcal{E}$ for some j'. Note that if we used another Markov partition $\mathcal{E}'$, constructed for instance via the classical proofs of existence of Markov partitions of (Sinai, 1968a, 1968b, 1994; Bowen, 1975), then the time reversal symmetry of our system would still imply that $i\mathcal{E}'$ is also a Markov partition (as $W^u_x = iW^s_{ix}$) so that the partition obtained by intersecting the two would still be a Markov partition, with the extra property of being time reversal invariant; hence by using $\mathcal{E}' \cap i\mathcal{E}'$ instead of $\mathcal{E}$ we could still carry the argument that follows which only depends on the time reversal invariance of the Markov partition $\mathcal{E}$ used for the construction of $\overline{\mu}_+$ and not on the particular Markov partition used. Note that if $\mathcal{E} = i\mathcal{E}$ then also the partition $\mathcal{E}_T$ in theorem II, $\mathcal{E}_T = \cap_{q=-T}^{T} s^q \mathcal{E}$, has the property of time reversal invariance $\mathcal{E}_T = i\mathcal{E}_T$.

Thus, for instance, up to the mentioned error of a factor (T,t)–independent:

$$\frac{\pi(p)}{\pi(-p)} = \frac{\sum_{j,a_t(x_j)=p} \overline{\Lambda}_{u,t}^{-1}(x_j)}{\sum_{j,a_t(x_j)=-p} \overline{\Lambda}_{u,t}^{-1}(x_j)} \tag{4.7}$$

where the sums run over the labels j of the prisms $E_j \in \mathcal{E}_T$ (with axes intersecting at x_j) verifying $a_t(x_j) \in (p, p+\delta)$.

By using the time reversal symmetry the above sums can be seen to consist of sums with the same number of addends. This can be done by pairing the contribution from $E_j \in \mathcal{E}_T$ with that of $iE_j \equiv E_{j'}$ that can be easily seen to give, if T is large (so that the prisms are really small), a value $-p$ to $a_t(x_{j'})$ if $a_t(x_j) = p$ and at the same time$\overline{\Lambda}_{u,t}(x_{j'}) = \overline{\Lambda}_{s,t}^{-1}(x_j)$. [13] Hence the addends can be paired so that the ratios of corresponding addends is:

$$\frac{\overline{\Lambda}_{u,t}^{-1}(x_j)}{\overline{\Lambda}_{s,t}(x_j)} = e^{(2N-1)\langle\alpha\rangle p t\, \tau_0} \frac{b(S^{t/2}x_j)}{b(S^{-t/2}x_j)} \tag{4.8}$$

where $\overline{\Lambda}_{s,t}(x)$ is defined as $\overline{\Lambda}_{u,t}(x) \equiv \prod_{j=-t/2\tau_0}^{t/2\tau_0 - 1} \Lambda_u(S^j x)$, with $\Lambda_c(x)$ being the absolute value of the determinant of the jacobian matrix of S as a map of W_x^s into W_{Sx}^s. Eq. (4.8) simply follows from the fact that a volume element around $S^{-t/2}x$ varies under the action of S^t by a factor $\overline{\Lambda}_{u,t}(x)\overline{\Lambda}_{s,t}(x) b(S^{t/2}x)/b(S^{-t/2}x)$ where $b(x)$ describes the transversality of the intersection of the stable and unstable manifolds at x, as defined in remark (iii) to definition 8: so that this equals $\exp -(2N-1)t\, \tau_0 \langle\alpha\rangle_t(x)$, (here τ_0 is considered extremely small, for simplicity).

---

[13] For instance, (Evans et al., 1993), from the two identities $S^{-\tau}(S^\tau x) = x$ and $S^{-\tau}(iS^{-\tau}x) = ix$ we deduce:

$$\begin{aligned} &\partial S^\tau(S^{-\tau}x)\, \partial S^{-\tau}(x) = 1 = \partial S^{-\tau}(S^\tau x)\, \partial S^\tau(x) \\ &\partial S^{-\tau}(iS^{-\tau}x)\, i\, \partial S^{-\tau}(x) = i \end{aligned} \tag{$*$}$$

We see that, by applying the first relation with x replaced by $\tilde{x}' \equiv iS^{-\tau}x$ and by applying successively the third (with $x \to \tilde{x}'$) and the second relations (with $x \to S^{-\tau}x$, using also $x = iS^{-\tau}\tilde{x}'$) in (*) one deduces:

$$\partial S^\tau(S^{-\tau}\tilde{x}') i \partial S^\tau(S^{-\tau}x) i = 1 \tag{$**$}$$

Therefore: $\big(\det \partial S^\tau(S^{-\tau}\tilde{x}')\big)^{-1} = \det \partial S^\tau(S^{-\tau}x) = e^{-(2N-1)\tau\tau_0\langle\alpha\rangle a_\tau(x)}$, where τ_0 is the value of the timing interval in real units (for us τ is an integer); *i.e.* , replacing x with $S^{-\tau/2}x$: $a_\tau(x) = -a_\tau(x')$.

As repeatedly mentioned above (4.6),(4.7) require a correction: but it has been remarked after (4.6) that the correction is a factor bounded by $e^{\pm B}$ for some B which is p-independent because of theorem II; and we can include in B also the correction due to the last ratio in (4.8) (bounded uniformly in t, by transversality property in the chaoticity assumption, see definition 8 in Section 3).

This means that $\pi(p)$ verifies (N large):

$$\begin{aligned} &\frac{1}{2Nt\,\tau_0 p}\log\frac{\pi(p)}{\pi(-p)}\xrightarrow[N\to\infty,\,t\to\infty]{}\langle\alpha\rangle = p-\text{independent} \\ &\pi(p)=e^{-\xi(p)+p\langle\alpha\rangle Nt\pm B} \end{aligned} \tag{4.9}$$

where $\xi(p)$ is an even function of p (over which we have no control) and B does not depend on p nor on t (but it may depend on N, E). The (4.9) can be considered as a *large deviation* result (both in N and t): its peculiarity is the p independence of the coefficient of p in the odd part othe argument of the exponential, to leading order in t.

The p, t, N independence of the first relation in (4.9), and the equality to $\langle\alpha\rangle$ can be quite easily tested even if N is moderately large (but not too large: as in such case the measurements simply could not be carried out): $N = 64$ seems feasible.

The idea of the above test, and its realization in a model different from (4.1), is in the basic paper (Evans et al., 1993) and it has been further developed in (Cohen and Gallavotti, 1994). The test *does not* require the measurement of the individual Lyapunov exponents: therefore it should be (relatively) easy to carry out. Particularly after its feasibility has been proved in similar models in (Evans et al., 1993).

Finally it is remarkable that one can, at all, find tests of the principle: in the case $E \neq 0$ there is no really well established non equilibrium thermodynamics with which one could compare the results of the principle. Such general results would constitute the real test of the principle from the point of view of Physics. A non controversial theory of non equilibrium thermodynamics could, actually, follow from the principle. For this, however, we must learn how to extract other consequences, just as we learnt to extract consequences from the Boltzmann Gibbs principle. And it is important to stress that whatever predictions it gives they should be as true as the statements derived in equilibrium are: *i.e.* essentially *exactly true*. Therefore the check above (and others that might be devised) will be satisfactory only if it gives exactly the expected results, within the experimental errors. This seems to be the case in (Evans et al., 1993) for the model considered there, see also (Cohen and Gallavotti, 1994).

5. Summary and Outlook

Here I regarded the Ruelle's principle as valid: to clarify its status I considered of some interest setting up a list of assumptions that would imply it rigorously. This essentially amounted to saying that the system behaves as an Anosov system (or more generally I could have supposed something like an axiom-A behavior) and then use the theory of Sinai (or, respectively, of Bowen and Ruelle) to "prove" the principle, see Section 3. The interest of course is not in the mathematical theorem as I just made enough assumptions to make it valid and checked its validity by adapting the ideas of the key papers (Sinai, 1968a, 1968b, 1994; Bowen, 1975; Ruelle, 1976), (while attempting, to avoid repetitions, at describing a more intuitive construction of Markov partitions), but rather the interest lies in the clarification of the meaning of the property (C) in Section 3. One cannot be too demanding on the matter of mathematical rigour: it should not be forgotten that even the ergodic hypothesis of Boltzmann is far from being proved, particularly in the generality one would want.

The "application" deviced in Section 4 is not very satisfactory as a test of validity of the principle because it can be performed only if N, τ are not too large. It would be, of course, nice to find a true thermodynamic property that could be computed and tested via the principle.

1) One should remark that the above analysis should improve with N, as the non chaotic phenomena should become less important: they are excluded by the strict chaoticity assumption, see Section 2, but they are possibly present if the assumption is made in the loose form of Section 3, (C). They would be certainly present if the hard core potential in (4.1) was replaced by a smooth finite range potential steeply diverging at zero distance (by general results from KAM theory).

2) Predictions analogous to the ones at the end of Section 4 could be made for other models, see (Cohen and Gallavotti, 1994): but except in the cases in (Evans et al., 1993) the numerical experiments do not seem to exist, yet.

3) The new principle appears to play the role that the ergodic hypothesis plays in equilibrium statistical mechanics: therefore one may be led to think that the time of approach to equilibrium should be of the order of the recurrence time on the attractor. Having set up a unified point of view for the equilibrium and the non equilibrium cases, we can adopt the classical explanation of Boltzmann, (Boltzmann, 1896, 1902), apparently relying also on an earlier suggestion by Thomson, (Thomson, 1874), intended to contradict such a hasty conclusion. The rate of approach to equilibrium is very short (essentially determined by the Boltzmann equation in the case of rarefied gases, as exemplified, for instance, by the Lorentz theory of conductivity in metals, (Becker), closely related to model 1) at least if one looks at the very few macroscopic observables relevant for equilibrium thermodynamics and for the transport coefficients: the reason

being that such observables have the same value on most of the surface W_O^u, see (Gallavotti, 1994).

6) Concerning the particularity of the gaussian thermostat, which could be called an "unphysical fiction", I think, see (Gallavotti, 1994), that there should be, also in non equilibrium, several equivalent ways of describing the same stationary distribution corresponding to different $\overline{\mu}$ and to different physical ways of reaching the stationary state. And it might well be that the gaussian thermostat turned out to be equivalent to other models of thermostats, which could be described by rather different attractors. For instance a stochastic thermostat, in which a particle colliding with the wall comes out with a maxwellian distribution at given temperature, will certainly be described by a statistics $\overline{\mu}$ which is absolutely continuous with respect to the Liouville measure.[14] In the thermodynamic limit this might just be the same as the result obtained with a statistics which, for finite N, is on a strange attractor. This mechanism is like the one realized by the microcanonical and the canonical ensembles (the first is concentrated on a set of configurations which has zero probability with respect to the second, as long as $N < \infty$). This is clearly a question that requires further investigations.

Aknowledgements: I am indebted to J. Lebowitz, G. Eyink and Y. Sinai for many helpful comments, and to G. Gentile for his help in the revision of the problem section. A special acknowledgement is for E. Cohen for stimulating my interest on the matter, and patiently explaining the details of his results and his intuition of their relation with the theory of chaotic systems: it was a great experience to profit of his advice. I have included here many ideas that grew up from our collaboration and that are included in our joint work (Cohen and Gallavotti, 1994). Support from Ministero della Pubblica Istruzione, (grants named 40 % and 60%), and from CNR-GNFM for travel is acknowledged together with support from the University of Granada, Spain, through the invitation to deliver the present lectures.

This paper is archived in *mp_ arc@math.utexas.edu* #94-333.

[14] Note that a stochastic model of thermostat is described by a stochastic differential equation and therefore our discussion does not apply without some major modification.

Appendix: Philosophical Questions; and a Few Concrete Ones

The following is a guided series of problems to the general theory of the Lyapunov exponents: it contains the ergodic commutative, subadditive and non commutative theorems and the Oseledec theorem. The theorems are mostly philosophical, i.e. they hold with essentially no assumptions. The problems are taken out of the preprint of the book Meccanica dei Fluidi, in italian, circulating in the form of a draft: the Oseledec theorem exposition is essentially taken from (Ruelle, 1979).

(1) If μ is a probability measure on the Borel sets of R^n and if Δ_n is a sequence of measurable sets such that $\sum_n \mu(\Delta_n) < +\infty$ then almost all points are contained in at most a finite number of sets in the sequence (*Borel Cantelli theorem*). (Idea: the set of the points in an infinite number of Δ_n's is $N = \cap_{k=1}^{\infty}(\cup_{h=k}^{\infty}\Delta_h)$, of course. Therefore $\mu(N) \le \sum_{h=k}^{\infty}\mu(\Delta_h)$ for all k's, hence $\mu(N) = 0$ because the series converges.)

(2) Let $(\mathcal{C}, S, \mu)$ be a triple formed by an invertible dynamical system and an invariant probability measure on (the Borel sets of) $\mathcal{C}$ (*i.e.* there is a zero measure set N such that S is invertible outside N and $\mu(E) = \mu(SE) = \mu(S^{-1}E)$ for all Borel sets $E \subset \mathcal{C}/N$). Let f be a (measurable) function bounded by a constant K almost everywhere (with respect to μ). Let D_n be the set of points x such that some average of f over a time $\le n$ is *non negative*: *i.e.* $m^{-1}\sum_{j=0}^{m-1} f(S^j x) \ge 0$ for some $1 \le m \le n$. Then $\int_{D_n} f(x)\mu(dx) \ge 0$. (*Garsia maximal averages theorem*). (Idea: if $n = 1$ the condition defining D_1 is simply $f(x) \ge 0$ and nothing has to be proved, besides the obvious. If $n = 2$ the condition defining D_2 is either $f(x) \ge 0$ or $f(x) + f(Sx) \ge 0$. Therefore the *new* points, *i.e.* those in D_2 but not in D_1, are points x where $f(x) < 0$ which can be paired with points $Sx \in D_1$ so that $f(x) + f(Sx) \ge 0$. Hence we can subdivide D_2 in the disjoint union of the sets $D_2/D_1 \cup S(D_2/D_1)$ and $D_1/S(D_2/D_1)$. On the last set it is $f(x) \ge 0$ while the integral over the first union can be written as $\int_{D_2/D_1}(f(x) + f(Sx))\mu(dx)$ because, by the invariance of μ the integral $\int_{S(D_2/D_1)} f(x)\mu(dx) \equiv \int_{D_2/D_1} f(Sx)\mu(dx)$. The case $n = 3$ is only slightly more involved and it is left to the reader (proceed in the "same way"), and, once understood, the general case becomes crystal clear.)

(3) The invertibility assumption is not necessary in (2): the invariance in the ordinary sense $\mu(E) = \mu(S^{-1}E)$ for all Borel sets E is sufficient. Prove this statement. (Idea: one has just to try to formulate what said by always using S^{-1}: for instance D_2 will consist of the points in D_1 and the inverse images of those in D_1 such that $f(x) + f(S^{-1}x) \ge 0$, etc.)

(4) Show that (2) implies the almost everywhere existence of the limit $\lim_{k\to\infty} \frac{1}{k}\sum_{j=0}^{k-1} f(S^j x)$. (Idea: let $f_{sup}(x) = \limsup_{k\to\infty} \frac{1}{k}\sum_{j=0}^{k-1} f(S^j x)$ and $f_{inf}(x) = \liminf_{k\to\infty} \frac{1}{k}\sum_{j=0}^{k-1} f(S^j x)$. The denial of the statement is that $f_{sup}(x) > \beta > \alpha > f_{inf}(x)$ on a set D which has *non zero measure* for a suitable pair $\beta > \alpha$. Quite absurd because on the (obviously) invariant set D the functions $f(x) - \beta$ and $\alpha - f(x)$ would have some non negative average. Hence by the previous theorem their integrals over D would have to be ≥ 0, but their sum would therefore be $(\alpha - \beta)\mu(D)$ which would be < 0, because we are (foolishly) thinking that $\mu(D) > 0$!)

(5) The boundedness assunption in (4) can be replaced by the summability assumption $f \in L_1(\mu)$. Furthermore if $\overline{f}(x)$ denotes the average of f, *i.e.* $\overline{f}(x) = \lim_{k\to\infty} \frac{1}{k}\sum_{j=0}^{k-1} f(S^j x)$, defined μ almost everywhere, show that:

$$|\overline{f}|_{L_1} \equiv \int_C \mu(dx)\,|\overline{f}(x)| \leq \int_C \mu(dx)|f(x)| \equiv |f|_{L_1}$$

$$\int_C \overline{f}(x)\mu(dx) = \int_C f(x)\mu(dx)$$

If there are no non trivial functions $f(x)$ which are constant of motion μ almost everywhere, *i.e.* if the system is *ergodic*, then the function $\overline{f}(x)$ is a constant (almost everywhere) and $\overline{f} = \int_C \mu(dy)f(y)$ for all $f \in L_1(\mu)$.

(6) Let (C, S, μ) be a dynamical system as in problem (4). And let $f_n(x)$ be a sequence of *subadditive* measurable functions, *i.e.* such that:

$$|f_1(x)| < K, \qquad f_{n+m}(x) \leq f_n(x) + f_m(S^n x) \qquad \mu \text{ almost everywhere}$$

Suppose μ ergodic; by applying the ergodic theorem (3),(4) above show the existence, μ almost everywhere, of the limit $\lim_{n\to\infty} \frac{1}{n} f_n(x) = \overline{f}(x)$, (*Kingman's subadditive ergodic theorem*). (Idea: Remark that the functions:

$$f_{sup}(x) = \limsup_{n\to\infty} \frac{1}{n} f_n(x), \qquad f_{inf}(x) = \liminf_{n\to\infty} \frac{1}{n} f_n(x)$$

verify $f_{sup}(Sx) \geq f_{sup}(x)$ and $f_{inf}(Sx) \geq f_{inf}(x)$ (because $f_n(x) \leq f_1(x) + f_{n-1}(Sx)$, then divide by n), so that the μ invariance of μ implies that they are constants of motion (because $\int (f_{sup}(Sx) - f_{sup}(x))d\mu = 0$). Therefore, μ almost everywhere, $f_{sup}(x) = \beta$ and $f_{inf}(x) = \alpha$ where $\alpha < \beta$ are suitable constants. Let us consider a number $\eta > 0$ such that $\alpha + \eta < \beta$.

Let Δ_n be the set of points where $\frac{1}{m} f_m(x) \leq \alpha + \eta$ for at least one value of $m \leq n$: hence $\lim_{n\to\infty} \mu(\Delta_n) = 1$. Given $\varepsilon > 0$ there is, therefore, a n_ε such that $\mu(\Delta^c_{n_\varepsilon}) < \varepsilon$ if ^{c} denotes the complementation operation on sets. If $x \in C$ one can suppose that the frequency of visit of x to $\Delta^c_{n_\varepsilon}$ is $< \varepsilon$, because such a frequency is the average value of $\chi_{\Delta^c_{n_\varepsilon}}(S^j x)$ (over j) if χ_Δ denotes the characteristic function of the set Δ (by the ergodicity).

Consider the sequence of times $j_1 < j_2 < \ldots$ when, instead, $S^{j_k} x \in \Delta^c_{n_\varepsilon}$: it follows that the number p of such j's with $j_k \leq T$ is such that $p/T < \varepsilon$ for T large enough (as p/T tends to the frequency of visit to $\Delta^c_{n_\varepsilon}$).

Let k_0 be the first time $\leq T$ in which $S^{k_0} x \in \Delta_{n_\varepsilon}$ and let k'_0 be the largest integer $\leq T$ such that $f_{k'_0 - k_0}(S^{k_0} x) \leq (k'_0 - k_0)(\alpha + \eta)$. One, at least, does exist with

$k'_o \le k_0 + n_\varepsilon$: the point $k'_0 + 1$ must be one of the j's, otherwise k'_0 could not be the largest integer k' such that $\frac{1}{k'-k_0} f_{k'-k_0}(S^{k_0}x) \le \alpha + \eta$ by the subadditivity of f, unless of course $k'_0 = T$.
Let $k_1 > k'_0$ be the first successive value, with $k_1 \le T$, not among the j's, *i.e.* such that $S^{k_1}x \in \Delta_{n_\varepsilon}$ and let $k'_1 > k_1$ be the largest value such that $f_{k'_1-k_1}(S^{k_1}x) \le (k'_1 - k_1)(\alpha + \eta)$, and so on.
In this way a sequence $[k_0, k'_0], \ldots, [k_s, k'_s]$ is constructed in the interval $[0, T]$. All the values $k < k'_s$ outside the intervals must be among the j's (hence their number is $\le p$). The last value k'_s will be, possibly, followed by a string of values among the j's but the first value k_{s+1} that does not have this property, if existing at all, *must* be within n_ε of the value T, otherwise we could form the interval $[k_{s+1}, k'_{s+1}]$. Therefore the number of values outside the intervals is bounded above by $p + n_\varepsilon$. The subadditivity then implies:

$$\frac{1}{T} f_T(x) \le \frac{1}{T}\Big(K(p+n_\varepsilon) + \sum_{i=0}^{s} f_{k'_i-k_i}(S^{k_i}x)\Big) \le$$
$$\le \frac{1}{T}\big(K(p+n_\varepsilon) + T(\alpha+\eta)\big) \xrightarrow[T\to\infty]{} \alpha + \eta$$

which shows the one *cannot* contemplate the case $\alpha < \beta$, by the contradiction it does provoke: hence $\alpha = \beta$.)

(7) Show that (6) implies that $\lim_{n\to\infty} \frac{1}{n} f(x) = \overline{f}(x)$ is a μ-almost everywhere constant, and:

$$\lim_{n\to\infty} \frac{1}{n} f_n(x) \equiv \overline{f} = \lim_{n\to\infty} \frac{1}{n} \int_C f_n(x)\mu(dx) = \inf_n \frac{1}{n} \int_C f_n(x)\mu(dx)$$

(Idea: subadditivity implies $f_n(x) \le f_1(x) + f_{n-1}(Sx)$; hence $\overline{f}(x) \le \overline{f}(Sx)$ and $0 \le \int_C (\overline{f}(Sx) - \overline{f}(x))d\mu = 0$ imply $\overline{f}(Sx) = \overline{f}(x)$ μ–almost everywhere. Then ergodicity yields that $\overline{f}$ is constant. By dominated convergence the first limit relation follows. The sequence $n \to \langle f_n \rangle \equiv \int_C f_n(x)\mu(dx)$ is subadditive and bounded by K; hence $\frac{1}{n}\langle f_n\rangle \xrightarrow[n\to\infty]{} \inf_n \frac{1}{n}\langle f_n\rangle$, by an elementary argument.)

(8) Show that the assumption $|f_1(x)| < K$ in (6) can be replaced by the summability of f^+, $f^+(x) \in L_1(\mu)$, if $f^+(x) = \max(0, f_1(x))$. (Idea: no idea is necessary; just a careful examination of the proofs in (6),(7).)

(9) Show that the ergodic theorem implies that if N has zero measure and $S^j N$ is measurable then $\mu(SN) = 0$. (Idea: the frequency of visit of the motion starting at x to the set SN, $\varphi_x(SN)$, is equal to that to N itself: $\varphi_x(N) = \varphi_x(SN)$; hence by the ergodic theorem $\mu(SN) = \int \varphi_x(SN)d\mu \equiv \int \varphi_x(N)d\mu = \mu(N) = 0$.)

(10) Define, via the ergodic theorem, the "future" and "past" averages of $f \in L_1(\mu)$ as the limits $f^\pm(x) = \lim \frac{1}{n}\sum_{j=0}^{n-1} f(S^{\pm j}x)$. Show that $f^+(x) = f^-(x)$ μ–almost everywhere. (Idea: Let D, α, β be such that $\mu(D) > 0$ and $f^+(x) > \beta > \alpha > f^-(x)$ for $x \in D$. Let D_n^+ be the set of points $x \in D$ such that

$\frac{1}{m}\sum_{j=0}^{n-1} f(S^j x) \equiv \langle f\rangle_m^+(x) > \beta$ for all $m \geq n$. Let D_n^- be the corresponding set where $\langle f\rangle_m^-(x) < \alpha$. Then, if n is large enough $S^{-(n-1)}D_n^- \cap D_n^+ \neq \emptyset$, because the quantities $\mu(D_n^-) \equiv \mu(S^{-(n-1)}D_n^-)$ and $\mu(D_n^+)$ are both very close to $\mu(D)$ for n large. Hence for x in the latter intersection it is $\alpha > \langle f\rangle_n^-(x) \equiv \langle f\rangle_n^+(S^{-(n-1)}x) > \beta$ by the properties defining D, D_n^+, D_n^-. This is impossible.)

The guided sequence of problems will now come "closer to Earth" by studying some finite matrix theorems.

(11) Let L be a real $d \times d$ non singular matrix. Consider the matrix $M^n \equiv ((L^*)^n L^n)^{1/2}$ and call $t_j^{(n)}$ the eigenvalues of M_n ordered by decreasing size. Show that the largest eigenvalue of M_n is such that the limit $\frac{1}{n}\log t_1^{(n)} \xrightarrow[n\to\infty]{} \lambda_1$ exists. Here * means transposition with respect to the scalar product $(u,v) = \sum_i u_i v_i$. (Idea: first show that the eigenvalues $t_j^{(n)}$ of $((L^*)^n L^n)^{1/2n}$, ordered according to decreasing size, do have a limit as $n \to \infty$. Note that $f_n = \log|L^n|$ verifies subadditivity $f_{n+m} \leq f_n + f_m$ and $f_n \leq n\log|L|$, if $|L|$ is the norm of the matrix L (*i.e.* $|L| = \max|Lv|/|v|$, with $|v| = (v,v)^{1/2}$). Hence the limit $n^{-1}\log|L^n| \xrightarrow[n\to\infty]{} t_1 = \inf_n \frac{1}{n}\log|L^n|$ exists, and $\frac{1}{n}f_n \geq \min_{1\leq j\leq d}\log|\mu_j|$ if μ_j are the eigenvalues of L. But $|L^n v| = (L^n v, L^n v)^{1/2} = (v,(L^*)^n L^n v)^{1/2} \leq \max_{1\leq j\leq d}(t_j^{(n)})^{1/n}$, if $|v| = 1$. Hence the largest eigenvalue has the appropriate convergence property.)

(12) Think of the vectors in R^d as functions $i \to u_i$ on the finite space $F \equiv (1,2,\ldots,d)$. Let $(R^d)^{\wedge q}$ be the space of the functions on F^q which are *antisymmetric*: these are the functions $u_{i-1\ldots i_q}$ which are antisymmetric. Define a scalar product on such functions by setting $(U,v) \equiv \sum u_{i_1\ldots i_q} v_{i_1\ldots i_q}$. Define the matrix $L^{\wedge q}$ by setting:

$$(L^{\wedge q}u)_{i_1\ldots i_q} = \sum_{j_1\ldots j_q}^{1,d} L_{i_1 j_1} L_{i_2 j_2}\ldots L_{i_q j_q} u_{j_1\ldots j_q}$$

Show that the result of (11) implies that the matrix $(((L^{\wedge q})^*)^n (L^{\wedge q})^n)^{1/2n}$, where the * denotes the adjunction operation with respect to the scalar product (u,v), is such that its largest eigenvalue logarithm divided by n has a limit as $n \to \infty$.

(13) Let $(t_j^{(n)})^2$ be the eigenvalues of $(L^*)^n L^n$ in decreasing order, repeated according to multiplicity. Prove that the eigenvalues of $((L^{\wedge q})^n)^*(L^{\wedge q})^n$ are just the products of the q-ples of eigenvalues $(t_{j_1}^{(n)})^2\ldots(t_{j_q}^{(n)})^2$ with $j_i \neq j_j$ (*i.e.* products of q–ples of pairwise distinct eigenvalues). (Idea: no idea is necessary.)

(14) Combine (12) and (13) to infer that the limits $\frac{1}{n}\log t_j^{(n)} = \lambda_j$ exist for all $j = 1,\ldots,d$. Let $\overline{\lambda}_1,\ldots,\overline{\lambda}_s$ be the *distinct* limits λ_j and let $m_1,\ldots,m_s$ be their multiplicities (*i.e.* the number of λ_i which are equal to $\overline{\lambda}_j$). Verify that $\sum_{i=1}^s m_i = d$ and define $r(i) = j$ if $\lambda_i = \overline{\lambda}_j$.

(15) Let $U_1^{(n)}$ be the linear space spanned by the first m_1 eigenvectors of $\Lambda_n = ((L^*)^n L^n)^{1/2}$; likewise $U_2^{(n)}$ will be the space spanned by the next m_2, and so on until $U_s^{(n)}$ is defined. Show that the notion of multiplicity introduced in (14) is even more justified by proving the following "orthogonality" property between unit vectors $u \in U_r^{(n)}$ and $u' \in U_{r'}^{(n+k)}$, $k \geq 0$. Given $\delta > 0$ there exists $C > 0$ such that:

$$|(u, u')| \leq Ce^{-(|\overline{\lambda}_r - \overline{\lambda}_{r'}| - \delta)n}$$

(Idea: nothing to prove if $r = r'$, of course. Let us suppose first $k = 1$. The "easy case" is $r' > r$. Let $\Lambda_{n+1} = \sum_{i=1}^d t_i^{(n+1)} P_i$ be a spectral decomposition for the matrix Λ_{n+1}. Then supposing n so large that for $m \geq n$ it is $|\frac{1}{m}\log t_j^{(m)} - \overline{\lambda}_i| < \delta_1$ for all $r(j) = i$ and all i, where for a given $\delta_1 > 0$, $\delta_1 < \frac{1}{2}\min(|\lambda_r - \lambda_{r'}|)$:

$$\begin{aligned}
|(u, u')| \leq& e^{-(n+1)\overline{\lambda}_{r'}} \max_{u' \in U_{r'}^{(n+1)}} |(u, \textstyle\sum_{r(i)=r'} e^{(n+1)\overline{\lambda}_{r'}} P_i u') \equiv \\
\equiv& e^{-(n+1)\overline{\lambda}_{r'}} \max_{u' \in U_{r'}^{(n+1)}} |(\textstyle\sum_{r(i)=r'} e^{(n+1)\overline{\lambda}_{r'}} P_i u, u') \leq \\
\leq& e^{-(n+1)\overline{\lambda}_{r'}} \left|\textstyle\sum_{r(i)=r'} e^{(n+1)\overline{\lambda}_{r'}} P_i u\right| \leq \\
\leq& e^{-(n+1)\overline{\lambda}_{r'} + (n+1)\delta_1} \left|\textstyle\sum_{r(i)=r'} t_i^{(n+1)} P_i u\right| \leq \\
\leq& e^{-(n+1)\overline{\lambda}_{r'} + (n+1)\delta_1} \left|((L^*)^{n+1} L^{n+1})^{1/2} u\right| \\
=& e^{-(n+1)\overline{\lambda}_{r'} + (n+1)\delta_1} (u, (L^{n+1})^* L^{n+1} u)^{1/2}
\end{aligned}$$

Thus, again by the spectral theorem and by $|TT'| \leq |T|\,|T'|$ we deduce:

$$\begin{aligned}
|(u, u')| \leq& e^{-(n+1)(\overline{\lambda}_{r'} - \delta_1)} \left|L^{n+1} u\right| \leq \\
&\leq e^{-(n+1)(\overline{\lambda}_{r'} - \delta_1)} |L| \left|L^n u\right| \leq e^{-(n+1)(\overline{\lambda}_{r'} - \overline{\lambda}_r - 2\delta_1)} |L|
\end{aligned}$$

because $u \in U_r^{(n)}$. This completes the proof in the case $r' > r$ and $k = 1$. The case $k > 1$ is simply obtained from the case $k = 1$ by applying the latter inequality k times, getting $\sum_{i=1}^k e^{-(n+i)(\overline{\lambda}_{r'} - \overline{\lambda}_r - 2\delta_1)} |L|$. One finds, since the series converges, the result with a C that can be taken $C_1 = |L|(1 - e^{-x})^{-1}$ if $0 < x < \min(|\overline{\lambda}_r - \overline{\lambda}_{r'}|) - 2\delta_1$ for all n large enough and all $k \geq 0$.

Consider the case $r > r'$. Let u_α be an orthonormal base with the first m_1 vectors spanning $U_1^{(n)}$, the next m_2 spanning $U_2^{(n)}$ and so on; let u'_α be the corresponding base for $n + k$. The orthogonal matrix $W_{\alpha\alpha'} = (u_\alpha, u'_{\alpha'})$ verifies the inequalities:

$$\begin{cases} |W_{\alpha,\alpha'}| \leq 1 & \text{for all } \alpha, \alpha' \\ |W_{\alpha,\alpha'}| \leq C_1 e^{-|\overline{\lambda}_{r(\alpha)} - \overline{\lambda}_{r(\alpha')} - 2\delta_1|n} & \text{if } r(\alpha') > r(\alpha) \end{cases}$$

If $r(\alpha') < r(\alpha)$, *however*, the orthogonality implies that $W_{\alpha\alpha'} \equiv (W^{-1})_{\alpha'\alpha}$. The latter quantity is "just" the determinant obtained by deleting the row α and the column α' from the matrix W. Such determinant consists in a sum of $(d-1)!$ products of $d - 1$ matrix elements W_{ij} picked up in pairwise distinct rows and columns (by Cramers' rule). Since we are interested in the non diagonal elements of $W_{\alpha\alpha'}$ we see that in each product there must be at least enough factors $W_{\beta\beta'}$ with $r(\beta') > r(\beta)$ so that $\sum \lambda_{r(\beta')} - \lambda_{r(\beta)} \geq \lambda_{r(\alpha)} - \lambda_{r(\alpha')}$ (hint: check this first when there are no degenracies, *i.e.* $r(\alpha) \equiv \alpha$). Hence we use for such factors the second inequality and just bound the others by 1. We finally get the result with $C = C_1(d-1)!$ if $2\delta_1(d-1) < \delta$.)

(16) Show that (15) implies that the planes $U_j^{(n)} \subset R^d$ have a limit U_j as $n \to \infty$ and the planes U_j are pairwise orthogonal. (Idea the planes $U_1^{(n+h)}$ and $U_1^{(n+k)}$ must form with the planes $\bigoplus_{r>1} U_r^{(n)}$ an angle closer to 90° by an amount prefixed arbitrarily if n is large enough. Hence they form a Cauchy sequence of planes converging to some U_1. The other planes are treated analogously.)

(17) Check that the previous problems (11)÷(16) imply that if L is a real $d \times d$ non singular matrix the limit $\lim_{n\to\infty}((L^*)^n L^n)^{1/2n} = D$ exists and it is a non singular positive definite matrix, with eigenvalues equal to the absolute values of those of L, counted according to multiplicity. The eigenspaces $U_j^{(n)}$ of $D_n \equiv ((L^*)^n L^n)^{1/2n}$ spanned by the $r(j)$ eigenvalues of D_n that converge to the j-th distinct limit value are planes that converge to limit planes U_j which are the eigenplanes of D which correspond to distinct eigenvalues. (Idea: the only thing still to check is the relation between the eigenvalues of D and those of L. Suppose for simplicity that the eigenvalues of L have pairwise distinct absolute values (hence they must be all real) and order them by decreasing absolute values $\mu_1, \ldots, \mu_d$. Let $L = \sum_j \mu_j v_j \times v_j^*$ be the spectral resolution of L where $Lv_j = \mu_j v_j$, $L^* v_j^* = \mu_j v_j^*$ and $(v^*, v) = 1$. Clearly $|L^n v_j| = |\mu_j|^n |v_j|$. Hence $|\mu_1| = \lambda_1$. The other equalities can be deduced for instance by the "trick" of considering the matrices $L^{\wedge q}$, as in (13).)

Going back to the realm of abstract thinking the following guided problems, combining the concrete theory (11)÷(17) and the philosophical results (1)÷(10), lead to the Oseledec theorem.

(18) Given an ergodic dynamical system (C, S, μ) with an invariant distribution μ such that there is a zero μ measure invariant set N outside which the transformation S is invertible and non singular, consider the matrix $\partial S^n(x) = T_n(x)$. Suppose that $|\det T(x)| \geq \varepsilon > 0$ and that $|T(x)| < E$ in C/N. Check that $T_n(x) = T(S^{n-1}x) \cdot \ldots \cdot T(Sx) \cdot T(x)$, and that $f_n(x) = \log |T_n(x)|$ verifies the subadditivity property of the subadditive ergodic theorem of (6). Therefore: $\lim_{n\to\infty} \frac{1}{n} \log |T_n(x)| = \lambda_1(x)$ exists μ almost everywhere.

(19) Define the matrices $T_n^{\wedge q}(x)$ on $R^{\wedge q}$ as in (12),(13) above, and by repeating the argument there (with (18) replacing (11)) prove that the limits $\lim_{n\to\infty} \frac{1}{n} \log t_j^{(n)}(x) = \lambda_j(x)$ exist almost everywhere for μ almost all x, if $T_n(x)$ replaces $L^n(x)$.

(20) Show that the analysis in (15) can be repeated word by word even when the matrices L^n are replaced by $T_n(x)$, in the points where the limits in (19) exist, *i.e.* almost everywhere. The matrix $D = \lim_{k\to\infty}((T_k^*(x) T_k(x))^{1/2k}$

has eigenplanes $U_1(x), \ldots, U_{s(x)}(x)$. Show that the spaces $V_j(x) = U_j(x) \oplus \ldots \oplus U_{s(x)}(x)$ can be identified with the system of scaling planes for S of definition 6, Section 2.

(21) The contraction exponents $\lambda_j(x)$ are defined almost everywhere and they are constants of motion together with their multiplicities $m_j(x)$. Therefore they are μ almost everywhere constants. (Idea: By (20) and an argument like that in (17).)

(22) Check that all the above results can be derived by just requiring that $T(x)$ is such that $\log^+ |T(x)|$, where $\log^+$ denotes the positive part of the logarithm function, is μ–summable. (Idea: just careful examination of the proofs.)

(23) If $T(x)$ is replaced by any $d' \times d'$–matrix valued function $x \to O(x)$ with $\log^+ |O(x)|$ μ–summable the "same results" can be proved. For instance if $O_n(x) = O(S^{n-1}x) \cdot \ldots \cdot O(x)$ then $D_n = (O_n^*(x) O_n(x))^{1/2n}$ has a limit D as $n \to \infty$ which is almost surely independent on x, and the eigenspaces, spanned by the eigenvalues of D_n whose eigenvalues converge to the same limit $\overline{\lambda}_j$, converge to the eigenspace of D with eigenvalue λ_j (see problem (4)). In other words the above theory extends trivially to a theory of products of random matrices $O(x)$ generated by selecting randomly by a distribution μ on $\mathcal{C}$ a point x and the corresponding random matrix $O(x)$ and by multiplying $O(S^{n-1}x) \ldots O(x)$.

(24) Let $\tilde{T}(x) = \partial S^{-1}(x)$, $\partial S(x) = T(x)$. Then $\tilde{T}(x) = T(S^{-1}x)^{-1}$ and $\tilde{T}_k(S^k x) = \partial S^{-k}(S^k x)$ can be written either $\tilde{T}(Sx) \ldots \tilde{T}(S^k x)$ or also as $T^{-1}(x) \ldots T^{-1}(S^{k-1}x)$. Show that if x admits a system of contracting planes, *i.e.* μ–almost everywhere, the eigenvalues $\tilde{t}_j^{(k)}$ (ordered by decreasing size) of $(\tilde{T}_k^*(S^k x) \tilde{T}_k(S^k x))^{1/2} \equiv \tilde{D}_k$ have the limit: $\lim \frac{1}{k} \log \tilde{t}_j^{(k)} = \tilde{\lambda}_j = -\lambda_{d-j}$, μ–almost everywhere. (Idea: note the following expression for $\tilde{T}_k$: $\tilde{T}_k^*(S^k x) \equiv \tilde{T}^*(S^k x) \ldots \tilde{T}^*(Sx)$ so that the matrices are multiplied in the correct order for the application of (23): hence $(\tilde{T}_k(S^k x) \tilde{T}_k^*(S^k x))^{1/2k}$ (identical to $(T_k^*(x) T_k(x))^{-1/2k}$) converges to a matrix $\tilde{D} \equiv D^{-1}$ (see problem (4)), μ–almost surely (or if x admits a scaling system of planes). Thus the limits exist, because the spectrum of T^*T and that of TT^* coincide in general, if T is non singular; their relation with the opposites of the contraction exponents λ_j in the forward direction is derived from $\tilde{T}_k(S^k x) \tilde{T}_k^*(S^k x) = (T_k^*(x) T_k(x))^{-1}$.)

(25) In the context of the problems (20),(24) consider the system of planes $\hat{V}_j(x)$, at x, defined by the eigenplanes $U_j(x)$ of the matrix D of problem (20) by $\hat{V}_j = U_{s-j+1} \oplus \ldots \oplus U_1(x)$, and existing μ–almost everywhere. Show that if $w_k \in V_j(S^k x)/V_{j+1}(S^k x)$ and $w_k = \partial S^k(x) u$ (hence $u \in \hat{V}_j(x)/\hat{V}_{j+1}(x)$) then $\lim_{k \to \infty} \frac{1}{k} \log |\tilde{T}(S^k) w_k| / |w_k| = -\overline{\lambda}_{s-j}$. Why one should not call $\hat{V}_j(x)$ an expanding plane when j is such that $\overline{\lambda}_{s-j'} < 0$ for all $j' < j$ and positive otherwise?

(26) Let us call the spectrum of a sequence of matrices O_n the d numbers $\omega_1, \ldots, \omega_d$ obtained by considering the logarithms of the eigenvalues of $(O_k^* O_k)^{1/2k}$, $o_1^{(k)} \geq \ldots o_d^{(k)}$, and setting, when they exist: $\lim_{k \to \infty} o_j^{(k)} = \omega_j$.

Show that the spectrum of the sequence $O_k(x) = T_k(S^{-k}x)$ coincides with that of $T_k(x)$ μ–almost surely. (Idea: if the functions $n \to f_n(x)$ and $n \to f_n(S^n x)$ are subadditive, in the sense of problem (6), the functions $n \to f_n(S^n x)$ and $n \to f_n(x)$ are such that $F = \lim_{n\to\infty} \frac{1}{n} f_n(S^n x)$ and $G = \lim_{n\to\infty} \frac{1}{n} f_n(x)$ exist and are constant μ–almost everywhere, by (6) above. Suppose that $F > G$. Consider the sets D_n and D_n' consisting in the points where $\frac{1}{m} f_m(S^m x) > F - \varepsilon > G$ for all $m \geq n$, or respectively where $\frac{1}{m} f_m(x) < F - \varepsilon$ for *all* $m \geq n$. Then $\mu(D_n), \mu(D_n') \to 1$. Hence for n large enough $S^{-n} D_n' \cap D_n \neq \emptyset$: if x is in this set then $\frac{1}{n} f(S^n x) < F - \varepsilon$ because $S^n x \in D_n'$ and $\frac{1}{n} f_n(S^n x) > F - \varepsilon$ because $x \in D_n$. Hence this contradiction shows that it is sufficient to check that the two functions are subadditive: this comes from the inequality $|AB| \leq |A||B|$.)

(27) The spectrum, see (26), of $\tilde{T}_{-k}(x) \equiv \tilde{T}(S^{-(k-1)}x) \ldots \tilde{T}(x)$ is the opposite of that of $T_k(x)$: $\omega_j = -\overline{\lambda}_{d-j+1}$. (Idea: $\tilde{T}_{-k}(x) = (T_k(S^{-k}x))^{-1}$; hence $\tilde{T}^*_{-k}(x)\tilde{T}_{-k}(x) = ((T_k(S^{-k}x)(T_k^*(S^{-k}x))^{-1}$; but the spectrum of the last matrix is the same as that of the matrix $(T_k^*(S^{-k}x)T_k(S^{-k}x))^{-1}$ which by (26) leads to the result.)

(28) The forward and backward systems of scaling planes exist μ–almost everywhere and have opposite corresponding exponents: $\lambda_j = -\omega_{d-j+1}$. (Idea: this is a corollary of what already discussed in the few preceding problems.)

(29) Show that if $V_{-j}(x)$ are the contracting planes for S^{-1}, defined μ–almost everywhere, then $V_r(x) \cap V_{-(s-r+2)}(x) = 0$ and $V_r(x) \oplus V_{-(s-r+2)} = R^d$, μ–almost everywhere. (Idea: since, by definition, the dimensions of V_r and of $V_{-(s-r+2)}$ are complementary to d one only has to check the first statement. Note that if $u \in V_r(x)$ it is $|T_k(x)u| \leq e^{k(\lambda_r+\delta)}|u|$ for any prefixed $\delta > 0$ and for k large enough and all $x \in D_k$ where D_k is a set such that $\mu(D_k) \to 1$. For the same reason $|T_{-k}(S^k x)v| \leq e^{-(\lambda_{r-1}+\delta)k}|v|$ for $v \in V_{-r}(S^k x)$ provided $S^k x \in D_k$. Since $T_{-k}(S^k x)V_{-(s-r+2)}(S^k x) = V_{-(s-r+2)}(x)$ we see that all the $u \in V_{-(s-r+2)}(x)$ have the form $u = T_{-k}(S^k x)v$ with $v \in V_{-r}(S^k x)$. So that using $T_k(x)T_{-k}(S^k x) = 1$ we get:

$$|T_{-k}(S^k x)v| \leq e^{-(\lambda_{r-1}-\delta)k}|T_k(x)T_{-k}(S^k x)v| \to$$
$$\to |T_k(x)u| \geq e^{(\lambda_{r-1}-\delta)k}|u| \qquad \text{for } x \in S^{-k}D_k,\ u \in V_{-(s-r+2)}(x)$$

and if $x \in S^{-k}D_k \cap D_k$ this is impossible because $\lambda_{r-1} > \lambda_r$ if also $u \in V_r(x)$ (so that $|T_k(x)u| \leq e^{(\lambda_r-\delta)k}|u|$) and δ is smaller than the twice the minimal difference between the scaling exponents.)

(30) The planes $W_j(x) = V_j(x) \cap V_{-(s-j+1)}(x)$ are such that

$$\frac{1}{k} \log |T_{\pm k}(x)u| \xrightarrow[k\to\infty]{} \lambda_j \qquad \text{for } u \in W_j(x)$$

and $V_j = W_j \oplus \ldots \oplus W_s$, $V_{-(s-j+2)} = W_1 \oplus \ldots \oplus W_{-(j-1)}$. Also $T(x)W_j(x) = W_j(Sx)$. This also implies the relation between the dynamical bases claimed in the theorem II. (Idea: this is just a corollary, or a summary, of the previous problems.)

(31) Show that if μ is non invariant the forward and backward Lyapunov exponents (and the relative systems of planes) may exist and be different. (Idea: consider the example on $\mathcal{C} = [-1,1] \times T^2$, where T^2 is the two dimensional torus. Let $x = (z, \varphi_1, \varphi_2)$ be a point in $\mathcal{C}$. Let $z \to f(z)$ be a smooth map such that $f^n(z) \xrightarrow[n\to\pm\infty]{} \pm 1$, and let $\nu(z) = 2$ if $z > 0$ and $\nu(z) = 1$ if $z < 0$; define:

$$x' = (z', \varphi_1' \varphi_2') = \begin{cases} f(z) \\ \varphi_1 + \nu(z)\varphi_2 \mod 2\pi \\ \nu(z)\varphi_1 + (\nu(z)^2 + 1)\varphi_2 \mod 2\pi \end{cases}$$

and check the statements using $\mu = dz\, d\varphi_1\, d\varphi_2 / 3(2\pi)^2$.)

(32) Find the positive and negative Lyapunov exponents in the example suggested in problem (31); find also the dynamical bases (of the points which admit them).

References

Artuso, Guarneri and Casati G. (1994): Talk at the conference, *Classical and Quantum Billiards*, Ascona, July.

Becker, R.: *Theory of electricity*, vol. II, Ch.36.

Boltzmann, L. (1896): *Entgegnung auf die wärmetheoretischen Betrachtungen des Hrn. E. Zermelo*, english translation in S. Brush, "Kinetic Theory", vol. 2, 218–, Pergamon Press, 1965.

Boltzmann, L. (1902): *Lectures on gas theory*, english edition annotated by S. Brush, University of California Press, Berkeley, 1964.

Bowen, R. (1975): *Equilibrium states and the ergodic theory of Anosov diffeomorphisms*, Lecture notes in mathematics, vol. **470**, Springer Verlag. See also (Gallavotti, 1981).

Bowen, R. (1978): *Markov partitions are not smooth*, Proceedings of the American Mathematical Society, **71**, 130–132.

Bunimovitch, L., Sinai, Y., Chernov, N.I. (1990): *Statistical properties of two dimensional hyperbolic billiards*, Russian Mathematical Surveys, **45**, n. 3, 105–152.

Chernov, N.I., Eyink, G.L., Lebowitz, J.L., Sinai, Y. (1993): *Steady state electric conductivity in the periodic Lorentz gas*, Communications in Mathematical Physics, **154**, 569–601.

Chernov, K., Lebowitz, J. (1994): *Stationary shear flow in boundary driven hamiltonian systems*, Rutgers University preprint, May 1994.

Cohen, E.G.D., Berlin, T.H. (1960): *Note on the derivation of the Boltzmann equation from the Liouville equation*, Physica, **26**, 717–729.

Cohen, E., Gallavotti, G. (1994): Non equilibrium ensembles, preprint 1994, in preparation.

Dressler, U. (1988): *Symmetry property of the Lyapunov exponents of a class of dissipative dynamical systems with viscous damping*, Physical Review, **38A**, 2103–2109.

Eckmann, J., Ruelle, D. (1985): *Ergodic theory of strange attractors*, Reviews of Modern Physics, **57**, 617–656.

Evans, D.J., Cohen, E.G.D., Morriss, G.P. (1990): *Viscosity of a simple fluid from its maximal Lyapunov exponents*, Physical Review, **42A**, 5990–5997.

Evans, D.J., Cohen, E.G.D., Morriss, G.P. (1993): *Probability of second law violations in shearing steady flows*, Physical Review Letters, **71**, 2401–2404.

Franceshini, V., Zironi, F (1985): *On constructing Markov partitions by computer*, Journal of Statistical Physics, **40**, 69–91.

Gallavotti, G. (1973): *Ising model and Bernoulli shifts*: Communications in Mathematical Physics, **32**, 183–190, 1973.

Gallavotti, G. (1981):*Aspetti della teoria ergodica, qualitativa e statistica del moto*, Quaderni dell' UMI (Unione Matematica Italiana), vol. 21, Bologna.

Gallavotti, G. (1994): *Ergodicity, ensembles, irreversibility in Boltzmann and beyond*, preprint Physics department, University of Wien, deposited in *mp_ arc@ math. utexas. edu*, #94-66.

Garrido, P., Gallavotti, G. (1994): *Billiards correlation functions*, in *mp_ arc@ math. utexas. edu*, #93-279, to appear in Journal of Statistical Physics.

Livi, R., Politi, A., Ruffo, S. (1986): *Distribution of characteristic exponents in the thermodynamic limit*, Journal of Physics, **19A**, 2033–2040.

Pesin, Y. (1992): *Dynamical systems with generalized hyperbolic attractors: hyperbolic, ergodic and topological properties*, Ergodic Theory and Dynamical Systems, **12**, pp.123-151.

Posch, H., Hoover, W. (1992): *Non equilibrium molecular dynamics of a classical fluid*, in "Molecular Liquids: new perspectives in Physics and chemistry", ed. J. Teixeira-Dias, Kluwier Academic Publishers, p. 527–547, 1992.

Ruelle. D. (1968): *Statistical mechanics of a one dimensional lattice gas*, Communications in Mathematical Physics, **9**, 267–278.

Ruelle, D. (1976): *A measure associated with axiom A attractors*, American Journal of Mathematics, **98**, 619–654.

Ruelle, D. (1978): *Thermodynamic formalism*, Encyclopedia of Mathematics, vol. 5, Addison-Wesley.

Ruelle, D. (1979):*Ergodic theory of differentiable dynamical systems*, Publications Mathematiques de l' IHES, **50**, 27–58.

Ruelle, D. (1980a): *Measures describing a turbulent flow*, Annals of the New York Academy of Sciences, **357**, 1–9.

Ruelle, D. (1980b): *Ergodic theory of differentiable dynamical systems*, Publications Mathémathiques de l' IHES, **50**, 275–306.

Ruelle, D. (1989): *Chaotic motions and strange attractors*, Lezioni Lincee, notes by S. Isola, Accademia Nazionale dei Lincei, Cambridge University Press.

Sarman, S., Evans, D., Morriss, G. (1992): *Conjugate pairing rule and thermal transport coefficients*, Physical Review, **45A**, 2233–2242.

Sinai, Y. (1968a): *Markov partitions and C-diffeomorphisms*, Functional Analysis and Applications, **2**, 64–89, n.1 (engl. p. 61).

Sinai, Y. (1968b): *Construction of Markov partitions*, Functional analysis and Applications, **2**, 70–80 n.2 (engl. p. 245).

Sinai, Y. (1970): *Dynamical systems with elastic reflections. Ergodic properties of dispersing billiards*, Russian Mathematical Surveys, **25**, 137–189.

Sinai, Y. (1994): *Topics in ergodic theory*, Princeton University Press, 1994.

Smale, D. (1965): *Differentiable dynamical systems*, Bulletin of the american mathematical society, **73**, 747–817.

Thomson, W. (1911): *The kinetic theory of the dissipation of energy*, in the collected "Mathematics and Physics papers", vol. V, p. 11-20, Cambridge University Press, from the 1874 original.

II

Contributions

Multistable Steady States and Oscillations in a Suface Reaction Model

M. C. Lemos, J. J. Luque, and F. Jiménez-Morales

Departamento de Física de la Materia Condensada,
Universidad de Sevilla, P.O. Box 1065,
41080-Sevilla, Spain

Multiplicity of stable steady states and oscillations of concentrations and temperature are phenomena often observed in reactions catalyzed on a surface. We describe a kinetic model based upon the reaction of carbon monoxide (CO) and oxygen (O) on a catalyst surface with the object of explaining some kinetic phase transitions which occur in this reaction.

Kinetic equations governing the time evolution of densities of monomers, n_{CO} and n_O, and pairs n_{CO-CO}, n_{CO-O}, and n_{O-O} are formulated from a doublet closure approximation (Luque, 1990) and they are completed with an equation for the surface temperature with the purpose of analyzing especially the oscillatory behaviour of the system. Kinetic phase transitions of the kind:

a) 1 single stable steady state (1 *sss*) $\Longleftrightarrow$ 2 stable steady states (2 *sss*).

b) 1 *sss* $\Longleftrightarrow$ 1 steady state and 1 oscillatory state, both stables (1 *sss* and 1 *sos*).

c) 2 *sss* $\Longleftrightarrow$ 1 *sss* and 1 *sos*

are found, depending on the reaction heat and the relaxation rate of the surface temperature.

These results are in agreement with those obtained by other models proposed to analyze the theory of oscillatory oxidation of carbon monoxide over catalytic surfaces (Lagos et al., 1979; Lemos and Córdoba, 1994).

- Lagos, R.E., Sales, B.C., and Suhl, H. (1979): Surf. Sci. **82** 525.
- Lemos, M.C., and Córdoba, A. (1994): Phys. Rev. B **49** 14648.
- Luque, J.J. (1990): Phys. Rev. A **42** 3319.

Comparation of Two Fokker-Planck Modelings for a Master Equation

M.A. Muñoz[1] and P.L. Garrido[2]

[1] Departamento de Física Aplicada, Facultad de Ciencias
Universidad de Granada, 18071-Granada, Spain.
[2] Instituto Carlos I de Física Teórica y Computacional and
Departamento de Física Moderna, Facultad de Ciencias
Universidad de Granada. 18071-Granada, Spain.

In a recent paper [Phys. Rev. E50, 2459, (1994)] we have introduce a new kind of Fokker Planck modeling for Master equations without detailed balance. In that modeling it is guaranteed by construction that the associated stationary probability distribution coincides with the exact one around the deterministic solutions, which are also exactly reproduced. This is an improvement with respect to previous approaches. Nevertheless, the validity of the new modeling far from the deterministic solutions has to be tested. In this communication we analyse the behavior of the stationary solution of the Fokker Planck modeling for a particular two variable Master equation that has previously appear in the literature. We compute explicitely its associated stationary potential in the weak noise limit, and we compare it with the exact one coming from the Master equation. We also consider an alternative, commonly used, Fokker Planck, obtained by truncating a Kramers Moyal expansion of the Master equation. We find for our Fokker-Planck equation an excellent agreement with the exact solution in the whole phase space, being in any event, far better than the other coming from the Kramers-Moyal expansion.

- Graham, R. (1978): "Stochastic Processes in Nonequilibrium Systems." in *Lecture Notes in Physics*, vol. 24, ed. L. Garrido, P. Seglar and P.J. Shephard, Springer, Berlin.
- Graham, R. and Tel, T. (1985): Phys. Rev. A**31** 1109.
- Graham, R. and Tel, T. (1986): Phys. Rev. A**33** 1322.
- Hanggi, P., Grabert, H., Talkner, P. and Thomas, H. (1984): Phys. Rev. A**29** 371.
- Muñoz, M.A. and Garrido, P.L. (1993): J. Phys. A **26** 309.
- Muñoz, M.A. and Garrido, P.L. (1994): Phys. Rev. E **50** 2459.
- van Kampen, N.G. (1981): *Stochastic Processes in Physics and Chemistry*, Noth-Holland, Amsterdam.

Computational Studies of the Complex Ginzburg-Landau Equation and Its Non-Equilibrium Potential

R. Montagne, E. Hernández-García, and M. San Miguel

Departament de Física, Universitat de les Illes Balears,
E-07071 Palma de Mallorca, Spain.

The Complex Ginzburg-Landau Equation (CGLE) is a model equation describing the dynamics of extended systems near a Hopf-bifurcation. Examples of physical context include amplitude waves in binary fluid convection, transversally extended lasers and electrohydrodynamic convection in liquid crystals. Suitable scaling of the complex amplitude A, space and time allows the CGLE to be written as a function of two real parameters b_1 and b_3

$$\partial_t A = A + (1 + ib_1)\nabla^2 A - (b_3 - i) \mid A \mid^2 A, \quad b_3 > 0,\, x \in [0, L]\,. \tag{1}$$

Recent work (Chaté, 1994) in the one-dimensional case has identified numerically regions of parameter $[b_1 - b_3]$-space displaying different kinds of regular and chaotic behavior. This numerical "phase diagram" and the "transition lines" contained on it deserve a better understanding as well as a theoretical justification. Graham and co-workers (Descalzi and Graham, 1992) have constructed an approximate nonequilibrium potential (NEP) for the 1-d CGLE that could be useful for describing the different phases. We have numerically (and analytically) checked the validity of this approximate nonequilibrium potential as a Lyapunov functional. Our results can be summarized as follows.

a) In the no-chaotic region of the parameter space this NEP monotonically decreases in time towards the plane wave attractors, as expected for a Lyapunov functional.

b) NEP also works for the strange attractor in the phase turbulence region near the Benjamin-Feir instability line.

c) NEP is not a correct Lyapunov functional for regions where the singularities of coherent structures make no longer valid the hypotheses for its approximate construction.

d) This NEP has been found to have local minima on spatially chaotic functions.

Thus it is of interest to study *per se* the statistical mechanics of such a new model with chaotic ground state.

- Chaté, H. (1994): Nonlinearity **7**, 185; *Spatiotemporal Patterns in Nonequilibrium Complex Systems* Addison-Wesley, New York.
- Descalzi, O., and Graham, R. (1992): Phys. Lett. A **170**, 84; (1994): Z. Phys. B **93**, 509.

Deterministic Simulation of Domain Pattern Formation Under Instantaneous Nucleation Conditions

C.H. Wörner

Instituto de Física, Universidad Católica de Valparaíso,
Avenida Brasil 2950, Valparaíso, Chile

In solid state phase transformations, the kinetics of domain formation is usually described by the well-known Mehl-Johnson-Avrami-Kolmorogov (MJAK) theory (Christian, 1975) or Monte Carlo modeling (Rollet et al., 1989). In the present work use is made of a deterministic simulation method that takes into account the two basic aspects of the problem, i.e., the velocities of the individual growing boundaries and the topological requirement that the final structure must fill the whole space. Also, as it is necessary to adopt an hypothesis on the nucleation conditions, we will assume that all nuclei appear at $t = 0$ (instantaneous nucleation). We establish a two-dimensional system in which we put at random some number of nuclei. In the present report, we use 10, 100 and 500 initial points. These nuclei grow at a constant rate until they touch one another. At this point the growth stops in this direction and continues to grow in other directions. Eventually, the whole structure fills the available space, and the process is finished. As a matter of fact, periodic boundary conditions are used. The final aspect of all patterns is the well known Voronoi tessellation, with right boundaries bisecting the lines formed for the original nuclei. This fact clearly reveals that our system shows scaling. A more interesting finding is the time dependence of the transformed area. The AMJK theory predicts, for two-dimensional phase transformations, a sigmoidal time dependence of the transformed area fraction, f. Also, a plot of $\ell og\{\ell n[1/(1-f)]\}$ against $\ell og(t)$ must appear as a straight line with slope n (the kinetic exponent), with n-values between 2 and 3. In the present work the simulated logarithmic plots appear to show a slope slightly greater that $n = 2$, apparently confirming the AMJK theory for instantaneous nucleation. This is a controversial result because the extended area concept laying behind the AMJK theory does not seem evident (Vandermeer and Jensen, 1994). Furthermore, we have used two simplified assumptions in this work; i.e., that the boundaries grow at constant velocity and that nucleation is instantaneous. Realistic results can be better achieved both with more complex migration rates, e.g. those found in secondary recrystallization (Wörner et al., 1991), and by using other nucleation hypothesis.

The support of a FONDECYT (Chile) grant No. 1941225, and the OEA-PMM Project are gratefully acknowledged.

- Christian, J.W. (1975): The Theory of Transformations in Metals and Alloys, Pergamon, Oxford.
- Rollet, A.D., Srolovitz, D.J., Doherty, R.D. and Anderson, M.P. (1989): Acta Metall., **37**, 627.
- Vandermeer, R.A. and Jensen, D.J. (1994): Acta Metall. Mater., **42**, 2427.
- Wörner, C.H., Romero, S.and Hazzledine, P.M. (1991): J. Mater Res., **6**, 1773.

Morphological Instabilities of a Nematic-Smectic B Interface

R. González-Cinca[1], L. Ramírez-Piscina[1], J. Casademunt[2]

and A. Hernández-Machado[2]

[1]Departament de Física Aplicada, Campus Nord - Módulo B4,
Universitat Politècnica de Catalunya, Jordi Girona Salgado s/n,
E-08034 Barcelona.
[2]Departament d'Estructura i Constituents de la Matèria,
Universitat de Barcelona, Av. Diagonal 647,
E-08028 Barcelona.

The dynamics of a nematic-smectic B interface under different growth conditions have been studied using a phase field model. This system has a strong anisotropy in the surface tension for both P(in P) and P(in H) interfaces. Their Wulff construction determines faceted equilibrium shapes. Simulations for different undercoolings give rise to different morphologies, which have been observed experimentally. At large undercoolings we get typical dendritic growth with a dinamically selected angle not present in the anisotropy of the model.

Dynamic Time Rescaled Monte Carlo Algorithm

L. Bitar, P.A. Serena, P. García-Mochales, and N. García

Física de Sistemas Pequeños, CSIC, Universidad Autónoma de Madrid
C-III, Cantoblanco, E–28049 Madrid, Spain

We present a new Monte Carlo (MC) algorithm designed to study systems of particles with several characteristic times associated with the presence of a broad range of energy barriers (ΔE), over which those particles attempt to jump. In standard MC calculations (Kehr and Binder 1984; Tringides and Gomer 1986), the jump over an energy barrier ΔE is decided by comparing a random number η ($0 < \eta < 1$) with the probability $p = \exp((\mathrm{E}_r - \Delta \mathrm{E})/k_\mathrm{B}\mathrm{T})$, being E_r a reference energy (lower than any energy barrier present in the system), k_B the Boltzmann constant, and T the temperature. If $\eta < p$ the jump is accepted, otherwise is rejected. A simple relation between physical and computational time is given by $\delta t = \nu^{-1} \exp(\mathrm{E}_r/k_\mathrm{B}\mathrm{T})$, where δt is the time elapsed per MC step (MCS) and ν is the vibrational frequency. In our MC algorithm we have considered that the reference energy E_r can be modified from a MCS to the next one (in order to optimize the total computational time) according to the following rules: (i) if there were successful jumps within the current MCS then E_r will become the value of the minimum possible energy barrier in the system, (ii) if no jumps were accept E_r increases up to $\mathrm{E}_r + \delta \mathrm{E}_r$, being $\delta \mathrm{E}_r$ a value which depends on the energy barriers distribution. The change of E_r determines a different physical time interval δt to be used in the next MCS. This MC algorithm has been applied to the study of the disintegration of nanostructures created on surfaces (Bitar et al. 1994) finding good agreement with experiments. We have noticed that within a simulation the principle of the detailed balance is satisfied overall except for very few events. However, a suitable choice of $\delta \mathrm{E}_r$ leads to very similar results to those of a standar MC simulation for the evolution of the nanostructure shape aswell as for the evaluation of the disintegration time of the nanostructure.

This work has been supported by the CICyT under contract MAT92-0129.

- Bitar, L., Serena, P.A., García-Mochales, P., García, N., (1994): Surf. Sci., in press.
- Kehr, K.W. and Binder, K. (1984): *Applications of the Monte Carlo Method in Statistical Physics*, ed. K. Binder, Springer, Berlin.
- Tringides, M.C. and Gomer, R. (1986): Surf. Sci. **166** 419.

Pattern Formation in Convection of Rotating Fluids

J. Millán-Rodriguez[1], C. Pérez-García[1] and M.Bestehorn[2]

[1]Departamento de Física y Matemática Aplicada,
Facultad de Ciencias, Universidad de Navarra,
E-31080 Pamplona, Navarra, Spain.
[2]Institut für Theoretische Physik und Synergetik,
Universität Stuttgart, Pfaffenwaldring 57/4,
70550 Stuttgart, Germany.

Convective patterns in laterally extended rotating cells with a vertically broken symmetry are analyzed. A simplified model allows one to determine the stability of different basic patterns. By means of a generalized Swift-Hohenberg equation (Millán et al., 1992a) one can study the pattern evolution for different rotation rates and different values of the symmetry-breaking term. In particular, a hexagonal pattern with oscillating amplitudes is analyzed and the role of the defects is discussed. We study defect motion in a rotating convective cell. We present numerical results of a generalized Swift-Hohenberg equation, which provides a model description of the vertically averaged three-dimensional (3-D) hydrodynamic equations. Our model includes non-Boussinesq effects. It also accounts for the effects of the Coriolis force induced by a rotation of the fluid layer around its vertical symmetry axis. We show that even a slow rotation well below the Küpers-Loortz instability causes defect motion perpendicular to the convective rolls. We derive an analytical estimation of that motion (Millán et al., 1992b). Finally, we perform a numerical integration of the 3D- hydrodynamic equations describing convection in a rotating fluid layer (Millán et al., 1994). We assume the validity of the Boussinesq approximation, an infinite Prandtl number and rigid, non-slip boundary conditions on the top and the bottom of the fluid. The role of the defect motion in pattern formation (gliding and climbing, defects annihilation, pinning effects and pattern reorientation) is analyzed when the fluid layer rotates slowly around its vertical axis. Climbing and gliding velocities of defects are determined and a mechanism leading to a reorientation of a parallel roll structure is suggested.

- Millán, J., Pérez-García, C., Bestehorn, M., Fantz, M. and Friedrich, R., (1992a): Phys. Rev. A **46**, 4729.
- Millán, J., Pérez-García, C., Bestehorn, M., Neufeld, M., and Friedrich, R. (1992b): Chaos **4** 369.

- Millán, J., Pérez-García, C., Bestehorn, M., Neufeld, M., and Friedrich, R. (1994): Phys. Rev. Lett., to appear.

Codimension Two Bifurcation in Unidimensional Heating

A.M. Mancho-Sánchez and C. Pérez-García

Departamento de Física y Matemática Aplicada,
Facultad de Ciencias, Universidad de Navarra,
31080 Pamplona, Navarra, Spain

Our purpose in this work is to explain the convective structures developed through unidimensional heating. The experiment consist on a long wire heating a fluid. We take a temporal sequence of the convective structure developed, along a direction parallel to the the wire. So we get a spatio-temporal pattern. Details about the experiment are in (Burguete, 1993).

The Fourier transform of the pattern reveals the presence of a standing wave, e^{i2kx} and travelling waves, $e^{i(kx+\omega t)}$, $e^{i(kx-\omega t)}$. Our explanation is only qualitative and it is valid around a codimension two point in which the modes $e^{i(kx+\omega t)}$, $e^{i(kx-\omega t)}$ and e^{i2kx} grow. In this point the convective structure developed by the physical system is described by a function:

$$U(x,t,T) = A_1 e^{i(\omega t-kx)} + A_2 e^{i(\omega t+kx)} + B e^{i2kx} + c.c$$

To describe the temporal evolution of the amplitudes we induce a set of ordinary differential equation from considerations on symmetries. So we search a description that preserves the invariances of the physical system, under space reflection, and under constant displacement in space and time. Similar models, for other kind of codimension two bifurcations are described in the references (Proctor and Jones, 1988; Knobloch and Proctor, 1988). We study the stability regions of the equations, for different values of the eigenvalues, ε and μ, of the linearized equation around the codimension two point. These eigenvalues are related with the experimental parameters ΔT and h. A simulation of the equations around the stable fixed points gives similar patterns than the ones obtained experimentally (Mancho-Sánchez, 1994).

- Burguete, J. (1993): Trabajo de Investigación, Universidad de Navarra.
- Knobloch, E. and Proctor, M.R.E. (1988): Proc. R. Soc. London, **61**.
- Mancho-Sánchez, A.M. (1994): Trabajo de Investigación Universidad de Navarra.
- Proctor, M.R.E. and Jones, J. (1988): J. Fluid Mech.,**188** 301.

Cylinders and Wheels Rolling into Chaos

C.M. Arizmendi[1], H.N. Núñez-Yépez[2], A.L. Salas-Brito[3]

[1]Departamento de Física, Facultad de Ingeniería,
Universidad Nacional de Mar del Plata, Av. J.B. Justo 4302,
7600 Mar del Plata, Argentina.
[2]Departamento de Física,
Universidad Autónoma Metropolitana-Iztapalapa,
Ap. Postal 55-534, Iztapalapa 09340, D.F. México.
[3]Departamento de Ciencias Básicas,
Universidad Autónoma Metropolitana-Atzcapotzalco,
Ap. Postal 21-726, Coyoacán 04000, D.F. México.

We study the chaotic dynamics of an expanding and contracting cylinder rolling on a curved cylindrical surface. This system is formally equivalent to an extensible or elastic pendulum. We describe the system using a two degree of freedom Hamiltonian which is formally analogous to an extensible pendulum Hamiltonian. As is the case in the last system the equations of motion are mildly singular. The transition to stochastic behavior with increasing energy is exhibited with the help of the Poincaré surface of section method. At the same time, the existence of a confined region with negative Gaussian curvature in the potential energy surface of the rolling cylinder, the so-called negative curvature criterion or NCC, is used to predict an order-chaos-order transition which indeed occurs in the Poincaré sections studied. A comparison of the numerical results of the Poincaré sections with the predictions of NCC shows that these are precise only when the system becomes parametrically unstable. In all the other cases NCC gives just rough estimates of the conditions for the transition to chaos.

On the Transition to Columnar Convection

A. Alonso and M. Net

Department de Física Aplicada, Universitat Politécnica de Catalunya, 08034 Barcelona, Spain.

Convection in a rotating annulus with stress-free ends, gravity radially inwards and side-walls outwards heating is considered. No-slip fixed temperature boundary conditions at side-walls and free-slip thermally insulating boundary conditions at the ends are adopted. The transition from fully three-dimensional convection cells to Taylor columns with increasing rotation rate is studied and its dependence on the annulus parameters is stablished.

Systems of this type admit solutions in the form of Taylor columns (Busse, 1970) which are expected for large rotation rates for which the Taylor-Proudman theorem applies. The resulting two-dimensional steady convection columns undergo precession once slanted ends are introduced, because they break the Taylor-Proudman constraint.

We consider horizontal ends and low enough rotation rates, so that the mode that first sets in does not take the form of a Taylor column. We numerically solve the linearized and nondimensionalized relative to the thermal difussion time across the gap form of the equations. The dimensionless parameters involved are the Rayleigh and Prandtl numbers, the angular velocity, the radius ratio (η) and the inverse aspect ratio of the annulus (β). The radial structure of the eigenfunctions is expanded in Chebyshev polynomials. Expansions are of the form: $\Theta(r,\phi,z,t) = e^{st}\sum_{lmn} T_{lmn}(r)e^{in\phi}cos\frac{m\pi z}{\beta}$

Taylor columns correspond to $m = 0$. As the rotation drops out from the equations, the bifurcation to Taylor columns will be a steady state one.

Results show that the dominant mode can precess even though the boundary conditions assumed admit solutions in the form of nonprecessing Taylor columns, provided the rotation rate is low enough. The transition between these two types of modes depends on the annulus aspect ratio and its radius ratio. The bifurcation to convection cells is a Hopf one.

Taylor columns are preferred even in the nonrotating system provided $\beta \leq 1$. For $\beta > 1$, the interval of rotation rates for which convection cells are found increases with β, although there is always a rotation rate above which Taylor columns are selected.

On the contrary, there is not a threshold η which limits states in which only Taylor columns are possible from those in which there is a transition from convection cells to Taylor columns as the rotation rate is increased. For some of them ($\eta = 0.2, 0.3, 0.5$) only Taylor columns are selected. For others ($\eta =$

0.1, 0.4) there is an interval of rotation rates in which the selected mode is an $m \neq 0$ mode. The origin of this behaviour is explored and it is found that the threshold for Taylor columns ($m = 0$ and different values of n) oscillates as a function of η. For certain radius ratios crosses the locus of $m = 1$ modes, either with $n = 1$ or $n = 2$. This trend continues and becomes more pronounced for larger β. For sufficiently low rotation rates the critical Rayleigh numbers for the $m \neq 0$ states are lower than that for the corresponding Taylor columns.

Busse, F.F. (1970): J. Fluid Mech. **44**, 441.

Finite Temperature dc Conductance of Random Dimer Superlattices

E. Díez[1], A. Sánchez[1], F. Domínguez-Adame[2]

[1]Escuela Politécnica Superior, Universidad Carlos III de Madrid, c/ Butarque, 15, E-28911 Leganés, Madrid, Spain
[2]Departamento de Física de Materiales, Facultad de Física, Universidad Complutense, E-28040 Madrid, Spain

Disordered systems in which disorder exhibits some kind of short range spatial correlation wave localization may be inhibited and bands of extended states appear (Dunlap et al., 1990; Sánchez et al., 1994; Maciá et al., 1994). We study how structural correlations manifest in experimentally measurable magnitudes, focusing on dc conductance of semiconductor superlattices with general potential profiles.

We show that bands of extended states give rise to very noticeable peaks in the finite temperature dc conductance as the chemical potential is moved through the bands or as the temperature is increased from zero. We discuss how dc conductance measurements can provide information on the location and width of the bands of extended states (Diez et al., 1994). The basis of our model is the Kronig-Penney one (Kronig and Penney, 1931) with a potential of the form $V(x) = \sum_n \lambda_n \delta(x - x_n)$. We are interested in the version called Continuous Random Dimer Model (CRDM) We introduce a paired correlated disorder which implies that λ_n takes only on two values, λ and λ', with the additional constraint that λ' appears only in pairs of neighboring sites (dimer). For a *single* dimer we show that there are an infinite number of resonant energies for which the reflection coefficient vanishes. We study this problem numerically from exact transfer-matrix expressions. The combination with multifractal analyses allowed us to establish that the single dimer resonances in the CRDM give rise to bands of finite width of truly extended states. We have studied the finite temperature conductance (Diez et al., 1994) of the CRDM, and we find a rapid increase in conductivity when the chemical potential is close to one of this bands. The marked peak in the dc conductance should be clearly observable at temperatures close to 5-10 K in most superlattices by sweeping a range of values of the chemical potential. Experiments could validate (or discard) the recent claims that correlation induces the appearance of bands of extended states in spite of the localization effects of disorder (Domínguez-Adame et al., 1994).

Work at Leganés is supported by DGICYT project PB92-0248, and by the European Union, contract ERBCHRXCT930413. Work at Madrid is supported by Universidad Complutense, project PR161/93-4811.

- Diez, E., Sánchez, A., Domínguez-Adame, F. (1994): Phys. Rev. B, in press.
- Domínguez-Adame, F., Sánchez, A. and Diez, E. (1994): Phys. Rev. B , unpublished.
- Dunlap, D.H., Wu, H.L.and Phillips, P. (1990): Phys. Rev. Lett. **65**, 88.
- Kronig, R. de L., and Penney, W.G. (1931): Proc. R. Soc. London A **130**, 499.
- Maciá, E., Domínguez-Adame, F. and Sánchez, A. (1994): Phys. Rev. B **49**, 9503.
- Sánchez, A., Maciá, E. and Domínguez-Adame, F. (1994): Phys. Rev. B **49**, 147.

A Multigrid Approach to Parallel Simulations in Nonlinear Optics

I. Martín[2], V. M. Pérez-García[1] , F. Tirado[2], L. Vázquez[1]

[1]Departamento de Física Teórica,
Universidad Complutense,
28040 Madrid. Spain.
[2]Departamento de Informática y Automática
Universidad Complutense, 28040 Madrid. Spain.

It has been a great interest in the transverse laser dynamics, e.g. to understand the complex patterns appearing when many transverse modes interact (Brambilla et al., 1994). The conventional approach cannot explain some complex features (Pérez-García and Guerra, 1994; Dangoisse et al., 1992). For this purpose, and for the purpose of studying the effects of symmetry changes on the dynamics it is necessary to integrate the complete laser equations. In a series of works (Fei et al., 1994; Martín et al., 1994a) we have developed a linearly implicit finite difference scheme for the nonlinear Schrodinger equation which is being successfully applied to a wide variety of physical situations. Thus, we have applied the method to the Maxwell-Bloch equations. As we use an implicit numerical scheme, it is necessary to solve a sparse system of equations at each time step. We need Conjugate Gradients methods with preconditioning and multigrid methods to solve these kind of linear equations. The linear system arising from the discretization on a finite reticulum was solved using a multigrid technique (Martín et al, 1994b; Briggs, 1987): relaxation methods are very effective at reducing the high frequency components of the error, but it is not so with the low frequency ones ("smooth" methods). However, the latter have higher frequency on a coarser grid, and they can be effectively reduced using a "smooth" method over such a grid. This can be done in a recursive way until a certain level, where the system will be easier to solve than the original one. If we define the problem on grids of various sizes, we can obtain the solution at each level, from the coarser grid to finer grids, with the multigrid method using as first approximation the interpolated one from coarser grids. We used a domain descomposition in physical space for all grids. The multigrid implementation was checked performing some direct simulations of the equation on some different cases.

This work has been partially supported by Spanish Goverment research grants CAM 264/93, CICyT TIC92-0088, DGICyT PB92-0226 and the Human Capital Mobility Network CT93-0331.

- Brambilla, M., et al., (1994): Phys. Rev. A, **49** 1427-1451.

- Briggs W. L, (1987): "Multigrid Tutorial", SIAM.
- Dangoisse, D., et al. (1992): Phys. Rev. A **46** 5955-5958.
- Fei, Z., Pérez-García, V.M. and Vázquez, L. (1994): unpublished.
- Martín, I., et al. (1994a): "Proceedings II Euromicro Workshop", IEEE Computer Society Press.
- Martín, I., et al. (1994b): "Proceedings of the 6th Int. Conf. on Physics Computing", European Physics Society Press.
- Pérez-García, M., Guerra, J.M. (1994): Phys. Rev. A **50** in press.

Surface Electromagnetic Waves on Index-Matched Interfaces

M. Lester[1] and R. Depine [2]

[1]Instituto de Física Arroyo Seco, Universidad Nacional del Centro, Provincia de Buenos Aires 7000 Tandil, Argentina.
[2]Grupo de Electromagnetismo Aplicado, Departamento de Física, Facultad de Ciencias Exactas y Naturales, Universidad de Buenos Aires, Ciudad Universitaria, Pabellón I, 1428 Buenos Aires, Argentina.

We report the singular behaviour of the reflectivity for grazing incident on a planar interface between permeable, isotropic dielectrics with equal refractive indices (index-matched interface). Consider an incident electromagnetic plane wave illuminating a flat interface at $y = 0$. This boundary separates two isotropic, magnetic media, each one characterized by two scalar constants, the permittivity ϵ_j and the permeability μ_j ($j = 1, 2$). The reflectivity on a flat interface between permeable, isotropic dielectrics with equal refractive indices (index-matched interface) is a constant, independent of angle of incidence and polarization of the incident wave, can be seen from the Fresnel coeficients (Chen, 1983; Depine et al., 1993; Lakhtakia et al., 1991). However, the question remains as to what happens at grazing incidence if both dielectrics have the same refractive indices. In this case, it is not possible to obtain te reflectivities R and transmition T coefficientes from de Maxwell equations, because the electric field with s polarization or the magnetic field with p polarization is everywhere perpendicular to the interface and then the continuity of its longitudinal component holds always for all values of R and T. This situation is completely equivalent to the propagation of an eigenmode along the flat surface, a solution to Maxwell equations plus boundary conditions without incident wave. However, it could be thought that a suitable corrugation would allow these modes to couple with a given incident wave. To investigate this posibility, we use a perturvative method to determine the dispersion relationship for surface polaritons on a grating. The method, based on the Rayleigh hypothesis, was originally developed for non magnetic media and p polarized surface waves (Maradudin, 1991) In order to asses this predictions a differential exact method (Lester and Depine, 1994) was implemented to calculate the efficiency as function of the incidence angle for an index-matched grating. Numerical results show the existence of anomalies in the specular reflectivity for angles close to those predicted by perturbative analysis. Therefore the realationship between the set of eigen-modes of the corrugated structure and the anomalus behaviour in the specular order is corroborated.

- Chen, H.C. (1983): Theory of electromagnetic waves: A coordinate-free approach, McGraw-Hill, New York.
- Depine, R.A., Fantino, A.N. and Lakhtakia, A. (1993): J. Mod. Optics. **40** 219-233.
- Lester, M. and Depine, R.A. (1994): J. Phys. D, to appear.
- Lakhtakia, A., Varadan, V.K., and Varadan, V.V. (1991): J. Mod. Optics **38** 649-657.
- Maradudin, A.A. (1991): Modern Problems in Cond. Matt. Sci., 405-508.

Conservative Numerical Schemes for Euler-Lagrange Equations

S. Jiménez

Universidad Alfonso X El Sabio,
28040-Villanueva de la Cañada, Madrid, Spain.

The lagrangian formulation is of great relevance in many fields of Physics. The idea is to construct numerical schemes in finite differences that integrate the Euler-Lagrange equations and have a discrete counterpart of the conservation/variation law of the energy. We consider a general lagrangian with explicit time dependence, and the associated physical quantities such as the energy and momenta. The idea from the numerical point of view is to consider a discrete equivalent of the variation law for the energy, and to deduce from there a suitable numerical representation of the derivative of the positions, the momenta and the Euler-Lagrange equation which define the numerical scheme. All the schemes that can be obtained in this way have, by construction, discrete counterparts of the variation law of the energy and all of them are conservatives whenever the lagrangian is time-independent. In general it is possible to obtain several schemes for the same lagrangian: further considerations on other symmetries may help to decide which one is best suited for a specific problem. There are families of conservative schemes that are widely and succesfully used (Strauss and Vázquez, 1978; Pascual and Vázquez, 1986; Jiménez and Vázquez, 1991; Jiménez, 1994). Nevertheless, these schemes are usually only applicable to problems with hamiltonians or lagrangians sepparable into a kinetic and a potential parts. The schemes we present here can be applied to more general problems. Applications: we have considered different applications of increasing complexity, obtaining schemes for a particle in a potential in one dimension with lagrangian $L = \frac{1}{2}\dot{q}^2 - V(q)$ (Malavasi, 1993); a forced oscillator: $L = \dot{q}^2/2 - aq^4/4 + q\cos(t)$; the Hénon system $L = \dot{x}^2/2 + \dot{y}^2/2 - (x^2+y^2)/2 - x^2y + y^3/3$; an application to a model for fusion accellerators, the Reverse Pinched Field (RPF) with $L(x,\dot{x};y,\dot{y};t) = F_1(x,y;t)\dot{x} + F_2(x,y;t)\dot{y} + F_3(x,y;t)$ (Zhang and Vázquez, 1991); and an application to an arbitrary large number of coupled oscillator, that is, to the finite difference representation of a partial differential equation, for instance the Sine-Gordon model. In all the cases, the scheme is locally of second order and four-level explicit, but in the RPF case, where it is two-level implicit. A discrete counterpart of the conservation/variation law exists and other symmetries of the continuous equations can also be taken into account.

Partially supported by the Human Capital and Mobility Network Grant CT93-0417.

- Strauss, W. and Vázquez, L. (1978): J. Comp. Phys. **28** 271-278.
- Pascual, P.J., and Vázquez, L. (1986): Hadronic Journal, **9** 307-310.
- Jiménez, S. and Vázquez, L. (1991): Nonlinear Problems in Future Particle Accelerators, proceedings, ed. World Scientific, 151-162.
- Jiménez, S. (1994): Applied Mathematics and Computation, **64** 13-45.
- Malavasi, M. (1993): PhD. Thesis, Università di Bologna, Bologna, Italy.
- Zhang, F., Vázquez, L (1991): Comp. Appl. Math. **10** 59.

Suppression of Chaotic Behaviour in Models of Chemical Chaos

J. Güémez[1], J.M. Gutierrez[2], A. Iglesias[2] and M.A. Matías[3]

[1]Departamento de Física Aplicada, Facultad de Ciencias, Universidad de Cantabria, E-39005 Santander, Spain.
[2]Departamento de Matemática Aplicada y Ciencias de la Computación, Facultad de Ciencias, Universidad de Cantabria, E-39005 Santander, Spain.
[3]Departamento de Química-Física, Facultad de Química, Universidad de Salamanca, E-37008 Salamanca, Spain.

In many practical situations one is interested in obtaining a regular behaviour in system with chaotic dynamics. In many cases it is possible to obtain periodic behaviour by making only small time-dependent perturbations in some accessible *system parameters* (Ott et al., 1990). In other cases weak periodic perturbations on some system parameters allows the system to perform regular behaviour (Lima and Pettini, 1991). But in those situations where finding an accessible parameter may be a difficult task, the case of chemical or biological systems, a method that performs changes in the *system variables* allows one to stabilize chaotic systems (Matías and Güémez, 1994). This method works by applying proportional instantaneous pulses spaced in time, every interval τ to the system variables (one, two or three) in the form $X_i(t) = X_i(t)(1+\lambda)$, where $X_i(t)$ represents the ith variable of the system and λ which can be positive or negative, regulates the strength of the pulse. In models of chemical chaos, with a virtually unlimited reservoir of periodic behaviour, this method is very symple to implement. We have considered four different chemical systems described by three-variable flows (differential equations) exhibiting deterministic chaos: the isothermal autocatalator (Peng et al., 1990), the Willamowski-Rössler model (Aguda and Clarke, 1988), and two Rössler chemical models, the first one proposed by Poland (Poland, 1993), and the second one proposed by Samardzija et al. (Samardzija et al., 1989). The main conclusion are that it is possible to achieve control of low-dimensional chaotic chemical systems by perturbing the system variables with a series of short unspecific pulses, whose function is to apply a proportional feedback to the system populations.

- Aguda, B.L. and Clarke, B.L. (1988): J. Chem. Phys. **89** 7428.
- Lima, R. and Pettini, M. (1991): Phys. Rev. A **41** 726.
- Matías, M.A. and Güémez, J. (1994): Phys. Rev. Lett. **72** 1455.
- Ott, E., Grebogi, C., and Yorke, J.A. (1990): Phys. Rev. Lett. **64** 1196.

- Peng, B., Scott, S.K. and Showalter, K. (1990): J. Chem. Phys. **94** 5243.
- Poland, D. (1993): Physica D **65** 86.
- Samardzija, N., Greller, L.D. and Wasserman, E. (1989): J. Chem. Phys. **90** 2296.

Perturbing Variables in Poincaré Sections to Suppress Chaos

J.M. Gutiérrez[1], A. Iglesias[1], J. Güémez[2] and M.A. Matías[3]

[1]Departamento de Matemática Aplicada y Ciencias de la Computación, Facultad de Ciencias, Universidad de Cantabria, E-39005 Santander, Spain.
[2]Departamento de Física Aplicada, Facultad de Ciencias, Universidad de Cantabria, E-39005 Santander, Spain.
[3]Departamento de Química-Física, Facultad de Química, Universidad de Salamanca, E-37008 Salamanca, Spain.

In the last few decades, investigators have studied some interesting chaotic effects in a very broad variety of nonlinear systems. Due to the rich variety of regimens embedded within chaos, chaotic systems can be stabilized, allowing the system to follow a regular orbit. So, controlling deterministic chaos represents one of the most interesting endeavours in nonlinear dynamics. Different chaos suppression methods have been proposed, based on the stabilization of unstable periodic orbits (Shinbrot et al., 1993). Recently a new method has been suggested, able to stabilize chaotic systems by applying proportional pulses to the system *variables*, and not in system parameters. This method has been successfully applied to many continuous and discrete systems (Matías and Güémez, 1994; Matías et al., 1994). In this work, we have considered the possibility of chaos suppression by performing small changes in certain system variables, just when the system is exactly located in a selected Poincaré section. In this way, continuous 3D chaotic flows are transformed in discrete 2D iterative maps, where effective chaos suppression algorithm is acting. Thus, the basic idea is to stabilize the dynamics of a system without introducing changes in the system parameters. We consider two different kinds of pulses: an additive constant pulse ($X_i' = X_i + \lambda$), that is independent of the system, and an additive proportional pulse ($X_i' = X_i(1 + \lambda)$), which depends on the system state. In both cases, the λ parameter regulates the strength of the pulse. Constant additive pulses have been applied to the three variables Rössler model, obtaining different periods (1,2,4,etc.) for stabilized orbits when varying λ. Thus, this method shows the usual transitions to chaos dependient upon λ, and it is possible to switch from period-1 to chaotic behaviour following the well-known Feigenbaum's route to chaos. Moreover, it is also possible to stabilize another kind of periods: period-3, 5, etc. These results imply that the present method works by creating a new dynamical system that has λ as a system parameter. Clearly, when $\lambda \to 0$, the new system is close to the original one. In the case of proportional pulses, a similar behaviour is observed for stabilized orbits.

- Matías, M.A. and Güémez, J. (1994): Phys. Rev. Lett. **72** 1455.
- Matías, M.A. et al. (1994): Phys. Lett. A **72** 1455.
- Shinbrot, T., et al. (1993): Nature (London) **361** 240.

Ergodic Theory for Unidimensional Jacobi Matrices

A.I. Alonso and C. Núñez

Departamento de Matemática Aplicada a la Ingeniería,
E.T.S. de Ingenieros Industriales,
Universidad de Valladolid, Spain

We study ergodic Jacobi matrices on $l^2(\mathbb{Z})$, and analize the directional differentiability of their Floquet coefficient.

Let Ω be a compact metric space and $T : \Omega \to \Omega$ a minimal homeomorphism. Let us fix an ergodic measure for T, m_0, and $V \in C(\Omega)$. We consider the family of Jacobi operators

$$-u_{n+1} - u_{n-1} + V(T^n(x))u_n = Eu_n \quad , \tag{1}$$

for $x \in \Omega$, which induces an action S_E of $\mathbb{Z}$ over the projective bundles $K_{\mathbb{R}} = \Omega \times P^1(\mathbb{R})$ and $K_{\mathbb{C}} = \Omega \times P^1(\mathbb{C})$. Our treatment is ergodic in nature. We analize properties verified by a whole family of operators with complete measure. Let A be the set of those $E \in \mathbb{C}$ with Lyapunov exponent $\gamma(E) = 0$. As a consequence of the Ishii-Pastur-Kotani theory it is known that the essential clousure of A agrees with the absolutely continuous spectrum of (1) for m_0-almost every $x \in \Omega$. Now we show that there exists a set $A_1 \subset A$ with Lebesgue measure $\rho(A - A_1) = 0$ such that $(K_{\mathbb{R}}, S_E)$ admits an absolutely continuous ergodic measure for all $E \in A_1$. At these points we prove the convergence in measure of the Weyl-Titsmarch functions, whose limit allows us to construct the above mentioned measures. This guarantees the directional differentiability of the Floquet coefficient. In particular we prove that a well-known inequality for the derivative of the rotation number due to Deift and Simon is in fact an equality. The derivative of the Floquet coefficient presents different properties according to the ergodic structure of the real projective flow. If $K_{\mathbb{R}}$ decomposes in ergodic sheets that derivative is not even linear. However, in any other case, the Floquet coefficient is Gateux differentiable. The formulas obtained are the starting point for a computational study, based on the family of Last periodic approximants, for the almost Mathieu operator

$$-u_{n+1} - u_{n-1} + \lambda \cos(2\pi\alpha\, n + \theta) = Eu_n \tag{2}$$

which gives us information about its spectral and ergodic properties.

Order in Binary Sequences and the Properties of the Routes to Chaos

R. López-Ruiz

Departamento de Física, Universidad de Navarra,
31080 Pamplona (Spain) and Departamento de Física,
Universidad Pública de Navarra, 31008 Pamplona (Spain).

The natural order in the space of binary sequences allows to recover the U-sequence and to explain the scaling laws of the period-doubling cascade, and the intermittency route to chaos as intrinsic properties of this ordered set, and independent from any consideration about dynamical systems.

- Metropolis, N., Stein, M.L. and Stein, P.R. (1973): J. Combinatorial Theory **15**, 25.
- Feigenbaum, M.J. (1978): J. Stat. Phys. **19** 25.
- Pomeau, Y., and Manneville, P. (1980): Comm. Math. Phys. **74** 189.
- Mindlin, G.B., López-Ruiz, R., Solari, H.G. and Gilmore, R. (1993): Phys. Rev. E **48** 4297.

List of Participants

Achahbar, A. *Instituto Carlos I de Física Teórica y Computacional, Universidad de Granada, E-18071 Granada.*

Alonso de M., A.I. *Departamento de Matemática Aplicada, E.T.S.I., Universidad de Valladolid, E-47011 Valladolid.*

Alonso, A. *Departamento de Física Aplicada, Universidad Politécnica de Cataluña, c/Jordi Girona Salgado s/n, módulo B4, E-08034 Barcelona.*

Alonso, J.J. *Laboratoire de P.M.M.H. Ecole Sup. de Physique et Chemie Ind., 10 rue Vauquelin, 75231 Paris 05, Francia.*

Ameziane, B. *Departamento de Física Moderna, Universidad de Granada, E-18071 Granada.*

Arizmendi, C.M. *Departamento de Física, Universidad Nacional de Mar del Plata, Avda. Juan B. Justo 4302, 7600 Mar del Plata, Argentina.*

Barceló, C. *Departamento de Física Aplicada, Universidad de Granada, E-18071 Granada.*

Burgos, D. *Servicio de Protección Radiológica, Hospital Universitario, S.A.S., Avda. Dr. Olóriz 16, E-18012 Granada.*

Calixto, M. *Instituto Carlos I de Física Teórica y Computacional, Universidad de Granada, E-18071 Granada.*

Caro, J. *Departamento de Física Moderna, Universidad de Granada, E-18071 Granada.*

Casas-Pérez, F. *Departamento de Matemáticas, Universidad Jaume I, Campus de Penyeta Roja, E-12071 Castellón.*

Cervantes de la T., F. *Departamento de Sistemas (CBI), Universidad Autónoma Metropolitana-Azcapotzalco, Avda. San Pablo 180, Col. Reynosa Tamaulipas, C.P. 02200 (Atdo. Postal 21-726), 0400 Distrito Federal, México*

Díaz-Calavia, E.J. *Departamento de Física y Matemática Aplicada, Universidad de Navarra, E-31080 Pamplona.*

Díaz-Guilera, A. *Departamento de Física Fundamental, Universidad de Barcelona, Avda. Diagonal 647, E-08028 Barcelona.*

Díez-Fernández, E. *E.P.S., Universidad Carlos III de Madrid, c/ Butarque 15, E-28911 Leganés, Madrid.*

Félix-Valdez, J. *Instituto de Física, Universidad de Guanajuato, Lomas de bosque 103, Lomas de campestre, León Gto. 37150, México.*

Fernández, J.F. *Instituto Carlos I de Física Teórica and Computacional y Departamento de Física Aplicada, Universidad de Granada, E-18071 Granada.*

Gallavotti, G. *Department of Physics, Universitá di Roma "La Sapienzia", 00185 Roma, Italia.*

García-Cortí, J.L. *Departamento de Matemáticas, Campus de Tafira baja, Universidad de Las Palmas de Gran Canaria, E-35017 Las Palmas de Gran Canaria.*

García-Mochales, P. *Física de Sistemas Pequeños-CSIC, Departamento de Física de la Materia Condensada, Universidad Autónoma de Madrid, Cantoblanco, E-28049 Madrid.*

Garrido, P.L. *Instituto Carlos I de Física Teórica y Computacional y Departamento de Física Moderna, Universidad de Granada, E-18071 Granada.*

González-Cinca, R. *Departamento de Física Aplicada, Universidad Politécnica de Cataluña, c/Jordi Girona Salgado s/n, módulo B4, E-08034 Barcelona.*

González-Martínez, J.R. *E.U.I.T. Industrial, Universidad de León, Campus de Vegazana, León*

Güemez, J. *Departamento de Física Aplicada, Universidad de Cantabria, Av. de los Castros s/n, E-39005 Santander.*

Gutierrez-Llorente, J.M. *Departamento de Matemática Aplicada, E.T.S.I., Universidad de Cantabria, E-39005 Santander.*

Herrero, H. *Departamento de Física y Matemática Aplicada, Universidad de Navarra, E-31080 Pamplona.*

Herrmann, H.J. *Laboratoire de P.M.M.H. Ecole Sup. de Physique et Chemie Ind., 10 rue Vauquelin, 75231 Paris 05, Francia.*

Iglesias, A. *Departamento de Matemática Aplicada, E.T.S.I., Universidad de Cantabria, E-39005 Santander.*

Jiménez, S. *Departamento de Matemática Aplicada, Facultad de Ciencias Químicas, Universidad Complutense de Madrid, E-28040 Madrid.*

Jiménez, J. *Departamento de Física Teórica y del Cosmos, Universidad de Granada, E-18071 Granada.*

Lemos, M.C. *Departamento de Física de la Materia Condensada, Universidad de Sevilla, Aptdo. 1065, E-41080 Sevilla.*

Lester, M.F. *Instituto de Física "Arroyo Seco", Universidad Ncnal. del Ctr. de la Pca. de Bns. Ars., Pinto 399, Tandil 7000 (Argentina).*

Linares, J.R. *Departamento de Física Aplicada, Universidad de Granada, E-18071 Granada.*

López-Lacomba, A.I. *Instituto Carlos I de Física Teórica y Computacional and Departamento de Física Aplicada, Universidad de Granada, E-18071 Granada.*

López-Ruiz, R. *Departamento de Física y Matemática Aplicada, Universidad de Navarra, E-31006 Pamplona.*

López-Sánchez, C. *Departamento de Física Moderna, Universidad de Granada, E-18071 Granada.*

Luévano, J.R. *Departamento de Sistemas (CBI), Universidad Autónoma Metropolitana-Azcapotzalco, Avda. San Pablo 180, Col. Reynosa Tamaulipas, C.P. 02200 (Atdo. Postal 21-726), 0400 Distrito Federal, México.*

Mancho, A.M. *Departamento de Física y Matemática Aplicada, Universidad de Navarra, E-31080 Pamplona.*

Marro, J. *Instituto Carlos I de Física Teórica y Computacional and Departamento de Física Aplicada, Universidad de Granada, E-18071 Granada.*

Matías, M.A. *Departamento de Química Física, Universidad de Salamanca, E-37008 Salamanca.*

Martín, I. *Departamento de Informática y Automática, Facultad de Ciencias Físicas, Universidad Complutense de Madrid, E-28040 Madrid.*

Millán, J. *Departamento de Física y Matemática Aplicada, Universidad de Navarra, E-31080 Pamplona.*

Montagne, H.R. *Departamento de Física, Universidad de las Islas Baleares, E-07071 Palma de Mallorca.*

Moya, A.A. *Departamento de Física Aplicada, Universidad de Jaén, E-23071 Jaén.*

Muñoz, M.A. *Instituto Carlos I de Física Teórica y Computacional and Departamento de Física Aplicada, Universidad de Granada, E-18071 Granada.*

Núñez, C. *Departamento de Matemática Aplicada, E.T.S.I., Universidad de Valladolid, E-47011 Valladolid.*

Plourabone, F. *Laboratoire de P.M.M.H. Ecole Sup. de Physique et Chemie Ind., 10 rue Vauquelin, 75231 Paris 05, Francia.*

Polo, I.M. *Departamento de Física Aplicada, Universidad de Granada, E-18071 Granada.*

Rojo, J. *Departamento de Matemática Aplicada, E.T.S.de Arquitectura, c/ Guzmán el Bueno 74, E-28015 Madrid.*

Romera, E. *Departamento de Física Moderna, Universidad de Granada, E-18071 Granada.*

Sánchez-Salcedo, F.J. *Departamento de Física Aplicada, Universidad de Granada, E-18071 Granada.*

Santos-Fernández, F. *Departamento de Física Moderna, Universidad de Granada, E-18071 Granada.*

Santos-Leal, E. *Escuela Politécnica Superior de Ingeniería, Universidad de Cantabria, Avda. los Castros s/n, E-39005 Santander.*

Sanz-Serna, J.M. *Departamento de Matemática Aplicada, Facultad de Ciencias, Universidad de Valladolid, E-47005 Valladolid.*

Sarsa, A. *Departamento de Electrónica, Universidad de Granada, E-18071 Granada.*

Sesma, J. *Departamento de Física Teórica, Universidad de Zaragoza, E-50009 Zaragoza.*

Tapia, R. *Escuela Politécnica Superior de Ingeniería, Universidad de Cantabria, Avda. los Castros s/n, E-39005 Santander.*

Toral, R. *Departamento de Física, Universidad de las Islas Baleares, E-07071 Palma de Mallorca.*

Torres-Agudo, J.J. *Instituto Carlos I de Física Teórica y Computacional andDepartamento de Física Aplicada, Universidad de Granada, E-18071 Granada.*

Vacas, J.A. *Instituto Carlos I de Física Teórica y Computacional and Departamento de Física Aplicada, Universidad de Granada, E-18071 Granada.*

Vázquez, L. *Departamento de Física Teórica, Facultad de Física, Universidad Complutense, E-28040 Madrid.*

Vidales, A.M. *Departamento de Física, Universidad Nacional de San Luis, Chacabuco 917, 5700 San Luis (Argentina).*

Vilches, M. *Servicio de Protección Radiológica, Hospital Universitario, S.A.S., Avda. Dr. Olóriz 16, E-18012 Granada.*

Wörner, C.H. *Instituto de Física, Universidad Católica de Valparaiso, Casilla 4059, Valparaiso 02, Chile.*

MIX
Papier aus verantwortungsvollen Quellen
Paper from responsible sources
FSC® C105338

If you have any concerns about our products,
you can contact us on
ProductSafety@springernature.com

In case Publisher is established outside the EU,
the EU authorized representative is:
Springer Nature Customer Service Center GmbH
Europaplatz 3, 69115 Heidelberg, Germany

Printed by Libri Plureos GmbH
in Hamburg, Germany